An Advanced Course in Nuclear Engineering

Volume 7

Series editors
Yoshiaki Oka, Tokyo, Japan
Haruki Madarame, Tokyo, Japan
Mitsuru Uesaka, Tokyo, Japan

This nuclear engineering textbook series aims to serve as global education standard for all universities that have a department of nuclear engineering and all research laboratories and companies related to nuclear engineering.

As global warming issues are strongly being advocated, nuclear power is examined again as a low-carbon and ecology-friendly energy source. Furthermore the accident at TEPCO Fukushima Daiichi Nuclear Power Plant on March 11, 2011, has intensified the attention to nuclear engineering. Industrial countries with nuclear power are not confined to domestic sales but are working toward global sales strategies of nuclear power plants, involving many nations. Against such a background, while nuclear power industry develops globally, the required human resources education should keep up. However, no English textbook with the potential to become the global standard has been published so far, although awaited for a long time. Recognizing the demand, Japanese nuclear power specialists concentrated their expertise and a Japanese nuclear engineering textbook series was published under the cooperation of the University of Tokyo, the Japan Atomic Energy Agency and other organizations related to nuclear engineering. The needs of such comprehensive textbooks are growing remarkably after the accident. Japan has steadily advanced the research and development of all fields of nuclear engineering. This is reflected in the broad coverage of the series, which includes not only nuclear reactors, but also the nuclear fuel cycle, radiation use, maintenance, etc.. This English version of the Japanese series serves domestic as well as foreign institutions. Although the Japanese version of the series was published before the accident, readers will find the in-depth coverage of diverse topics in nuclear engineering still relevant.

More information about this series at http://www.springer.com/series/10746

Hisaaki Kudo
Editor

Radiation Applications

Springer

Editor
Hisaaki Kudo
The University of Tokyo
Tokai, Ibaraki, Japan

Original Japanese edition, Genshiryoku Kyokasho: Houshasen Riyou, edited by Hisaaki KudoCopyright © 2011 Published by Ohmsha, Ltd. 3-1 Kanda Nishikicho, Chiyoda-ku, Tokyo, Japan

ISSN 2195-3708 ISSN 2195-3716 (electronic)
An Advanced Course in Nuclear Engineering
ISBN 978-981-10-7349-6 ISBN 978-981-10-7350-2 (eBook)
https://doi.org/10.1007/978-981-10-7350-2

Library of Congress Control Number: 2017963644

Printed on acid-free paper

This Springer imprint is published by Springer Nature
The registered company is Springer Nature Singapore Pte Ltd.
The registered company address is: 152 Beach Road, #21-01/04 Gateway East, Singapore 189721, Singapore

Preface

To use radiation "intentionally" for something may not be considered desirable, particularly taking into account the feelings of Japanese people in this regard. For example, for some commercial products that use radiation during their manufacturing process, producers tend not to advertise this fact. However, in reality, radiation is extensively utilized in many diverse fields such as industry, conservation of the environment, agriculture, and medicine.

This book addresses part of the developing advanced subjects that make up the nuclear engineering textbook series.

In the first part of the book, which is built on the fact that most radiation application is based on the academic philosophy of radiation chemistry, the fundamentals of radiation chemistry are described. There are other subjects related to radiation in this series, but they are focused on radiation safety, protection, measurements, and shielding, and not much has been written about radiation chemistry. This book therefore addresses this gap. However, a book should provide an overall view, and thus the fundamentals of radiation also need to be described to a limited extent. Therefore, while there may be some overlap with other textbooks, the editor and authors hope that readers recognize that the overlap is important and necessary.

The latter half of this textbook is very application-oriented. As the content shows, radiation application has long depended on the contribution of the Japan Atomic Energy Agency (JAEA), Quantum Beam Directorate (formerly Takasaki Radiation Chemistry Research Establishment (TRCRE), Japan Atomic Energy Research Institute (JAERI)), which has been reorganized as the National Institutes for Quantum and Radiological Science and Technology (QST). Experts from the JAEA (QST) provide the state-of-the-art in their specific field. The evolution of JAERI (TRCRE) – JAEA, is also described, from the institution of radiation polymer chemistry from semi-conductors, to conservation of the environment and production of inorganic materials and biotechnology.

References are listed at the end of each chapter. This reflects the recent rise in publications that introduce radiation application to the public. This is a welcome development, and although this book takes a similar direction, its structure as a textbook is maintained. Compared with other books in this series, the introduction of current research activities is often described, and readers will understand the necessity of this due to the nature of such an advanced subject. Simultaneously, the introduction of current research activities means that this research is at a cross-section at the moment, on the path of progress, which is ongoing.

In the latter half of the book, in particular the chapters on biological, agricultural, and medical fields, many illustrations, graphics, and photographs are provided. Some of these were provided in color, but in the main text, illustrations and photographs are published in black and white for readers' convenience.

This book targets senior undergraduates, students of professional schools, and those doing masters' courses, as well as young researchers and engineers of research organizations. If this book can be of help in implementing radiation application, which is sometimes compared to another wing or another wheel, it would be to our, the editor and authors, great pleasure.

Finally, it took an unexpectedly long time to complete this book, from inception to publication, and the editor and authors are most grateful and appreciative of the generous encouragements from the University of Tokyo, JAEA, and others. Also, the editor would like to thank and acknowledge his respect for all the authors who were involved in the creation of this book.

Tokai, Ibaraki, Japan
January, 2011

Hisaaki Kudo

Contents

Chapter 1
Radiation: Types and Sources

Yosuke Katsumura and Hisaaki Kudo

1.1 Definitions of Radiation

You may find the definition of radiation to be somewhat more complicated than expected. According to one definition, visible light and infrared rays can be described as radiation because they are energy fluxes, but this may seem surprising. Radiation is also described as causing ionization in a medium, so it is frequently called "ionizing radiation." Scientifically, radiation is assumed to be a flux of energy traversing space or a medium. Japanese regulations define radiation as electromagnetic waves or corpuscular beams that can ionize air directly or indirectly. By this definition, visible rays and infrared rays are not classified as radiation, even though they are energy fluxes. They are sometimes called non-ionizing radiation. Therefore, radiation is, in a broad sense, an energy flux traversing space or a medium and, in a narrow sense, an energy flux that can ionize materials. Many terms such as "quantum beam," "energy beam," and "quantum energy beam" were introduced recently; they have almost the same meaning as "radiation."

For details, refer to books such as "Radiation Safety" and "Nucleus and Radiation Measurements" in this textbook series.

Y. Katsumura · H. Kudo (✉)
The University of Tokyo, Tokai, Ibaraki, Japan
e-mail: hkudo@tokai.t.u-tokyo.ac.jp

H. Kudo (ed.), *Radiation Applications*, An Advanced Course in Nuclear Engineering 7, https://doi.org/10.1007/978-981-10-7350-2_1

1.2 Chemical Effects of Radiation

Radiation loses its energy through interactions with materials; this energy is transferred to the materials, causing ionization and excitation in them. As a result of ionization and excitation, ions (M^+), electrons (e^-), excited states (M^*, where the symbol * indicates that species M is in an electronically high-energy state), and radicals (denoted as $R^•$; see Chap. 4) are formed. They are normally chemically active and react with each other, or with surrounding molecules, to form new reactive species, but they have short lifetimes and are generally called reactive intermediates or short-lived chemical species. The chemical reactions of these species produce stable final products. This process is described in Fig. 1.1, where the symbol ⟿ represents irradiation (exposure to radiation).

For example, when liquid water is irradiated, the irradiation evokes ionization and excitation, forming H_2O^+, e^-, and H_2O^* within a very short time (less than 10^{-12} s). Subsequently, 10^{-6} s after the moment of irradiation, radicals such as •H and •OH (a water molecule after undergoing scission), "electrons in water" (called hydrated electrons), and acids such as H_3O^+ are formed. They may, in the presence of some solutes, react or be decomposed or converted to other species. However, in pure water, they will form H_2, H_2O_2, and as stable final products. This field of study is called radiation chemistry; it identifies and/or evaluates the yields of short-lived chemical species and determines their chemical reactivity, identifies and/or evaluates the yields of final products, elucidates radiation-induced reaction processes, and applies the characteristic reactions to commercial products and/or environmental conservation. Radiation chemistry is closely related to radical chemistry and photochemistry, which focus on similar chemical species. However, radiation chemistry differs from radiochemistry and nuclear chemistry, although they have similar names, as the latter two deal with the chemical characteristics and/or behavior of radioisotopes (RIs) themselves, and aim to understand chemical reactions through nuclear reactions. This textbook series on nuclear engineering does not include a book that systematically describes radiochemistry.

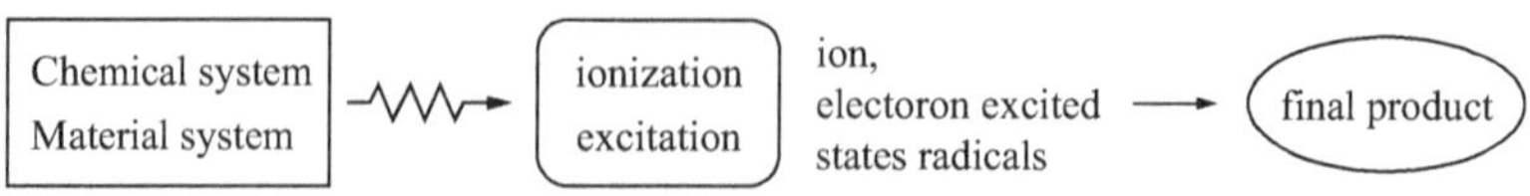

Fig. 1.1 Scheme of radiation-related chemical reactions

1.3 History of Research in Chemical Action of Radiation

In 1895, W. C. Roentgen of Germany discovered X-rays, and in the following year, A. Becquerel of France discovered radioactivity; these discoveries led to research on the structure of the nucleus. The photosensitizing effect on photographic plates was critical evidence in the discovery of X-rays. Accordingly, the chemical action of X-rays was already recognized from their discovery. In 1898, radium, a type of RI, was separated by M. Curie, a Polish scientist working in France. Radium emits X-rays, and radiation-based decomposition of water was observed by observing colorization and gas evolution in an aqueous solution containing radium in the early twentieth century. The gas was composed of hydrogen and oxygen, which indicates the decomposition of water. At that time, the X-ray emitters radium and polonium were frequently used as radiation sources. However, after the 1920s, intense X-ray sources were developed and have since been used for research. Then an interesting phenomenon was observed; gas evolution was not observed from aqueous solutions upon X-ray irradiation, while gas evolution was observed with the same amount of radiation of alpha-rays, gas evolution was not observed. This was the first observation of the effect of the radiation quality (different chemical action under different types of radiation) with current knowledge. This finding was not presented until after World War II. In the meantime, radiolysis of water and the oxidative action of sonic species in water had been observed. The Manhattan Project in the USA during World War II dramatically advanced the understanding of radiation chemistry. This project aimed to develop atomic bombs, but the development of nuclear reactors and research on the physical and chemical actions of radiation were simultaneously conducted by E. Fermi, an Italian scientist working in the USA, with a view to the use of nuclear energy. These were classified as military secrets during and after the war but were disclosed upon US President Eisenhower's "Atoms for Peace" address in 1953. Subsequently, radiation sources such as nuclear reactors became available with the development and use of nuclear energy. Industrial technologies using radiation were developed, and in the 1960s, cables made of material cross-linked by an electron beam were produced and are still used. The most well-known source is the Co-60 γ-ray source. In addition, electron beam accelerators have frequently been used.

1.4 Types and Sources of Radiation

1. Radiation and radioactivity
   Radiation can be defined as a flow of energy or particles that can ionize or encircle materials. Examples are α-, β-, γ-, and X-rays; beams of electrons,

positrons, ions, and mesons; and fission fragments. The sources of radiation are RIs, accelerators, and nuclear reactors. Radioactivity is the ability to emit radiation and is measured in terms of the number of disintegrations per unit time.

2. Radiation from radioactive materials

Materials containing radioactive elements are called radioactive materials, and the radiation they emit is classified as α-, β-, or γ-rays, which are assigned to high-energy helium ions, electrons, and electromagnetic waves, respectively. They are emitted upon nuclear reactions or disintegration.

(a) α-rays

The α-ray is a naked helium atom, namely, the nucleus of helium. The nuclear reaction that emits α-rays is called α-decay, and it is observed in nuclides heavier than Pb. The energy spectrum of α-rays is monoenergetic in the range of 4–6 MeV. Thus, sometimes this high energy can have important biological effects.

(b) β-rays

The β-ray is a high-energy electron (or positron, the antiparticle to the electron) beam that appears during a nuclear reaction called β-decay. β-Rays have a continuous energy spectrum, unlike α-rays; the energy spectrum comes from the energy of neutrinos, which are emitted simultaneously upon β-decay.

(c) γ-rays

The γ-ray is an electromagnetic wave emitted during a nuclear transition from one energy level to another; these levels are similar to the electronic energy levels of an atom or a molecule. X-rays are also a type of electromagnetic wave. In practice, electromagnetic waves from the nucleus are called γ-rays. The energy of γ-rays is expressed as $E = h\nu$, where h is Planck's constant, 6.6×10^{-34} J•s, and ν is the frequency of the electromagnetic wave in s^{-1}. For example, Co-60, which is the most commonly used RI in radiation applications, emits γ-rays having energies of 1.17 and 1.33 MeV, and has a half-life of 5.27 years.

(d) Neutron beams

Some radioactive nuclides such as californium-252 (Cf-252) emit neutrons spontaneously. Neutrons are also emitted during nuclear fission reactions. Neutrons have no electrical charge, so their interaction with matter differs from that of α-, β-, γ-, and X-rays.

3. Radiation from accelerators

An accelerator is an apparatus for accelerating a beam of electrons or ions and increasing their energy. Historically it was recognized as an apparatus in the field of physics, but, with increasing use of accelerators in radiation science and technology and radiation medicine, a wide variety of accelerators is currently in use. An accelerator is not a special apparatus; a (traditional) television monitor is an accelerator. The Brown tube in a television monitor has an electron gun behind it, and electrons called thermoelectrons are emitted by a heating filament. The electric potential of the electron gun is set 26 kV lower than that of the surface of the Brown tube, and electrons are accelerated toward the surface because they have a negative electronic charge. When the electrons reach the surface, their energy is 26 keV, and this energy is dispersed in the emission of light from the fluorescence plate when the electrons collide with its surface. The tracks of the electrons are controlled by a scanning magnet to obtain the image.

The electron beam accelerators used in industry also have an electron gun and an acceleration tube. The electron beam is extracted from the acceleration tube, the inside of which is under vacuum, through a thin film (roughly dozens of micrometers thick) of titanium or stainless steel, and the materials to be irradiated are placed under air. Many types of industrial accelerators are available, but in principle, they are the same as the television monitor. In addition to electron beams, ion beams have recently become very widely used; they are produced by ion beam accelerators, which are based on the same principal as the electron beam accelerator.

4. Radiation from nuclear reactors

Currently, it may not be typical to use a nuclear reactor as a radiation source for applications. In the early days of radiation chemistry, research on radiation chemistry was conducted, and radiation applications were tested using neutron beams from nuclear reactors. Today, however, as described in later chapters, some applications, such as neutron activation analysis in scientific fields and boron neutron capture therapy in medicine, are observed to benefit from the use of a nuclear reactor. The characteristic is that much higher intense flux (the number of neutrons per unit cross section per unit time) is available compared to RIs.

Exercises

1. Describe the scheme of radiation chemistry.
2. Consult the economic scale of radiation application.

Further Readings

On Radiation Chemistry

1. Shida S (1960) Houshasen kagaku. Nikkan kogyo shinbunsha, Tokyo (in Japanese); Shida S (1960) Radiation chemistry. Nikkan Kogyo Shimbun Ltd., Tokyo (in Japanese)
2. Kuri Z (1975) Houshasen kagaku. Kyoritsu shuppan, Tokyo (in Japanese); Kuri Z (1975) Radiation chemistry. Kyoritsu Shuppan Co., Ltd., Tokyo (in Japanese)
3. Tabata Y (1978) Houshasen kagaku. Tokyo daigaku shuppankai, Tokyo (in Japanese); Tabata Y (1978) Radiation chemistry. University of Tokyo Press, Tokyo (in Japanese)
4. Kondo M (1980) Shinozaki Yoshiharu, Houshasen kagaku, Korona sha, Tokyo (in Japanese); Kondo M (1980) Shinozaki Yoshiharu, Radiation chemistry, CORONA PUBLISHING CO., LTD., Tokyo (in Japanese)

5. Makuuchi K (2000) Porima no houshasen kako. Raba daijesuto sha (in Japanese); Makuuchi K (2000) Radiation processing of polymers. Rubber-Digest Inc. (in Japanese)
6. Houshasen kagakukai (hen) (2006) Houshasen kagaku no susume. Gakkai shuppan senta (in Japanese); Japanese Society of Radiation Chemistry (ed) (2006) Invitation to radiation chemistry. Gakkai Shuppan Center (in Japanese)

On Radiation Application

7. Zaidanhojin Genshiryoku bunka sinko zaidan (hen) (1998) Ett konna tokoro ni houshasen. Zaidanhojin Genshiryoku bunka sinko zaidan, Tokyo (in Japanese); Japan Atomic Energy Relations Organization (ed) (1998) Oh, here is radiation. Japan Atomic Energy Relations Organization, Tokyo (in Japanese)
8. "Genshiryoku no subete" henshu iinkai (hen) (2003) Genshiryoku no subete (Chapter 8). Kokuritsu insatsukyoku, Tokyo (in Japanese); Editorial Committee on "All about nuclear" (ed) (2003) All about nuclear (Chapter 8). National Printing Bureau, Tokyo (in Japanese)
9. Iwasaki T (2003) Shitte imasuka? Houshasen no riyo, Maruzen, Tokyo (in Japanese); Iwasaki T (2003) Do you know it? Radiation Application Maruzen Company, Limited, Tokyo (in Japanese)
10. Shadanhojin Nihon genshiryoku gakkai (hen) (2004) Genshiryoku ga hiraku seiki (Chapter 5). Shadanhojin Nihon genshiryoku gakkai, Tokyo (in Japanese); Atomic Energy Society of Japan (ed) (2004) Century opened by Nuclear (Chapter 5). Atomic Energy Society of Japan, Tokyo (in Japanese)
11. Iida T (2005) Senshin houshasen riyo. Osaka daigaku shuppankai, Osaka (in Japanese); Iida T (2005) Advanced radiation application. Osaka University Press, Osaka (in Japanese)
12. Tojima W (2006) Houshasen riyo no kiso chishiki. Kodan sha, Tokyo (in Japanese); Tojima W (2006) Fundamental knowledge of radiation application. Kodansha Ltd., Tokyo (in Japanese)
13. Hayashi T (hen) (2008) Shokuhin· nogyo bunya no houshasen riyo. Saiwai shobo, Tokyo (in Japanese); Hayashi T (ed) (2008) Radiation Application in food and agricultural field. Saiwai shobo, Tokyo (in Japanese)
14. Minna no kurashi to houshasen chishiki fukyu jikko iinkai (hen) (2008) Minna no kurashi to houshasen. Osaka koritsu daigaku kyodo shuppankai, Osaka (in Japanese); Executive on knowledge dissemination of "Your life and radiation" (ed) (2008) Your life and radiation. Osaka Municipal Universities Press, Osaka (in Japanese)

Data Books

15. Shadanhojin Nihon aisotopu kyokai, Houshasen riyo tokei (in Japanese); Japan Radio Isotopes Association, Statistics on the use of radiation in Japan (in Japanese)
16. Zaidanhojin Nihon genshiryoku sangyo kyokai, Genshiryoku nenkan (in Japanese); Japan Atomic Industrial Forum Inc., Almanac of Atomic Industry (in Japanese)
17. Shadanhojin Nihon denki kyokai shinbunbu, Genshiryoku poketto Bukku (in Japanese); The Japan Electric Association News Paper Division, Nuclear Pockets Book (in Japanese)

Chapter 2
Interactions Between Radiation and Matter

Yosuke Katsumura and Hisaaki Kudo

2.1 Introduction

Radiation deposits energy in water and induces ionization, excitation, and reactions. Energy is deposited through interactions between radiation and matter; the larger the interaction is, the more energy is deposited and vice versa. The interaction differs depending on the radiation and is classified roughly into three cases: high-energy electromagnetic waves such as γ-rays and X-rays, charged particles including electrons and ions, and neutrons.

2.2 γ-Rays and X-Rays

Both are electromagnetic waves and can be treated as photons. Here, the interactions are described from macroscopic and microscopic viewpoints.

2.2.1 Macroscopic Behavior

The intensity of γ-rays and X-rays is reduced (attenuated) in matter; because they are electromagnetic waves, their absorption in matter is similar to that of ultraviolet rays and visible light. Let us consider X-rays of intensity I_0 penetrating matter of thickness x. When the intensity of the light after transmission is I, the following relationship holds:

Y. Katsumura · H. Kudo (✉)
The University of Tokyo, Tokai, Ibaraki, Japan
e-mail: hkudo@tokai.t.u-tokyo.ac.jp

H. Kudo (ed.), *Radiation Applications*, An Advanced Course in Nuclear Engineering 7, https://doi.org/10.1007/978-981-10-7350-2_2

$$I = I_0 \exp(-\mu x) \tag{2.1}$$

The intensity decays exponentially as a function of the thickness x, as in the absorption of light (see Sect. 5.2), because $\exp(-\mu x)$ can be written as $\exp(-\mu x) = 10^{-(\mu x/2.303)}$, where μ is the absorption coefficient and depends on the energy of the X-rays and the material. If x is measured in centimeters, μ has units of cm^{-1}. A more common formula is

$$I = I_0 \exp\left[-\left(\frac{\mu}{\rho}\right)(\rho x)\right] \tag{2.2}$$

where ρ is the destiny ($g{\cdot}cm^{-3}$) and μ/ρ is the mass absorption coefficient. The mass absorption coefficients have been tabulated. The units of ρx, which is the mass of the medium within a thickness x of cross section 1 cm^2, are $g{\cdot}cm^{-2}$. As described later, radiation interacts with material through electrons as described by the absorption coefficient, which is sometimes written as cm^2/electron (when considering the value per electron) or in cm^2/atom (when considering the value per atom).

The absorption coefficients for simple elements or typical compounds have been tabulated; for example, the International Commission on Radiation Measurements and Units tabulates them with their energy dependence. Even if the material of interest is not listed and only the elemental composition of the material is known, the mass absorption coefficient $(\mu/e)_T$ can be obtained as the summation of $(\mu/e)_i$, each multiplied by the weight fraction W_i for the ith element:

$$\left(\frac{\mu}{\rho}\right)_T = \sum_i \left(\frac{\mu}{\rho}\right)_i W_i \tag{2.3}$$

The simple exponential decay described above occurs for collimated (spatially aligned) X-rays. The degree of collimation is given by a simple exponent. In actual cases, not only collimated X-rays but also “radiative” radiation or curtain-shaped beam can exist, where scattering from a nonlinear geometry must be considered. In such cases, the build-up factor B must be considered as follows:

$$I = I_0 B \exp\left[-\left(\frac{\mu}{\rho}\right)(\rho x)\right] \tag{2.4}$$

This consideration is important in designing the shielding in radiation facilities. Readers should refer to “Radiation Shielding” in this textbook series.

2.2.2 *Microscopic Behavior*

The matter–radiation interaction can also be described microscopically. The dominant process varies depending on the energy of the X- and γ-rays. The

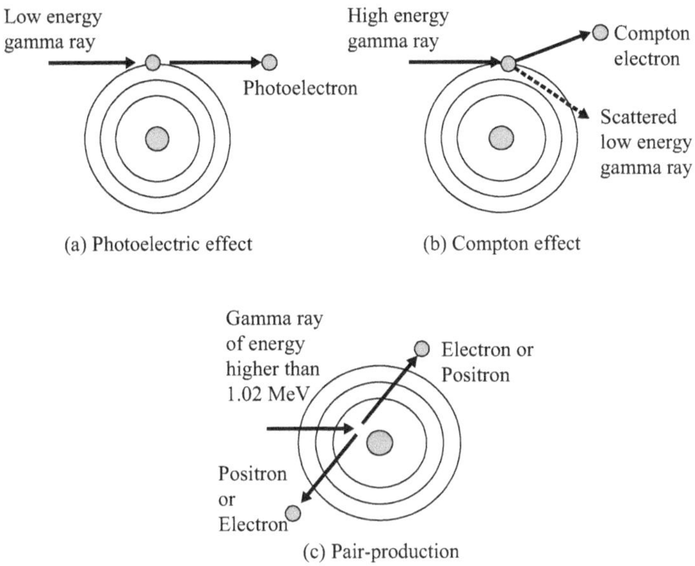

Fig. 2.1 Interaction between photon and material

photoelectric effect is dominant in the lowest region (less than hundreds of kiloelectron volts), followed by the Compton effect, which is dominant at hundreds of kiloelectron volts to several megaelectron volts. Pair production starts at 1.022 MeV and dominates above several megaelectron volts. These three processes are schematically shown in Fig. 2.1.

Photoelectric Effect

X-rays and γ-rays are photons, and their energy is expressed as $h\nu$, where h is Planck's constant, 6.63×10^{-34} J·s, and ν is the frequency in s^{-1}. Under electron bombardment in a medium, a low-energy photon emits an electron, which is called a photoelectron. Because electrons are bonded to atoms, the kinetic energy of the emitted electron, E_k, is the different between $h\nu$ and the binding energy I_B, i.e.:

$$E_k = h\nu - I_B = h\nu - \phi \quad (2.5)$$

where ϕ is the work function necessary for an electron to be emitted from the surface of the medium. This process experimentally underlies the photoquantum hypothesis presented by A. Einstein in 1905.

An example of the application of the photoelectric effect is the photomultiplier. It uses a material having a low work function and emits photoelectrons. The photoelectrons are accelerated by a high voltage and collide with another material called the dynode, which then emits multiple electrons; this multiplies the electron into many electrons. It is difficult to detect the charge of an electron emitted by a photon. However, multiplying the electron makes it possible to detect the electrons emitted by a photon. This method is indispensable to light detection in photochemistry and related fields; it is known as a single photon counting method, which detects a photon as a pulse-shaped electronic signal.

Another example of the application of the photoelectric effect to material analysis is X-ray photoelectron spectroscopy. The material is exposed to monoenergetic X-rays (e.g., the characteristic X-rays of Cu at an energy of 8.9 keV or those of Mg at 1.3 keV) and emits photoelectrons; the energy of the photoelectrons is the difference between the binding energy and that of the irradiated X-rays. Consequently, analyzing the energy of the photoelectrons makes it possible to obtain the binding energy and identify the elements in the target material. The photoelectron energy varies slightly depending on the electrons' environment. This is called the chemical shift, and it enables the chemical status (such as the oxidation status) to be obtained. This is a powerful tool in material sciences.

Compton Effect

When the energy of the incident X-rays is increased, the Compton effect becomes dominant. In Compton scattering, a photon collides with an electron (assumed to be a free electron), and the photon loses energy and is simultaneously scattered (where the photon's energy becomes $h\nu'$). An electron having an energy of E_e is emitted.

As shown in Fig. 2.2, a photon having energy $h\nu$ comes from the left, collides with an electron, and is scattered in the direction θ; its energy becomes $h\nu'$. The electron is scattered in the direction ϕ, and its energy is E_e. The following is obtained:

$$h\nu' = \frac{h\nu}{\{1 + (h\nu/m_0c^2)(1 - \cos\theta)\}} \tag{2.6}$$

$$E_e = h\nu\frac{(h\nu/m_0c^2)(1 - \cos\theta)}{\{1 + (h\nu)/(m_0c^2)(1 - \cos\theta)\}} \tag{2.7}$$

where m_0c^2 is the static mass of an electron and equals 0.511 MeV. When $\cos\theta = -1$, $\theta = 180°$, $h\nu'$ is minimum and E_e is maximum:

$$h\nu'_{\text{min}} = \frac{h\nu}{1 + 2(h\nu)/(m_0c^2)} \tag{2.8}$$

$$E_{\text{max}} = \frac{2h\nu(h\nu/m_0c^2)}{1 + 2(h\nu)/(m_0c^2)} \tag{2.9}$$

The energies of γ-rays from Co-60 are 1.33 and 1.17 MeV. The minimum $h\nu'$ values are 0.22 and 0.21 MeV, and the corresponding energies of the recoiled

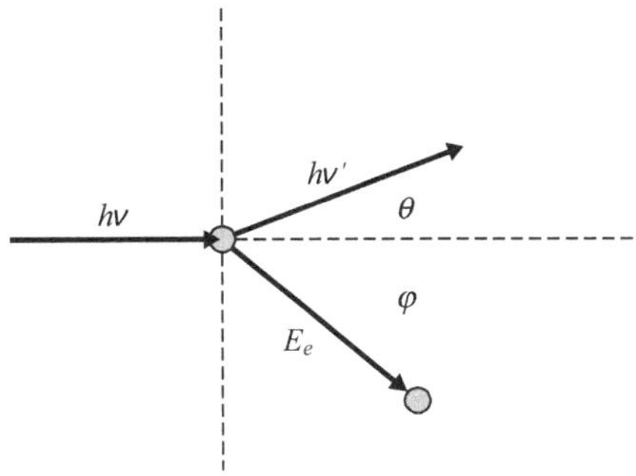

Fig. 2.2 Compton scattering

electrons are 1.11 and 0.96 MeV, respectively. Klein and Nishina calculated the magnitude of the interaction (the cross section) of the Compton effect for an electron. Readers should refer to "Nucleus and Radiation Detection" in this series.

Pair Production

If the energy of the incident γ-rays exceeds $2m_0c^2 = 1.022$ MeV, the γ-rays are annihilated in the electric field of the nucleus, and an electron/positron (antielectron) pair is produced. This process is called pair production, and it satisfies the following relation:

$$h\nu - 2m_0c^2 = E^- + E^+$$

where E^- and E^+ are the energies of the produced electron and positron, respectively. E^- and E^+ can vary from 0 to $h\nu$-$2m_0c^2$. The cross section (probability) has a threshold energy of 1.022 MeV for ions coming from γ-rays, and with further energy it increases gradually, eventually becoming constant. The positron gives excess energy to the medium, reacts with the electron, and is annihilated. The annihilation yields two γ-rays having energies of 0.511 MeV each. They are emitted in opposite directions. This phenomenon is called positron annihilation and is applied in medicine, in particular for diagnosis as the positron in positron emission tomography, as described later.

As mentioned above, when a γ-ray enters a medium, it causes the photoelectric effect, Compton effect, or pair production, depending on γ-ray's energy. As a result, electrons (secondary electrons) are generated. Subsequently, these electrons interact with matter physically, chemically, and biologically. Considering this fact, γ-ray irradiation is equivalent to electron beam irradiation, as a rough approximation, and yields identical effects.

2.3 Electron and Charged Particles

When charged particles enter a medium, they lose energy through interactions with atoms and molecules in the medium along the track of the charged particles. The amount of energy deposited by a charged particle of energy E per unit length is called the stopping power and is indicated by dE/dx. This parameter represents the energy required to stop the charged particles from penetrating the medium. The stopping power depends on the type of charged particles, their energy, and the material of interest. The stopping power is expressed as the summation of three processes:

$$-\frac{dE}{dx} = \left(-\frac{dE}{dx}\right)_I + \left(-\frac{dE}{dx}\right)_k + \left(-\frac{dE}{dx}\right)_S \tag{2.10}$$

The first term represents ionization and excitation. When swift ions travel through a material, they interact with electrons in the material, causing ionization and excitation. The second term becomes dominant when the energy of the charged particles is low. On the other hand, the third term becomes dominant when the energy of the charged particles is high; for an electron beam, if the energy of the electron beam is 90 MeV, the bremsstrahlung fraction is 50%. For an ion beam, it is sufficient to consider the ionization and excitation represented by the first term.

The formula that expresses the ionization and excitation is called Bethe's formula, and it is given below for an electron beam:

$$\begin{aligned}\left(-\tfrac{dE}{dx}\right)_I = \frac{2\pi e^4 NZ}{m_0 v^2}\Bigg[\ln\frac{m_0 v^2}{2I^2(1-\beta^2)} - \left(2\sqrt{1-\beta^2} - 1 + \beta^2\right)\ln 2 + 1 \\ -\beta^2 + \frac{1}{8}\left(1-\sqrt{1-\beta^2}\right)^2\Bigg] \ [\mathrm{erg/cm}]\end{aligned} \tag{2.11}$$

where ν is the speed of the electron, c is the speed of light, and $\beta = \nu/c$; N, Z, and I are the number of molecules in 1 cm^3 of the material, the atomic number of the material, and the excitation potential unique to the material, respectively. I is 75 eV for liquid water. M_0 is the rest mass of an electron.

For charged particles, the following formula is used instead:

$$\begin{aligned}\left(-\tfrac{dE}{dx}\right)_I &= \frac{4\pi e^4 z^2 nZ}{m_0 v^2}\left[\ln\frac{2m_0 v^2}{I} - \beta^2 - \ln\left(1-\beta^2\right)\right] \\ &= 8\pi e^4 z^2 nZ\left(\frac{\mu}{m_0}\right)\frac{1}{E}\left[\ln\frac{4(m_0/\mu)E}{I} - \beta^2 - \ln\left(1-\beta^2\right)\right]\end{aligned} \tag{2.12}$$

where ν, z, μ, and E are the speed, charge, rest mass, and energy of the charged particles, respectively.

Table 2.1 Stopping Power of water to radiations

Radiation	Average stopping power, LET [eV/nm]
Co-60 gamma ray	0.2~0.3
1~2 MeV electron	0.2~0.3
250 kV X-ray	1
10 kV X-ray	2
2 MeV H^+	17
Polonium α-ray	90
Fission fragment (70–80 MeV)	7000

By using the stopping power, the range $R(E)$ (i.e., the depth to which particles having energy E penetrate the target medium) can be calculated as follows:

$$R(E) = \int_E^\varepsilon \frac{dx}{dE} dE + R(\varepsilon) = \int_E^\varepsilon \left(-\frac{dx}{dE}\right) dE + R(\varepsilon) \qquad (2.13)$$

where $R(\varepsilon)$ is a correction term used when the energy is very low; for typical values of E, $R(\varepsilon)$ can be assumed to be negligible.

The stopping power is the energy loss of the charged particles. If all the energy is dissipated into the medium, it represents the energy deposition density per path. This is called the linear energy transfer (LET). Bremsstrahlung is the emission of electromagnetic waves, and it is generated not only around the path. Thus, the LET is generally lower than the stopping power. The stopping power and the LET change with the energy. When the range is short, the average LET, i.e., the energy loss divided by the range, is often used. Table 2.1 shows the average stopping powers of different types of radiation in water. High-energy electrons and γ-rays have a stopping power (LET) of 0.2–0.3 eV/nm^{-1}, and low-energy X-rays have a somewhat higher stopping power. However, these values are very low compared to those of ion beams. Therefore, high-energy electrons, X-rays, and γ-rays are classified as low-LET radiation, and other types of radiation such as ion beams, mesons, neutrons, and fission products are classified as high-LET radiation. High-LET radiation yields high-energy deposition, and events that cannot be observed for low-LET radiation can occur. This is usually called the LET effect or radiation quality effect. It affects, for example, the survivability of cells and the evolution of hydrogen gas upon radiolysis of water.

2.4 Neutrons

Neutrons interact with matter through absorption (captive), elastic scattering, inelastic scattering, and nuclear reactions (e.g., charged particle production, fission). In radiation chemistry, the important interactions are elastic and inelastic scattering. When water on polymer compounds is irradiated, neutrons and atoms consisting of water or polymer compounds recoil. In particular, light elements such as hydrogen are readily scattered, and they can cause secondary irradiation effects.

The neutron reaction of nuclear spallation can be important when the energy of the neutrons is in the resonance region (where the probability of nuclear reactions increases abruptly) or when the energy of the neutron is very high.

Exercises

1. Why is the Compton effect the dominant process for the core of Co-60 gamma rays and water?
2. Draw a figure of charge in stopping power of a carbon ion of 6 Mev per nuclear (expressed as 6 MeV/u, where u means a nuclear) subjected to water.

Further Readings

1. Iida H (hen) Houshasen gairon. Tsusho sangyo kenkyu sha, Tokyo (in Japanese); Iida H (ed) Introduction of radiation. TSUSHO-SANGYO-KENKYU-SYA, Tokyo (in Japanese)
2. Ito N (1981) Houshasen sonsho. Morikita shuppan, Tokyo (in Japanese); Ito N (1981) Radiation damage. Morikita Publishing Co., Ltd, Tokyo (in Japanese)
3. Ishino S (1979) Shosha sonsho. Tokyo daigaku shuppankai, Tokyo (in Japanese); Ishino S (1979) Radiation damage. University of Tokyo Press, Tokyo (in Japanese)
4. Sekiguchi A (1979) Houshasen keisoku gairon. Tokyo daigaku shuppankai, Tokyo (in Japanese); Sekiguchi A (1979) Introduction to radiation measurements. University of Tokyo Press, Tokyo (in Japanese)
5. Kosako T (hen) (2010), Houshasen shahei. Omu sha, Tokyo (in Japanese); Kosako T (ed) (2010) Radiation shielding. Ohmsha, Ltd., Tokyo (in Japanese)
6. Ziegler JF, Biersack JP (2008) Ziegler M D. SRIM-The stopping and Range of Ions in Matter, SRIM Co

Data Books

7. ICRU Report 37 (1984) Stopping Powers for Electrons and Positrons
8. ICRU Report 49 (1993) Stopping Powers and Ranges of Protons and Alpha Particles
9. ICRU Report 73 (2005) Stopping of Ions Heavier than Helium

Chapter 3
Dosimetry

Yosuke Katsumura and Hisaaki Kudo

3.1 Dosimetry: Measuring the Quantity of the Dose

The quantity of radiation is called the radiation dose. It is important to evaluate the radiation dose not only for safe handling of radiation but also for quality control in radiation therapy and radiation sterilization. This chapter describes the types, definition, measurements, and methods of radiation dosimetry.

3.2 Three Types of Dose

In practice, the frequently used types of radiation dose are the exposure dose, absorbed dose, and dose equivalent.

1. Exposure dose

The common characteristic of radiation is that it causes ionization in matter. The exposure dose is determined from the quantity of ionization (the number of ion pairs) in air; 1 R (roentgen) is defined as producing 2.58×10^{-4} C (1 cgs electrostatic unit) in 1 cc of dried air at 0 °C and 1 atm. This effectively expresses the radiation level in the environment, as the exposure dose uses air as the standard. The exposure dose is typically used for X-rays and γ-rays. Because electrons and ion beams exhibit low penetration, the exposure dose varies by positions, and it may be difficult and misleading to use the exposure dose for electrons and ion beams.

Y. Katsumura · H. Kudo (✉)
The University of Tokyo, Tokai, Ibaraki, Japan
e-mail: hkudo@tokai.t.u-tokyo.ac.jp

H. Kudo (ed.), *Radiation Applications*, An Advanced Course in Nuclear Engineering 7, https://doi.org/10.1007/978-981-10-7350-2_3

Table 3.1 W value of various gases

Gas	W value/γ-ray or electron beam	W value/α-ray	Gas	W value/γ-ray or electron beam	W value/α-ray
He	41.5	46.0	CO_2	32.9	34.1
Ne	36.2	35.7	H_2O	30.1	37.6
Ar	26.2	26.3	NH_3	35	–
Kr	24.3	24.0	CCl_4	25.3	26.7
Xe	21.9	22.8	$CHCl_3$	26.1	–
H_2	36.6	36.2	SF_6	34.9	35.7
N_2	34.6	36.39	CH_4	27.3	29.1
O_2	31.8	32.3	C_2H_2	25.7	27.3
Air	33.73	34.98	C_2H_4	26.3	28.03
Cl_2	23.6	25.0	C_2H_6	24.6	26.6
Br_2	27.9	–	C_3H_8	27.8	–
HCl	24.8	27.0	C_4H_{10}	–	24.8
HBr	24.4	27.0	C_6H_{14}	22.4	–

2. Absorbed dose

The absorbed dose expresses the amount of radiation regardless of the material, and 1 Gy (gray) means that 1 J of energy is dissipated in 1 kg of material. This amount is based on the unit mass, and it is not specific to the material, e.g., water or air. The unit of the rad (an abbreviation for "radiation-absorbed dose") was used previously; it was defined as 100 $erg{\cdot}g^{-1}$. Note that $1\ Gy = 1\ J{\cdot}kg^{-1} = 10^7\ erg{\cdot}kg^{-1} = 10^4\ erg{\cdot}g^{-1} = 100$ rad.

An exposure dose of 1 R is defined with respect to air; here we consider the absorbed dose. The energy necessary to produce one non-pair (a cation and an anion) is called the W value and is 33.7 eV in air. From the definition of 1 R, the total amount of ion pairs is $2.58 \times 10^{-4}/1.6 \times 10^{-19} = 1.61 \times 10^{15}$ pairs. To produce a pair in air, 33.7 eV is necessary; thus, $[(1.61 \times 10^{15}) \times 33.7\ eV] \times 1.6 \times 10^{-19}\ J{\cdot}eV^{-1} = 0.0087$ J. Table 3.1 shows the W values for many gases, including air.

3. Equivalent dose

This dose is used for radiation protection and is obtained as the product of the absorbed dose and a radiation weighting factor that depends on the type (quality) of radiation and its biological effect. Namely, the equivalent dose is defined as the radiation weighting factor times the absorbed dose; the unit is the sievert (Sv). When the unit of rads was used for the absorbed dose, the unit of rem (an abbreviation for "radiation equivalent to man") was used for the equivalent dose. The radiation weighting factors recommended by the International Commission on Radiological Protection are shown in Table 3.2.

Table 3.2 Radiation weighting factor

Type and energy of radiation	Radiation weighting factor
Photon	1
Electron, μ-meson	1
Neutron E < 10 keV	5
10 keV < E < 100 keV	10
100 keV < E < 2 MeV	20
2 MeV < E < 20 MeV	10
20 MeV < E	5
Proton (E < 2 MeV)	5
Fission fragment, heavy nucleus	20

3.3 W Value and Ionization Potential

As mentioned previously, the w value is the energy necessary to ionize gaseous matter by absorbing energy from radiation. The ionization potential (IP) is a similar parameter. It is the minimum energy necessary to ionize an atom or a molecule of interest; the values for various materials are listed in Table 3.3. A comparison of the w values and the IPs reveals that the w value is always higher than the IP. Both parameters have to do with ionization. However, the IP is the minimum energy required for ionization, whereas the W value includes other processes in addition to ionization. In other words, when the radiation energy is absorbed by atoms in molecules, the energy is used not only for ionization but also for excitation. Consequently, when an ion pair is produced, the total energy necessary is the w value.

3.4 Dose Measurement and Method

Evaluation of the dose is called dose measurement, and the apparatuses and devices used for dose measurement are called dosimeters.

Ionization chambers are widely used to evaluate the exposure dose; they do this by collecting the electric charges produced in a gas in a vessel under a radiation environment and determining the amount of ionization. The proportional tube and Geiger–Müller tube, which both operate on a principle similar to that of the ionization chamber, are also available.

On the other hand, the absorbed dose can be determined if the amount of deposited energy per unit mass is known. The direct application is a calorimeter. If all the deposited energy can be converted to thermal energy, the energy appears as a temperature increase. From this temperature increase, one can determine the absorbed dose. The calorimeter is efficient for a high dose and high dose rate, but

Table 3.3 Ionization potential

Chemical species	IP [eV]	Chemical species	IP [eV]
H	13.60	HCl	10.4
D	13.60	HBr	11.6
He	24.59	CO_2	13.8
Ne	21.56	NH_3	10.2
Ar	15.76	CH_4	12.7
Kr	14.00	C_2H_2	11.4
Xe	12.13	C_2H_4	10.5
H_2	15.4	C_2H_6	11.5
D_2	15.5	C_3H_8	11.1
N_2	15.6	C_4H_{10}	10.6
O_2	12.1	c-C_6H_{12}	9.9

its sensitivity is generally low. Irradiation can cause various effects that can be clearly correlated one to one with the absorbed dose; thus, one can determine the dose from the degree of radiation-induced changes. The irradiation effects that can be used are (1) oxidation and reduction reactions in aqueous solution (e.g., the Fricke dosimeter and ceric-cerous dosimeter), (2) coloration [the Radcolor Film™, Co-glass dosimeter, and poly(methyl)methacrylate dosimeter], (3) fluorescence [the thermoluminescence dosimeter (TLD)], (4) discoloration (blue cellophane), (5) radical formation (alanine, Teflon™), and (6) biological effects (the index virus).

Coloration and discoloration exploit changes in the color of materials upon irradiation toward darker (coloration) or lighter (discoloration) colors. For example, a TLD consisting of crystallites such as LiF and $Li_2B_4O_7$ forms an electron and a hole (positive ion) through ionization upon irradiation. At room temperature, some of the formed electrons and holes are trapped in the crystal, and the absorbed energy is accumulated. When this material is heated, the trapped electrons and holes become mobile and recombine, emitting light. Radicals, which are described in a later chapter, are formed by dissociation of bonds inside a molecule; they are reactive and unstable. However, in alanine (a type of protein) and Teflon™ (a type of polymer material), radicals are rather stable over the long term, and the correlation between the radical concentration and absorbed dose can be obtained. The ionization chamber and calorimeter directly observe the dose as it is defined, and they can be described as primary dosimeters. Those that correlate the radiation effects to the absorbed dose can be described as secondary dosimeters. Dosimeters can be characterized in terms of the coverage dose range, usable dose rate accuracy, reproducibility, procedure, stability, operability, and cost. Dosimeters are selected according to the intended usage.

The dose is an integrated quantity, and the dose per unit time is called the dose rate, in the units of, e.g., $Gy{\cdot}s^{-1}$ or $Sv{\cdot}year^{-1}$.

3.5 Chemical Dosimeters (Fricke Dosimeter and Ceric-Cerous Dosimeter)

Dosimeters based on oxidation and reduction reactions are called chemical dosimeters. Many chemical dosimeters have been proposed or developed. Among them, the most widely used and established ones for which the mechanisms have been studied are the Fricke dosimeter and ceric-cerous dosimeter.

First, the G value, which is indispensable to quantitative discussion in radiation chemistry, is described. As a result of radiation-induced chemical reactions, molecules and ion may be consumed or converted, decreasing their concentrations. On the other hand, other new molecules and ions may be formed. The number of decreases or formations per energy absorption of 100 eV is defined as the G value. The recommended unit recently became mole·J^{-1}, and the G value is no longer recommended. Conversion of the ordinary G value to the recommended unit should be considered.

One joule is $(1.60 \times 10^{-19})^{-1}$ eV/100 eV = 6.25×10^{16}; to convert this to a molar concentration, it is divided by Avogadro's number, 6.02×10^{23}, which yields 1.037×10^{-7} mol. Therefore, the ordinary G value of 1 is 1.037×10^{-7} mol in the new units. In this textbook, the ordinary unit system is adapted.

Further, it is necessary to know the effect of radiolysis of water in acid aqueous solutions to discuss the G(Fe^{3+}) value of the Fricke dosimeter. Water radiolysis can generally be written as follows (it is discussed in more detail in another chapter):

$$H_2O \rightsquigarrow e^-_{aq}, \bullet H, \bullet OH, H_2, H_2O_2 \tag{3.1}$$

where e^-_{aq} is called a hydrated electron and •H and •OH are the hydrogen atom radical and hydroxyl radical, respectively. In an acid aqueous solution, a proton (H_3O^+) reacts with a hydrated electron to form a hydrogen atom:

$$e^-_{aq} + H_3O^+ \rightarrow H + H_2O \tag{3.2}$$

Namely, in an acid aqueous solution, it is difficult to distinguish the yields of hydrated electrons and hydrogen atoms because hydrated electrons are converted to hydrogen atoms. If oxygen is present, a hydrogen atom reacts with oxygen to form molecules: $\bullet H + O_2 \rightarrow \bullet HO_2$. Table 3.4 shows the G values of the radiolysis products of water in acid aqueous solutions.

3.5.1 *Fricke Dosimeter*

This dosimeter is named for H. Fricke of the U.K., who proposed this system and also studied the effect of radiation on iron ions in aqueous solution and on hemoglobin. The dosimeter consists of a 0.8 N (normal) H_2SO_4 aqueous solution

Table 3.4 Water decomposition in acid aqueous solution

Product	e^-_{aq} + •H	•OH	H_2	H_2O_2
G value	3.7	2.9	0.4	0.8

dissolving 1 mM Fe^{2+} ($M = mol{\cdot}l^{-1}$); because of its composition, it is also called an iron dosimeter. This system undergoes the oxidation reaction $Fe^{2+} \rightarrow Fe^{3+}$ upon irradiation. The absorbed dose is determined from the degree of this conversion. Below are the reactions that occur in acid aqueous solutions containing Fe^{2+}:

$$Fe^{2+} + \bullet OH \rightarrow Fe^{3+} + OH^- \quad (3.3)$$

$$Fe^{2+} + \bullet H + H^+ \rightarrow \left(FeH_2^{3+}\right) Fe^{3+} + H_2 \quad (3.4)$$

$$Fe^{2+} + \bullet HO_2 \rightarrow Fe^{3+} + HO_2^-, HO_2^- + H^+ \rightarrow H_2O_2 \quad (3.5)$$

$$Fe^{2+} + H_2O_2 \rightarrow Fe^{3+} + \bullet OH + OH^- \quad (3.6)$$

Note that if oxygen is absent, the hydrogen atom radical (•H) remains unchanged, but if oxygen is present, •H is converted to $\bullet HO_2$. The G(Fe^{3+}) value of the Fricke dosimeter is considered differently in aerated (air- or oxygen-saturated) and deaerated cases.

1. G value in the presence of oxygen (aerated): $G(Fe^{3+})_A$

In an acid aqueous solution, hydrated electrons are covered to hydrogen atoms, which react with O_2 to form $\bullet HO_2$. Reaction (3.5) oxidizes Fe^{2+} once. As a product, H_2O_2 is formed, and the H_2O_2 oxidizes Fe^{2+} through reaction (3.6) and also forms •OH. The •OH oxidizes Fe^{2+} through reaction (3.3). Therefore, the hydrated electron and hydrogen atom oxidize Fe^{2+} three times. H_2O_2 oxidizes Fe^{2+} twice through (3.6) and (3.3). The •OH oxidizes Fe^{2+} once through reaction (3.3). Considering these reactions, the total Fe^{3+} formed is given by

$$G\left(Fe^{3+}\right)_A = 3\left(G_H + G_{e^-_{aq}}\right) + G_{OH} + 2G_{H_2O_2} \quad (3.7)$$

2. G value in the absence of oxygen (deaerated): $G(Fe^{3+})_D$

In the absence of oxygen, hydrogen atoms oxidize Fe^{2+} only once through reaction (3.4). Therefore, the G value is obtained as follows:

$$G\left(Fe^{3+}\right)_D = \left(G_H + G_{e^-_{aq}}\right) + G_{OH} + 2G_{H_2O_2} \quad (3.8)$$

By using the G values of the radiolysis products of water in acid solution in Table 3.4, one can obtain $G(Fe^{3+})_A = 15.6$ and $G(Fe^{3+})_D = 0.2$. When one determines the absorbed dose, it is necessary to measure the Fe^{3+} concentration. The molar extinction coefficient ε at 304, 2164 M^{-1} cm^{-1}, is used. A spectroscopic method using the Lambert–Beer law is adopted: Absorbance $= \log^{10} I_0/I = \varepsilon cl$, where I_0 and I are the intensity of the incoming and outgoing light, respectively, and

l is the optical path length of the quartz (glass) cell containing the aqueous solution. The Fricke dosimeter is well known to be usable in the dose range of dozens to several hundreds of grays under a low dose rate source such as γ-rays but unsuitable for high dose rate irradiation such as electron beam pulses.

3.5.2 *Cerium Dosimeter (Ceric-Cerous Dosimeter)*

The cerium dosimeter is also a widely used aqueous solution dosimeter. It consists of 0.8 N H_2SO_4 dissolving $Ce(SO_4)_2$ (0.5 mM) and $Ce_2(SO_3)_3$ (0.5 mM). This acid aqueous solution is similar to the Fricke solution. Upon irradiation, the reduction reaction $Ce^{4+} \rightarrow Ce^{3+}$ occurs. The reaction mechanism is as follows:

$$Ce^{3+} + \bullet OH \rightarrow Ce_{4+} + OH^- \tag{3.9}$$
$$Ce^{4+} + \bullet H \rightarrow Ce^{3+} + H^+ \tag{3.10}$$
$$Ce^{4+} + \bullet HO_2 \rightarrow Ce^{3+} + H^+ + O_2 \tag{3.11}$$
$$Ce^{4+} + H_2O_2 \rightarrow Ce^{3+} + \bullet HO_2 + H^+ \tag{3.12}$$

The •OH radical oxidizes Ce^{3+}, and •H, •HO_2, and H_2O_2 reduce Ce^{4+}. Note that the reductions occur regardless of the presence or absence of oxygen, which is different from the Fricke dosimeter. By using the G values of the radiolysis products of water, $G(Ce^{3+}) = 2.5$ is obtained.

The cerium dosimeter is usable in the wide dose range of $1–10^6$ Gy. It is widely used in industrial radiation applications because of its accuracy. The purity of the water used affects the reproducibility of measurements by Fricke or cerium dosimeters. Ultrapure water and well-washed glassware are necessary. The use of triple-distilled water and quartz glassware used to be recommended. Currently, however, apparatuses for obtaining water of equivalent purity are available.

Other dosimetry systems have been proposed, but they are less useful than the Fricke or cerium dosimeter.

Exercises

1. Aqueous solution of density of 1(g/cm^2) as irradiated to absorbed dose of D [Gy], and the concentration of product is C[M]. Obtain the formula to express the G value of the product. Draw a figure of charge in stopping power of a carbon ion of 6 Mev per nuclear (expressed as 6 MeV/u, where u means a nuclear) subjected to water.
2. Derive the equivalent equations for the case of ceric-cerous dosimeter.

Further Readings

1. ISO/ASTM standard 51205 (2009) Practice for use of a ceric-cerous sulfate dosimetry system
2. ISO/ASTM standard 51401 (2003) Practice for use of a dichromate dosimetry system

Chapter 4
Radicals and ESR

Yosuke Katsumura and Hisaaki Kudo

4.1 Free Radicals

Free radicals, or simply radicals, are atoms having unpaired electrons, which can form bonds. For example, when the hydrogen and oxygen forming a water molecule are dissociated, •O and •OH are formed, and they are called the hydrogen atom radical and hydroxyl radical, respectively. The dissociated bond was a covalent bond, in which an electron was donated and received. In the radical, an electron is localized to either fragment. To emphasize the unpaired electron, a minus sign is added on the right or left side of the chemical symbol. Radicals play an important role in chemical reactions.

4.2 The Production of Radicals

4.2.1 Pyrolysis

Some compounds decompose at elevated temperature in a process called pyrolysis. For example, iodinate forms a molecule from two iodinate atoms and is decomposed at 700 °C as follows:

$$I_2 \leftrightarrow 2 \bullet I$$

When •I becomes detectable, equilibrium is reached between I_2 and •I. Above 1700 °C, this equilibrium shifts toward the right.

Y. Katsumura · H. Kudo (✉)
The University of Tokyo, Tokai, Ibaraki, Japan
e-mail: hkudo@tokai.t.u-tokyo.ac.jp

H. Kudo (ed.), *Radiation Applications*, An Advanced Course in Nuclear Engineering 7, https://doi.org/10.1007/978-981-10-7350-2_4

As another example, di-tert-butyl peroxide decomposes at a relatively low temperature of 120–200 °C and is often used as a radical source.

$$(CH_3)_3CO\text{-}OC(CH_3)_3 \rightarrow 2\,(CH_3)_3CO\bullet$$
$$\rightarrow 2\bullet CH_3 + 2\,(CH_3)_2CO$$

Both O–O bonds dissociate, and the tert-butoxy radical [$(CO_3)_3CO\bullet$] is formed. At higher temperatures, the tert-butoxy radical decomposes further into the methyl radical ($\bullet CH_3$) and acetone [$(CH_3)_2CO$].

Normal saturated hydrocarbon (n-paraffin) has the form $CH_3CH_2CH_2\bullet\bullet\bullet CH_2CH_3$; carbon atoms are combined linearly, and each carbon atom is combined with two hydrogen atoms. The bonds between atoms decompose above 350 °C, and many types of radicals having unpaired electrons at different carbon positions are formed. This radical reaction process is used to purify petroleum.

4.2.2 *Photolysis*

Molecules absorb light with wavelengths in the ultraviolet region. Ultraviolet light absorption dissociates chemical bonds and forms radicals. For example, hydrogen peroxide (H_2O_2) absorbs light at wavelengths shorter than 300 nm, forming two $\bullet OH$ radicals.

$$H_2O_2 + h\nu \rightarrow 2\bullet OH$$

The HO–OH bonds are dissociated. This process is often used to form the $\bullet OH$ radical. Di-tert-butyl peroxide also exhibits a reaction similar to pyrolysis upon light absorption. The chemical reaction triggered by light absorption is called a photoreaction; it plays an important role in reactions in the atmosphere and dominates the formation and behavior of atmospheric pollutants.

4.2.3 *Oxidation and Reduction*

Radical formation is often observed upon oxidation and reduction. For example, hydroperoxide in acidic aqueous solution reacts with Fe^{2+}, forming the $\bullet OH$ radical.

$$H_2O_2 + Fe^{2+} \rightarrow Fe^{3+} + \bullet OH + H^-$$

Fe^{2+} is oxidized to Fe^{3+}. This is called the Fenton reaction and is a well-known $\bullet OH$ radical formation procedure.

4.2.4 *Radiolysis*

When radiation energy is absorbed by molecules, because the energy dissociates bonds, radicals are generally formed.

$$H_2O \rightsquigarrow \bullet H + \bullet OH$$
$$CH_4 \rightsquigarrow \bullet H + \bullet CH_3$$

4.3 Characteristics and Reactions of Radicals

4.3.1 *Scission and Rearrangement*

Radicals are unstable and are decomposed or rearranged into more stable formations.

$$\begin{array}{lll} CH_3CH_2CH_2CH_2 & \rightarrow CH_3CH_2CH = CH_2 + \bullet H & \\ & \rightarrow CH_2 = CH_2 + \bullet C_2H_5 & \text{(scission reaction)} \\ (C_6H_5)_3CCH_2 & \rightarrow (C_6H_5)_2C - CH_2(C_6H_5) & \text{(rearrangement reaction)} \end{array}$$

where C_6H_5 is the phenyl group and the benzene ring group is moved (rearranged) in a molecule toward a more stable formation.

4.3.2 *Addition to Double Bond*

Electrons are concentrated around a double bond, which has two conjugating bonds. The electrons of the double bond and unpaired electrons in radicals can form a new bond, resulting in the formation of a new high-molecular-weight radical.

$$\bullet OH + H_2C = CH_2 \rightarrow HOCH_2\text{-}CH_2\bullet$$

If an ethylene molecule exists around this radical, the radical reacts with the surrounding ethylene molecules, and the carbon atom branch develops.

$$HOCH_2 - CH_2\bullet + H_2C = CH_2 \rightarrow HOCH_2 - CH_2 - CH_2 - CH_2\bullet$$

This molecule reacts with another ethylene. If this reaction continues consecutively, a macromolecule having a long carbon chain develops. This is called the polymerization reaction. Further, if a radical is present, the process is called the radical polymerization reaction.

Another example of a chain reaction is the reactions of chlorine atoms formed by decomposition of chlorine gas with organics, which is called chlorination.

$$Cl_2 \rightarrow 2 \bullet Cl$$
$$RH + \bullet Cl \rightarrow R \bullet + HCl$$
$$Cl_2 + R \bullet \rightarrow \bullet Cl + RCl$$

Hydrogen radicals react with the benzene ring and form cyclohexadienyl radicals, which have two double bonds in a hexagonal ring.

$$\bullet H + C_6H_6 \rightarrow \bullet C_6H_7$$

4.3.3 Hydrogen Absorption Reaction

•OH and •H radicals abstract a hydrogen from organic compounds and are converted to water molecules and hydrogen molecules, respectively; this forms organic radicals.

$$\bullet OH + CH_3OH \rightarrow \bullet CH_2OH + H_2O$$
$$\bullet H + CH_2OH \rightarrow \bullet CH_2OH + H_2$$

By these reactions, the •OH radical formed by radiative decomposition of an aqueous solution attacks the organic solutes and eventually decomposes the organics. This is the principle of water purification by radiation. The initial process of radiation damage in biological systems is similar. The living body consists of up to 70% water; •OH radicals formed by decomposition of water in the living body under irradiation attack biological macromolecules such as DNA.

4.3.4 Recombination and Disproportion

Because they have unpaired electrons, radicals react with each other.

$$\bullet OH + \bullet OH \rightarrow H_2O_2$$

Hydrogen peroxide, formed by radiolysis of water, is attributed to the recombination of two •OH radicals. If •H and •OH react, a molecule of water is regenerated.

$$\bullet H + \bullet OH \rightarrow H_2O$$

Radicals from ethanol react with each other, forming ethylene glycol. On the other hand, if a H atom is present, an ethanol molecule is generated along with formaldehyde. This is called the disproportion reaction because the two reactants produce two different products. In aqueous solution, the ratio of recombination and disproportion is around 1:1.

$\bullet CH_2OH + \bullet CH_2OH \rightarrow HOH_2C - CH_2OH$ Ethylene glycol (recombination) $\rightarrow CH_3OH + CH_2OH$ Methanol and formaldehyde (disproportion)

4.3.5 Reaction with Oxygen

Oxygen molecules, which have two unpaired electrons, are diradicals (also called biradicals). Radicals often react easily with oxygen.

$$\bullet CH_2OH + O_2 \rightarrow \bullet O - O - CH_2OH$$

The •O-O-R type of radical is called the peroxy radical. It takes in oxygen and forms oxygen-containing groups such as peroxide, alcohol, and ketene in subsequent reactions. Accordingly, this reaction is related to oxidative degradation, for example, deterioration of polymer materials or reactions in biological systems.

As described above, radicals are formed through various processes and produce various reactions. These reactions include petroleum purification, photoinduced reactions, atmospheric reactions, radiation-induced reactions, and reactions in biological systems.

4.4 Electron Spin Resonance (ESR)

The ESR method detects unpaired electrons using the interaction between the spin of unpaired electrons in radicals and an external magnetic field. Unpaired electrons in a magnetic field can take two quantum states: alpha spin and beta spin. They exhibit an energy difference proportional to the intensity of the magnetic field. This is called Zeeman splitting. The energy difference ΔE is $\Delta E = g\beta H$ (g is around 2.002), where g, β, and H are the g-factor, Bohr magneton (0.927×10^{-16} erg·T^{-1}), and intensity of the magnetic field, respectively (T, Tesla, 1 T = 10^4 gauss). By measuring the absorption of electromagnetic waves corresponding to the energy difference, the radical can be detected. Accordingly, as in ordinary light absorption, the condition $h\upsilon = g\beta H = \Delta E$ is met.

Many commercial apparatuses use a frequency of 9 GHz (1 GHz = 10^9 Hz) and a wavelength of 3 cm, which is in the microwave or X band. The magnetic field H corresponding to the energy of microwaves at this wavelength is around 3.3 T. The energy of light having a wavelength of 300 nm is 4 eV. On the other hand, the frequency of microwaves having a wavelength of 3 cm is one-[3 cm/(300 nm)]th

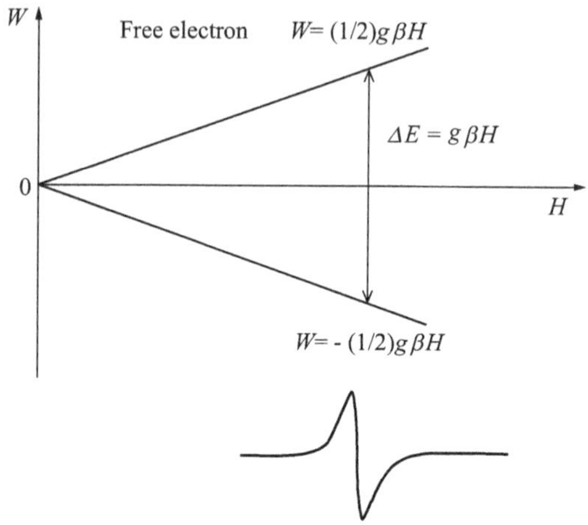

Fig. 4.1 Zeeman effect of free electron

that of light having a wavelength of 300 nm. Because this is one-hundred-thousandth, the difference in energy is also one-hundred-thousandth or 4×10^5 eV.

As shown in Fig. 4.1, when the applied magnetic field H is taken as the horizontal axis, and the energy is taken as the vertical axis, the energy levels of the electron split into two states, where the difference is proportional to the intensity of the magnetic field. Externally applied microwaves will be absorbed when their energy equals the energy difference, and a single absorption line will be observed. To detect it more sensitively, the magnetic field can be modulated, which often yields a differential absorption spectrum. Further, secondary differentiation may be adopted.

The ESR method uses the fact that the spectrum shows further fine structure depending on the characteristics of the radical. For example, the hydrogen atom has a nucleus having a nuclear spin with two spin states, α_N and β_N. The electron spin interacts with the nuclear spin, and both the α and β spins of the electron split into two levels. Accordingly, the hydrogen atom shows two absorption lines separated by 50 mT, as shown in Fig. 4.2, plus a constant that is unique to the radical.

The spectra of radicals having more hydrogen, such as the methyl radical ($\bullet CH_3$), which forms from methanol, exhibit more complex splits depending on factors such as the number of hydrogen atoms and the distance between the unpaired electrons and the H nucleus.

The ESR spectrum shows an apparently complicated change due to interactions between unpaired electrons and the nucleus of the radical. However, analyzing the spectrum yields information on the interactions between the nucleus and the unpaired electrons, and ultimately the structure of the radical can be assigned, which makes the ESR method a useful technique. The separation width, i.e., the coupling constant, is unique to the radical, and if a list of coupling constants and radicals is available, the radical can be assigned using the list. Further, with the recent development of molecular dynamics based on quantum mechanics, the

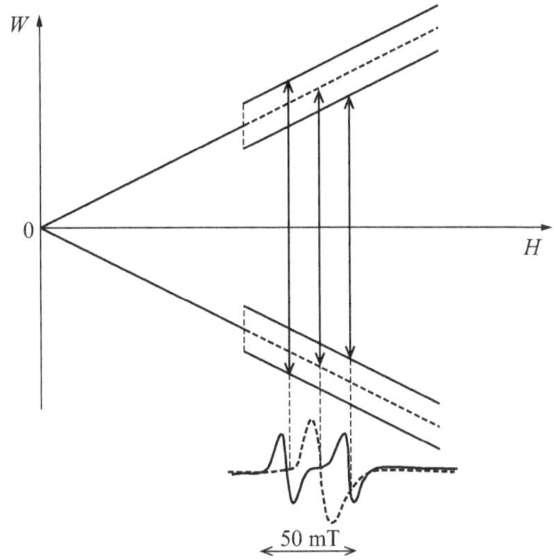

Fig. 4.2 Zeeman effect of hydrogen electron

coupling constants can be evaluated. It is helpful to use a theoretical calculation to analyze the spectrum. Historically, experimental ESR data have contributed to the development of molecular dynamics.

The ESR method is applicable to not only radical assignment but also quantification of the radical concentration, because the spectral signal's intensity is proportional to the radical concentration. For example, a time-resolved method is also available, and the spin dynamics can be discussed. More sophisticated methods have also been developed and used.

Exercises

1. Explain the difference between hydroxyl radical (•OH) and hydroxyl ion (OH^-).
2. How many lines does the ethyl radical (•CH_2-CH_3) have in the ESR spectrum?

Chapter 5
Pulse Radiolysis

Yosuke Katsumura and Hisaaki Kudo

5.1 Pulse Radiolysis Method

When matter is irradiated, the radiation energy is deposited in the irradiated matter, inducing ionization and excitation; subsequently, reactive ions, excited states, and radicals are formed. They interact with each other and with surrounding molecules, eventually producing stable final products. A precise analysis of the final product reveals the effects of the radiation. However, the final product is the result of reactions, and even though it may be possible to speculate about the reaction process, this provides only a conjecture. Thus, researchers wish to know what types of intermediates are really formed and how they react. To answer this need, the pulse radiolysis method was developed. After pulsed irradiation, the behavior of short-lived intermediates is observed using time-resolved light absorption spectroscopy.

The pulse radiolysis setup is shown in Fig. 5.1. The sample call (reaction cell) is placed in front of the radiation source, typically a high-energy electron beam accelerator. To analyze the light absorption of intermediates formed in the cell, light is passed through the cell and collected by optical lenses. A monochromator selects light of a certain wavelength after it passes through the cell, and a light detector, such as a photomultiplier in a photodiode, is placed at the exit of the monochromator to convert the light intensity with time to an electrical signal, which is recorded by an oscilloscope, digitized, and transmitted to a computer for storage. The system converts the electrical signal to the light absorption.

"Lysis" means decomposition, as in photolysis, electrolysis, and pyrolysis.

The pulse radiolysis method uses the same principle as laser flash photolysis, which uses laser light and is well known in research on photoinduced reactions.

Y. Katsumura · H. Kudo (✉)
The University of Tokyo, Tokai, Ibaraki, Japan
e-mail: hkudo@tokai.t.u-tokyo.ac.jp

H. Kudo (ed.), *Radiation Applications*, An Advanced Course in Nuclear Engineering 7, https://doi.org/10.1007/978-981-10-7350-2_5

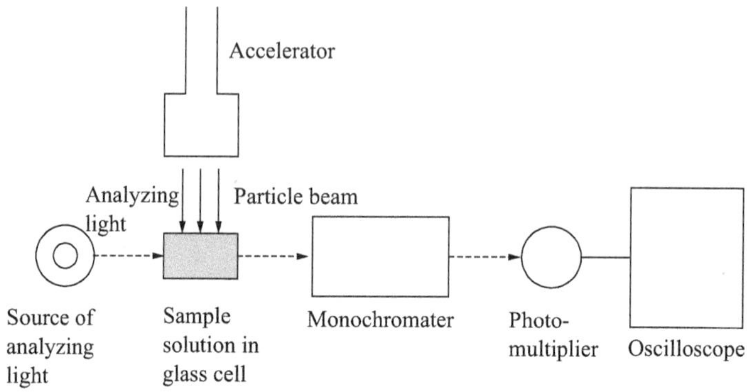

Fig. 5.1 Setup of pulse radiolysis system

Flash photolysis was established by R. G. W. Norrish and G. Porter. G. Porter was a British chemist with electronics skills who worked on the development of radar during World War II. After the war, at the University of Cambridge, he developed a 10 μs pulsed light using an electrical circuit. Chemical species formed under the pulsed light; the pulsed light used to analyze the chemical species is synchronized with the pulsed light that initiates the reaction and then recorded on photographic film. He established this method of observing short-lived species directly and was awarded the Nobel Prize in chemistry in 1967. This method is also called the pump and probe method because it triggers reactions using a pulse and measures the light absorption using another pulse. The time resolution was about 10 μs. With subsequent technological developments and the desire to observe much faster reactions, the time resolution has been improved to the nanosecond and picosecond range, and currently a time resolution as high as 10 fs is possible. The 1999 Nobel laureate in chemistry, A. H. Zewail, developed a new optical spectroscopic method by using a femtosecond laser pump-probe system. It can be assumed to follow the same principle as flash photolysis and pulse radiolysis.

The development of the pulse radiolysis method was begun in the early 1960s, and its usefulness was greatly enhanced by the discovery of hydrated electrons in 1962 (this is described in another chapter). Since then, many systems have been developed and used, and several systems are available even in Japan. The time resolution of the pulse radiolysis system was improved from the initial 10 μs to the picosecond and sub-picosecond range. The Nuclear Engineering Research Laboratory at the University of Tokyo in Tokai-mura recently developed and began operating a pulse radiolysis system having the highest time resolution in the world.

5.2 Lambert–Beer Law

Light absorption measurement is based on the Lambert–Beer law. As shown in Fig. 5.2, when light is absorbed by solutes in a solution, the absorbance can be expressed as follows:

$$I = I_0\,10^{-\varepsilon cl} = I_0\exp(-2.303\varepsilon cl)$$

where I_0 is the intensity of the injected light, and I is the intensity of the transmitted light. This formula, which is called the Lambert–Beer law, shows that the light intensity decreases exponentially as a function of the thickness of the sample cell when the solution has absorption bands appropriate for the incoming light. It may be reasonable to use the natural logarithm to express the decay, but for historical reasons, it is customary in chemistry to use the base-10 logarithm. Both formulas have the same meaning and can be converted using a conversion factor of 2.303 (= $\log_{10}$). In this formula, ε is the molar absorption coefficient (mol•l^{-1}•cm^{-1}), c is the concentration (mol•l^{-1}, which is sometimes expressed as M), and l is the optical path length. Here, it may be readily understood that ε has units of mol•l^{-1}•cm^{-1} because the natural logarithm is dimensionless.

Taking the logarithms of the terms on both sides yields $\varepsilon cl = \log_{10}(I_0/I)$. This is called the absorbance or optical density (OD). I_0 and I are the experimentally measured values, and the ratio is proportional to the solute concentration. The proportional constant is a molar absorption coefficient. When the absorption is strong, the coefficient is high. Optical measurements are based on the Lambert–Beer law.

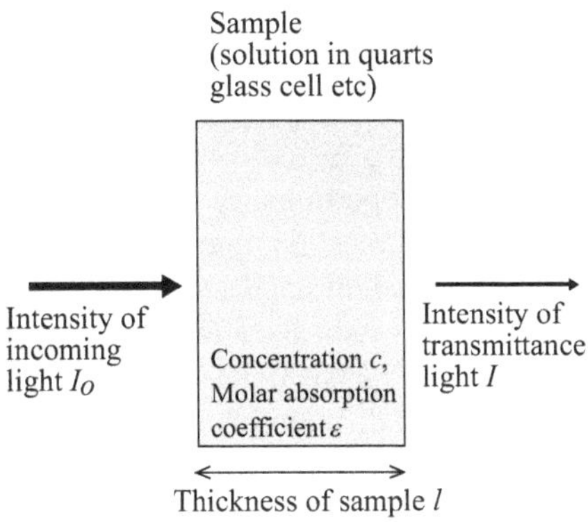

Fig. 5.2 Aqueous solution of thickness l absorbing light

5.3 Elementary Processes of Chemical Reactions and Rate Constants

Here the three types of chemical reactions involving radical A• will be considered.

5.3.1 *First-Order Reaction*

Radical A• changes (decays) with time into B, A• → B, which is expressed by the differential equation $d[\mathrm{A}\bullet]/dt = -k[\mathrm{A}\bullet]$, where k is the reaction rate constant and has the units of s^{-1}. Solving this yields $[\mathrm{A}\bullet](t) = [\mathrm{A}\bullet](0)\exp.(-kt)$, where [A•](0) represents [A•] at time zero. The semilogarithmic plot of [A•](t) yields the lifetime. The time to kt in the experiment becomes unity; namely, $t = 1/k$, which has the dimension of time, gives the time scale to 0.37 or $1/e$ (37%) of the initial concentration. [B] is assumed to be zero at time zero, which is expressed as $[\mathrm{B}](t) = [\mathrm{A}\bullet](0)[1 - \exp.(-kt)]$. After sufficient time has elapsed, all of radical A• is changed to B.

5.3.2 *Recombination (Second-Order) Reaction*

The scheme in which the radical A• reacts with other A•, especially by recombination, is expressed as A• + A• → A-A. Here, the change in the concentration of A• is expressed by the following differential equation:

$$dt[\mathrm{A}\bullet]/dt = -2k[\mathrm{A}\bullet]^2$$

Because two A• radicals react with each other, the rate would be proportional to $[\mathrm{A}\bullet]^2$. The proportionality coefficient k, which describes the rate of disappearance, is called the reaction rate constant. To denote that [A•] decreases through the reaction, the right-hand term has a minus sign. Here, because one reaction consumes two A• radicals, a factor of 2 is applied. The concentration [A•] has units of $\mathrm{M\text{-}mol{\bullet}l^{-1}}$, so k has units of $\mathrm{l{\bullet}mol^{-1}{\bullet}s^{-1}}$.

The above differential equation can be rewritten as follows:

$$(d[\mathrm{A}\bullet]/dt)/\left([\mathrm{A}\bullet]^2\right) = -2k$$

[A•] as a function of time is written as [A•](t). Integrating the above equation yields the following expression:

$$-1/[\mathrm{A}\bullet](t) = -2kt + C$$

[A•](t) at $t = 0$ is written as [A•](0); consequently, $C = -1/[\mathrm{A}\bullet](0)$:

$$1/[\mathrm{A}\bullet](t) = 2kt + 1/[\mathrm{A}\bullet](0)$$

If the reciprocal of the concentration is plotted versus time, the slope is 2 k, and the intersection is 1/[A•(0)].

5.3.3 Pseudo First-Order Reaction

Consider the reaction between radical A• and solute B in a system where [A•] « [B], and reaction products such as P or A′ and B′ are produced. This reaction is expressed as follows:

$$\mathrm{A}\bullet + \mathrm{B} \rightarrow \mathrm{P} \text{ or } \mathrm{A}' + \mathrm{B}'$$

Here, the reaction rate is proportional to the concentration of A• and also to the concentration of B•. Accordingly, the time profile of the concentration of A• is expressed by the following differential equation:

$$d[\mathrm{A}\bullet]/dt = -k[\mathrm{A}\bullet][\mathrm{B}]$$

As in the previous case, the minus sign in the right-hand term indicates that A• decreases, and k is the proportionality coefficient. In this reaction, a factor of 2 is not necessary because one reaction consumes one A• radical. Again, k has units of $\mathrm{l{\bullet}mol^{-1}{\bullet}s^{-1}}$. In many cases, the condition of [A•] « [B] is satisfied, and the change in [B] is negligible, so [B] can be approximated as a constant. This situation is called the pseudo first-order reaction.

By rewriting the above equation, we obtain

$$(d[\mathrm{A}\bullet]/dt)/[\mathrm{A}\bullet] = -k[\mathrm{B}]$$

By integrating, we obtain

$$\ln([\mathrm{A}\bullet](t)) = -k[\mathrm{B}]t + C$$

This can be rewritten as

$$[\mathrm{A}\bullet](t) = \exp(C)\exp(-k[\mathrm{B}]t) = C'\exp(-k[\mathrm{B}]t)$$

If [A•](t) at $t = 0$ is written as [A•](0), $C' = [\mathrm{A}\bullet](0)$. Consequently,

$$[\mathrm{A}\bullet](t) = [\mathrm{A}\bullet](0)\exp(-k[\mathrm{B}]t)$$

When the logarithm of the concentration is plotted versus time, a straight line of slope k[B] is obtained.

The temporal behavior of chemical species can be expressed if the reaction rate constants are known and the initial conditions are given. As described in this chapter, when one or two reactions are of interest, an analytical solution is possible. However, when the number of reactions of interest increases, an analytical solution becomes very difficult or impossible. For example, the ground atmosphere can be described using more than 200 reactions, and the radiation chemical reaction in water can be simulated with more than 40 reactions. In such cases, an analytical solution is difficult to obtain, and numerical evaluation is feasible. Many chemical software packages for this purpose are available on the market.

Exercises

1. In pulse radiolysis method, why is the pulsed radiation necessary?
2. Explain why the synchronization of timing of linear accelerator, analyzing light, and oscilloscope is important in pulse radiolysis method.

Chapter 6
Radiation Chemistry of Aqueous Solutions

Yosuke Katsumura and Hisaaki Kudo

6.1 History and Concept of Radiation Chemistry of Aqueous Solutions

Radiation chemistry of aqueous solutions has a long history; it dates back to almost the same time as the discovery of radiation. The observation of gas evolution from an aqueous solution containing radioisotopes (RIs) was discovered at almost the same time as radiation. However, substantial research and development was performed during the Manhattan project. The currently accepted fundamental idea was developed during the project, i.e., radical theory and the concept that a H atom and OH radical are formed after ionization and excitation of the water molecule. In the 1950s, experiments demonstrated that two types of reducing species appear after the decomposition of water. On the other hand, in 1953, R. L. Platzman predicted that the electron emitted during ionization would exist for some time in a hydrated state after thermalization. These suggested phenomena were expected to be observed directly. In the 1960s, J. P. Keene, E. J. Hart, and J. W. Boag independently observed intense light absorption in the visible region using the pulse radiolysis method and confirmed that this absorption is due to hydrated electrons. The pulse radiolysis method uses pulsed radiation (usually an electron beam pulse) for momentary irradiation of the samples. The species formed in the sample are analyzed using the time-resolved light absorption, and the behaviors of the species are monitored dynamically. It is based on the same principle as the flash photolysis method developed by R. G. W. Norrish and G. Porter (which was successfully developed in 1950 and received the Nobel prize in 1967). Observations of hydrated electron confirmed the usefulness of the pulse radiolysis method; many apparatuses have been developed, and pulse

Y. Katsumura · H. Kudo (✉)
The University of Tokyo, Tokai, Ibaraki, Japan
e-mail: hkudo@tokai.t.u-tokyo.ac.jp

H. Kudo (ed.), *Radiation Applications*, An Advanced Course in Nuclear Engineering 7, https://doi.org/10.1007/978-981-10-7350-2_6

radiolysis has become an essential tool for research on fundamental processes. Since then, many data have been accumulated, and direct observation of intermediate species became possible. On the basis of these achievements, modeling of radiolysis was attempted, and a diffusion kinetic model was proposed by A. Kupperman in 1962. This model successfully reproduced many experimental results semiquantitatively. This idea was further developed by H. A. Schwarz in 1969, and it is currently the standard model of radiolysis of aqueous solutions. In this chapter, the standard concept of the reaction is introduced, where the G values are defined as the yield per 100 eV of energy absorption.

6.2 Radiolysis of Water

When a water molecule absorbs radiation energy, it undergoes ionization and excitation. As shown in Fig. 6.1, ionization produces H_2O^+ and e^-. This occurs within as little as 10^{-16} s. H_2O^+ reacts with the surrounding water molecules to form H_3O^+ and •OH. The reaction of $H_2O^+ + H_2O \rightarrow H_3O^+ + \bullet OH$ is well known as the ion-molecule reaction. Its time scale is evaluated as 10^{-14} s from the reaction rate constant in the gas phase. On the other hand, excess energy from the electron formed during ionization further ionizes or excites the water molecules, and the electron loses its energy and eventually becomes thermalized (reaches equilibrium with its surroundings). The electron rearranges the water molecule around itself, forming a potential well in which the electron is stabilized. This state is called hydration, and the electron is called a hydrated electron and denoted as e^-_{aq}. A femtosecond laser flash photolysis study recently revealed that the time scale of hydration is hundreds of femtoseconds. Excited states decompose to •OH, •H (H_2, O). Consequently, 1 ps after the radiation energy is absorbed; e^-_{aq}, H_3O^+, •H, •OH, and (H_2) are formed; and the yield at 1 ps is conventionally called the initial yield. These decomposed species of water are speculated to exist within a local spatial region of hundreds of nanometers, which is called a spur. These spurs are considered to be separated from each other by thousands of angstroms for radiation with low linear energy transfer (LET) such as electrons and γ-rays. The decomposed species of water are formed at locally high concentrations and then tend to diffuse and become distributed homogenously in the system. Simultaneously, because these species are reactive, they react with each other in so-called spur reactions (see Table 6.1). Diffusion and reaction proceed simultaneously, and after 1 μs, the distribution in the system becomes homogeneous. The yield of decomposed species of water at 1 μs is called the primary yield. During the spur reactions and diffusion, molecular products of H_2O_2 and H_2 are considered to form through the spur reactions listed in Table 6.1. Because the spur reactions include the regeneration of water, the decomposition yield is less at 1 μs than at 1 ps. The scenario above is derived from experimental results and generally accepted. The diffusion model simulates this scenario. The model assumes the initial distribution of decomposed species at 1 ps and reproduces the spur reactions and diffusion.

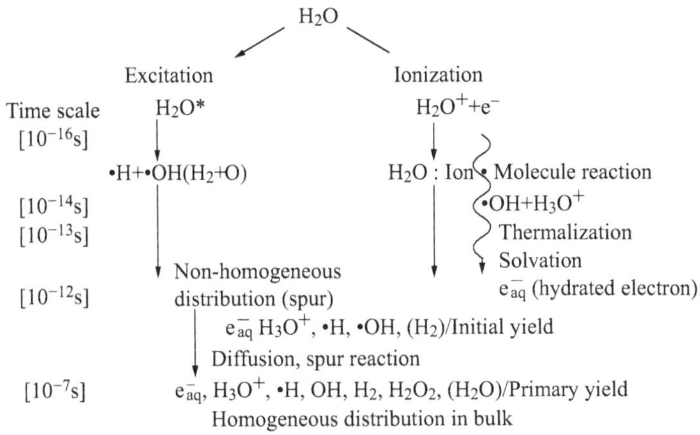

Fig. 6.1 Scheme of radiolysis of water

Table 6.1 Spur reactions in water

Reaction	Rate constant k $[10^{10}\ M^{-1}\ s^{-1}]$
$e^-_{aq} + e^-_{aq} + 2H_2O \rightarrow H_2 + 2OH^-$	0.54
$e^-_{aq} + \bullet OH \rightarrow OH^-$	3.0
$e^-_{aq} + H_3O^+ \rightarrow \bullet H + H_2O$	2.3
$e^-_{aq} + \bullet H + H_2O \rightarrow H_2 + OH$	2.5
$\bullet H + \bullet H \rightarrow H_2$	1.3
$\bullet OH + \bullet OH \rightarrow H_2O_2$	0.53
$\bullet OH + \bullet H \rightarrow H_2O$	3.2
$H_3O^+ + OH^- \rightarrow 2H_2O$	14.3

The reaction in pure water is thought to proceed as described above, but an aqueous solution generally contains solutes. When the solute concentration is low, the fraction of the solute in the spur is low, so intraspur reactions proceed without being affected by the solute, even if the solute has high reactivity with hydrated electrons, hydrogen atoms, and OH radicals. Most reactions with the solute occur when the decomposition products are distributed homogeneously in the system after the spur reactions. So far as this condition is met, even when the solute concentration is changed, the yield of molecular products is increased further and becomes equal to that of radicals. However, if the solute concentration is increased further, the spur reactions would be affected; for example, a solute having high reactivity with OH radicals (in this case, the solute is called a •OH scavenger because it reacts selectively with •OH) reacts with •OH radicals in competition with the spur reactions. The yield of the reaction product becomes higher than the yield of the primary •OH, and because some fraction of the •OH involved in the spur reaction is removed, the H_2O_2 yield becomes low. Note that, in reality, the concept of the spur has been developed on the basis of these historical observations.

The radiation chemistry of dilute aqueous solution is well established under the condition that the solute has little effect on the spur reactions.

6.3 Reactivity of e^-_{aq} (Hydrated Electrons), Hydrogen Atoms, and •OH Radicals

As described so far, the radiolysis products of water are distributed homogeneously in the system through local spur reactions within a rapid process. The main products of water radiolysis are e^-_{aq}, •H, •OH, H_2, and H_2O_2. Among these products, H_2 and H_2O_2 are stable compounds, and their chemical characteristics are well known. On the other hand, e^-_{aq}, •H, and •OH are specific reactive intermediates in the radiation chemistry of water. Reactions between these species and solutes in the solution appear as effects of radiation. The characteristics of these reactive intermediates are given below.

6.3.1 Absorption Spectra of e^-_{aq} (Hydrated Electrons), Hydrogen Atoms, and •OH Radicals

These intermediates are produced by pulsed radiation on a spot at high concentration, and in most cases, they can be monitored by time-resolved light absorption. Hydrated electrons produce intense absorption with a peak at 720 nm (molar absorption coefficient $\varepsilon = 18{,}500$ l•mol^{-1}•cm^{-1}), but other species exhibit rather weak absorption at shorter wavelengths, which makes them difficult to detect.

Characteristics and Reactivity of e^-_{aq} (Hydrated Electrons)

Figure 6.2 shows the physicochemical properties of hydrated electrons and the hydrated structure.

Hydrated electrons are a very active reduction agent with an oxidation reduction potential of −2.9 V; they react readily with many metal ions, as described below.

$$e^-_{aq} + M^{n+} \rightarrow M^{(n-1)+}$$

The reaction rate is around 10^{10} l•mol^{-1}•s^{-1}, and the reaction is often diffusion limited.

Further, hydrated electrons react with compounds having high electron affinity and form anions. An example is oxygen and N_2O; the reaction with oxygen is written as a capture reaction:

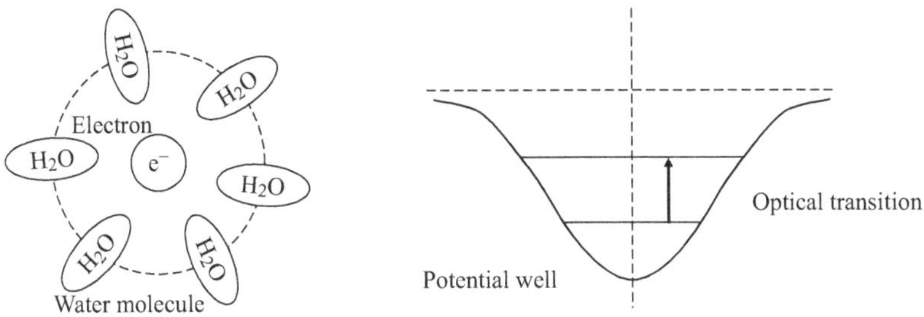

Fig. 6.2 Structure of hydrated electron in water

$$O_2 + e^-{}_{aq} \rightarrow O_2{}^-$$

The other reaction is

$$N_2O + e^-{}_{aq} \rightarrow (N_2O) \rightarrow N_2 + O^-$$
$$O^- + H_2O \rightarrow OH + \bullet OH^-$$

and the formed anion decomposes rapidly; therefore, this reaction is called dissociative electron capture. Dissociative electron capture is often observed for halogen compounds. The reaction between N_2O and hydrated electrons changes the hydrated electrons into oxidative O^- and •OH; therefore, this reaction is often used in radiation chemistry. Table 6.2 shows examples of reactions involving hydrated electrons.

Characteristics and Reactivity of Hydrogen Atoms

The hydrogen atom, like the hydrated electron, is also a reduction magnet. Its oxidation reduction potential is −2.3 V, which is somewhat higher than that of the hydrated electron (i.e., it is a less powerful reducing agent). Its high absorption spectra appear at short wavelengths but are not intense (molar absorption coefficient, $\varepsilon = 1620\ l{\bullet}mol^{-1}{\bullet}cm^{-1}$ at 188 nm). Hydrogen atoms often reduce metal ion. Hydrated electron reacts with proton in acid solution and changes to hydrogen atom. The hydrogen atom is the simplest radical and reacts with other radicals; it also undergoes reactions that form double bonds, abstract hydrogen from organic compounds, and form organic radical and hydrogen molecules. These reactions are shown below.

(a) Reducing reagent

$$\bullet H + M^{n+} \rightarrow M^{(n-1)+}$$
$$\bullet H + Cu^{2+} \rightarrow Cu^+ + H^+ \quad k = 9.1 \times 10^7\,M^{-1}s^{-1}$$

Table 6.2 Rate constants of hydrated electron with various species (in bracket is reaction product)

Inorganic	Rate constant $10^{-10}\,k$ $[M^{-1}\,s^{-1}]$	Organic	Rate constant $10^{-10}\,k$ $[M^{-1}\,s^{-1}]$
O_2 ($O_2^- \Leftrightarrow HO_2$)	1.9	C_6H_6 ($C_6H_7 + OH^-$)	1.2×10^{-3}
H_3O^+ ($H + H_2O$)	2.3	C_6H_5Cl ($C_6H_5 + Cl^-$)	1×10^{-2}
NH_4^+ ($H + NH_3$)	$< 2 \times 10^{-4}$	C_6H_5I ($C_6H_5 + I^-$)	1.2
$Ag + (Ag^0)$	3.6	$CH_2 = CH_2$	$<2.5 \times 10^{-4}$
Cd^{2+} (Cd^+)	5.0	$CH_2 = CCl_2$	2.3
In^{3+} (In^{2+})	5.6	CH_4	$<10^{-3}$
$Fe(CN)_6^{3-}$ ($Fe(CN)_6^{4-}$)	0.3	CH_3I ($CH_3 + I^-$)	1.7
NO_3^- ($NO_3^{2-} \rightarrow NO_2 + 2OH^-$)	0.87	CH_3OH	$<10^{-6}$
UO_2^{2+} (UO_2^+)	1.7		

(b) Reaction with radicals

$$\bullet H + \bullet OH \rightarrow H_2O \quad k = 3.2 \times 10^{10}\,M^{-1}s^{-1}$$
$$\bullet H + \bullet H \rightarrow H_2 \quad k = 1.3 \times 10^{10}\,M^{-1}s^{-1}$$
$$\bullet H + O_2 \rightarrow \bullet HO_2 \quad k = 2.1 \times 10^{10}\,M^{-1}s^{-1}$$

(c) Addition to unsaturated bonds

$$\bullet H + CH_3C \equiv N \rightarrow CH_3CH_2 = N\bullet \quad k = 2 \times 10^{6}\,M^{-1}s^{-1}$$
$$\bullet H + C_6H_6 \rightarrow \bullet C_6H_7 \quad k = 9.1 \times 10^{8}\,M^{-1}s^{-1}$$

(d) Abstraction of hydrogen

$$\bullet H + CH_3OH \rightarrow \bullet CH_2OH \quad k = 2.6 \times 10^{6}\,M^{-1}s^{-1}$$
$$\bullet H + c\text{-}C_6H_{12} \rightarrow \bullet c\text{-}C_6H_{11} \quad k = 7.8 \times 10^{7}\,M^{-1}s^{-1}$$

Characteristics and Reactivity of •OH Radical

The •OH radical is an oxidation reagent, and its oxidation-reduction potential is 2.72 V; thus, it oxidizes many ions. Its light absorption spectra are in the short wavelength region but are not intense (molar absorption coefficient, $\varepsilon = 400\ l\bullet mol^{-1}\bullet cm^{-1}$ at 260 nm). In a basic solution, •OH dissociates and behaves as O^-. The •OH radical reacts with other radicals, undergoing addition to double bonds and abstracting hydrogen from organic compounds. The reactions of •OH are shown below.

(a) Oxidizing reagent

$$\bullet OH + M^{n+} \rightarrow M^{(n+1)+}$$
$$\bullet OH + Fe^{2+} \rightarrow Fe^{3+} + OH^- \quad k = 4.3 \times 10^8\, M^{-1}s^{-1}$$

(b) Reaction with radicals

$$\bullet OH + \bullet H \rightarrow H_2O \quad k = 3.2 \times 10^{10}\, M^{-1}s^{-1}$$
$$\bullet OH + \bullet OH \rightarrow H_2O_2 \quad k = 1.1 \times 10^{10}\, M^{-1}s^{-1}$$

(c) Addition to double bonds

$$\bullet OH + CH_2 = CH_2 \rightarrow CH_2OHCH_2\bullet \quad k = 4.4 \times 10^9\, M^{-1}s^{-1}$$
$$\bullet OH + C_6H_6 \rightarrow \bullet C_6H_6OH \quad k = 7.8 \times 10^9\, M^{-1}s^{-1}$$

(d) Abstraction of hydrogen

$$\bullet OH + CH_3OH \rightarrow \bullet CH_2OH + H_2O \quad k = 1.9 \times 10^9\, M^{-1}s^{-1}$$
$$\bullet OH + c\text{-}C_6H_{12} \rightarrow \bullet c\text{-}C_6H_{11} + H_2O \quad k = 6.1 \times 10^9\, M^{-1}s^{-1}$$

As mentioned, $e^-{}_{aq}$ (hydrated electrons), hydrogen atoms, and •OH radicals, which are formed in water radiolysis, are reactive and react with solutes in solution. Consequently, oxidation or reduction in solution, decomposition, and an increase or decrease in molecular weight can occur. Regarding the decomposition products of $e^-{}_{aq}$, H, and •OH (except for H_2 and H_2O_2), many experiments have been conducted using the pulse radiolysis method. The reactivity between them and various solutes has been evaluated. Many results were recently compiled in a data book, and the reactivities with metal ions, inorganics, organics, and bio-related compounds have been well studied. The study of radiation chemical reactions in aqueous solutions has a long history, and its achievements are summarized well in textbooks and review articles. Readers are advised to refer to them.

6.4 Dependence of Decomposition of Water on pH

Upon decomposition through spur reactions, the species shown below are distributed homogeneously 10^{-7} s after the radiation energy is deposited.

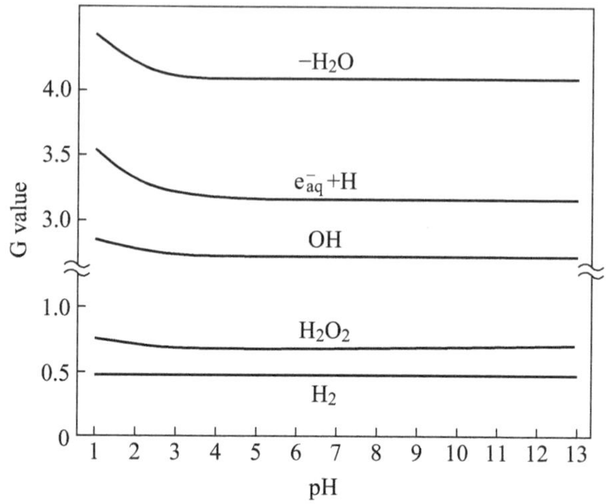

Fig. 6.3 pH depends of G value of decomposition species of water

$$H_2O \rightsquigarrow e^-{}_{aq}, \bullet H, \bullet OH, H_2O_2, H_2, (\bullet HO_2) \quad (6.1)$$

It is known that the yields of these species do not depend greatly on the pH of the solution. The yields are almost constant unless the solution has a very high or low pH (Fig. 6.3). Accordingly, under typical conditions, the yield of water decomposition can be assumed to be constant.

6.5 Dependence of Decomposition of Water on LET

It is well known that the G values of radiolysis products and the decomposition pattern vary with the type of radiation. Table 6.3 shows the experimentally determined yields of water radiolysis products as a function of the energy deposition density along the path of radiation for LET from Co-60 γ-rays, 18 MeV D^+ ions, 23 MeV He^{2+}, and 12 MeV He^{2+} ions.

The G value of water decomposition is expressed as G_{-H_2O}, and the G values of formation are expressed as G_H, G_{OH}, and so forth. The table reveals the following trends.

1. G_{-H_2O} (decomposition of water) decreases with increasing LET.
2. $G_{e^-{}_{aq}}$, G_H, and G_{OH} (which are called the radical yields) decrease with increasing LET.
3. G_{H_2}, $G_{H_2O_2}$, and G_{HO_2} (which are called the molecule yields) increase with increasing LET.

These observations can be explained in terms of spur reactions. For low-LET radiation, spurs, in which radiolysis products are formed in a cluster, are distributed

Table 6.3 LET dependence of radiolysis product in neutral water

	G value						
LET [eV/nm]	$-H_2O$	e_{aq}^-	•OH	•H	H_2	H_2O_2	HO_2
0.23	4.08	2.63	2.72	0.55	0.45	0.68	0.008
12.3	3.46	1.48	1.78	0.62	0.68	0.84	
61	3.01	0.72	0.91	0.42	0.96	1.00	0.05
108	2.84	0.42	0.54	0.27	1.11	1.08	0.07

and well separated from each other. However, as the LET increases, the distance between spurs decreases. When the distance is sufficiently short, they can be assumed to be linked, and the structure resembles a long cylinder. These structures are called track structures. In the initial distribution of water radiolysis products, when the spur reactions are of interest, the rate of spur reactions increases with decreasing spur distance and increasing overlap between spurs. The spur reactions occur between the water radiolysis products and include regeneration of water. If the molecular products such as H_2 and H_2O_2 increase as a result of radical reactions, the yields of radicals (e^-_{aq}, •H, •OH) would decrease accordingly.

These dependencies on the LET or type of radiation are called the LET effect and radiation quality effect, respectively. Historically, such experimental results gave birth to the concept of the spatial distribution of the radiolysis of water. Furthermore, the spatially nonhomogeneous energy deposition inspired the concept of microdosimetry, which is dosimetry in a microscopically small region.

6.6 Relationship Between G Values of Water Radiolysis Products: Material Balance

Radiolysis products are formed after ionization of the water molecule and bond dissociation. Therefore, the amount of all of the radiolysis products together is equal to the original amount of water. Between them, the material balance relationship is satisfied.

$$H_2O \rightsquigarrow e^-{}_{aq}, \bullet H, \bullet OH, H_2O_2, H_2, (\bullet HO_2) \tag{6.2}$$

Because H_2O is electrically neutral, $e^-{}_{aq}$ and H_3O^+ are counterparts and are formed with the same yields. The yield of $e^-{}_{aq}$ corresponds to that of H_3O^+, and regarding the decomposition of water, they can be assumed to originate as hydrogen atoms.

By counting the number of hydrogen atoms, the number of hydrogen atoms is

$$G_{e^-_{aq}} + G_H + G_{OH} + 2G_{H_2O_2} + 2G_{H_2} \tag{6.3}$$

The G values of species having two hydrogen atoms are doubled.

On the other hand, by counting the number of oxygen atoms,

$$\text{The number of oxygen atoms} = G_{OH} + 2G_{H_2O_2}.$$

Because the number of hydrogen atoms is twice the number of oxygen atoms,

$$\begin{aligned} 2(G_{OH} + 2G_{H_2O_2}) &= G_{e^-_{aq}} + G_H + G_{OH} + 2G_{H_2O_2} + 2G_{H_2} \\ G_{OH} + 2G_{H_2O_2} &= G_{e^-_{aq}} + G_H + 2G_{H_2} \end{aligned} \tag{6.4}$$

The number of decomposed water molecules and oxygen atoms in products is

$$G_{\text{-}H_2O} = G_{OH} + 2G_{H_2O_2} = G_{e^-_{aq}} + G_H + 2G_{H_2} \tag{6.5}$$

This relationship indicates that all the G values are not necessary for calculating the yields of specific species of interest.

When the formation of HO_2 is considered, the relationship becomes

$$G_{\text{-}H_2O} = G_{OH} + 2G_{H_2O_2} + 2G_{HO_2} = G_{e^-_{aq}} + G_H + 2G_{H_2} - G_{HO_2} \tag{6.6}$$

6.7 Examples of Radiation Chemical Reactions in Aqueous Solution

Consider the radiation chemical reactions of water dissolving alcohol to dozens of m mol/e$^-$. Thus, 1 l of this aqueous solution contains 55 mol of water and mM of alcohol, so most of the molecules by number are water. Accordingly, most of the energy is deposited to water, so the reaction dynamics begin with the decomposition of water. Consequently, hydrated electrons, •OH radicals, hydrogen atoms (H), hydrogen molecules (H_2), and hydrated peroxide are produced. If this aqueous solution is acidic, the hydrated electrons react with protons (acid, H_3O^+), forming hydrogen atoms. Hydrated electrons have low reactivity with alcohol (the rate constant is lower than 10^4 l•mol^{-1}•s^{-1}).

For example, consider an aqueous solution containing methanol. In this system, •H and •OH react with methanol as follows:

$$\bullet H + CH_3OH \rightarrow H_2 + \bullet CH_2OH \quad \text{rate constant} : 2.6 \times 10^6 M^{-1}s^{-1} \tag{6.7}$$

$$\bullet OH + CH_3OH \rightarrow H_2O + \bullet CH_2OH \quad \text{rate constant} : 9.7 \times 10^8 M^{-1}s^{-1} \tag{6.8}$$

The rate constants are reported above; when the alcohol concentration is 20 mM, the calculated decay rate $k[CH_3OH]$ is $(2.6 \times 10^6) \times 0.02 = 5.2 \times 10^4\ s^{-1}$ for •H and $(9.7 \times 10^8) \times 0.02 = 1.94 \times 10^7\ s^{-1}$ for •OH, so the lifetimes are 19 μs and 51 ns for •H and •OH, respectively.

The formed radical •CH_3OH undergoes the following reaction in the absence of oxygen.

$$2\bullet CH_2OH \rightarrow CH_3OH + HCHO \quad \text{methanol and formaldehyde} \tag{6.9}$$
$$\rightarrow (CH_2OH)_2 \quad \text{ethylene glycol} \tag{6.10}$$

The rate constant of this reaction is 2.4×10^9 $l{\bullet}mol^{-1}{\bullet}s^{-1}$, and the branching ratio is 1:1.

$$\bullet CH_2OH + H_2O_2 \rightarrow HCHO + H_2O + \bullet OH \tag{6.11}$$

The G value of hydrogen formation is

$$G(H_2) = G_{H_2} + G_{e^-_{aq}} + G_H \tag{6.12}$$

The scheme changes depending on the absence or presence of oxygen. When oxygen is present in the system, the reactions proceed as follows:

$$\bullet CH_2OH + O_2 \rightarrow \bullet O_2CH_2OH(\text{peroxy radical}) \quad 4.2 \times 10^9 M^{-1}s^{-1} \tag{6.13}$$
$$2\bullet O_2CH_2OH \rightarrow 2HCHO + H_2O_2 + O_2 \quad 2.1 \times 10^9 M^{-1}s^{-1} \tag{6.14}$$
$$\bullet H + O_2 \rightarrow \bullet HO_2 \quad 2.1 \times 10^{10} M^{-1}s^{-1} \tag{6.15}$$
$$\bullet HO_2 + \bullet HO_2 \rightarrow H_2O_2 + O_2 \quad 8.3 \times 10^5 M^{-1}s^{-1} \tag{6.16}$$

The reaction scheme does not result in the decomposition of alcohol, and its characteristic is that the yield of H_2O_2 is high.

Table 6.4 summarizes the experimental results for aqueous solutions containing up to 34 mM methanol at pH 1.2.

The G values of radiolysis in an acidic solution are considered using this table. In an acidic system, $G_{\text{-}H_2O}$, $G_{e^-_{aq}} + G_H$, G_{OH}, G_{H_2}, and $G_{H_2O_2}$ are 4.5, 3.7, 2.9, 0.4, and 0.8, respectively. (They are described in the section on the Fricke dosimeter in the chapter on dosimetry.)

In the absence of oxygen, hydrated electrons, hydrogen atoms, and •OH radicals abstract hydrogen atoms from alcohol and form alcohol radicals. From the equations above, the sum of aldehyde and alcohol should be half the sum of hydrated electrons, hydrogen atoms, and •OH radicals. This condition is roughly satisfied. The sum of aldehyde and glycol is 1.90 + 1.6 = 3.55, and the sum of hydrated electrons, hydrogen atoms, and •OH radicals is 3.7 + 2.9 = 6.6, which is nearly twice that of aldehyde and glycol.

On the other hand, in the presence of oxygen, reaction (6.15) replaces (6.8), and most of the hydrogen formation occurs at $G_{H2} = 0.4$. Aldehyde is formed through (6.14), and peroxy radicals are formed through the •OH radical only; $G_{OH} = 2.9$

Table 6.4 G values in methanol aqueous solution

	G value	
Product	In absence of oxygen	In presence of oxygen
H_2	4.2	~0.6
H_2O_2	~0.6	4.15
CH_3CHO	1.90	2.6
Glycol	1.65	0

coincides well with the experimental value G(GH_3CHO, methyl aldehyde) =2.6. Hydrogen peroxide is formed by (6.14), (6.15), and (6.16) and direct decomposition of water; the yield should be 2.9/2 + 3.7/2 + 0.8 = 4.1, which agrees with the experimental value, G(H_2O_2) = 4.15. It is characteristic that oxygen is incorporated as peroxy radicals and that oxidative decomposition proceeds.

6.8 Direct and Indirect Effects

The reactions described above follow water radiolysis; the intermediates subsequently react with and decompose the solute alcohol. This process is called an indirect effect because the radiation energy is absorbed by the water, not the solute, and decomposes the water. If the alcohol concentration is increased and becomes comparable to the water concentration, the energy is directly absorbed by the solute alcohol. This process in called a direct effect.

6.9 Application of Radiolysis of Aqueous Solutions: Water Purification by Radiation

Environmental conservation is one of the applications of water radiolysis. For purification and sterilization, the radiation technique has been attempted at a fundamental or pilot scale. The object is the decomposition or removal of contaminants, and the possible processes include (1) oxidation and reduction; (2) decomposition of organics in fertilizer, pesticide, and so forth; (3) decolorization of dyes; (4) sterilization of microorganisms; and (5) deodorizing. The radiolysis products of hydrated electrons and •OH radicals react with contaminants in the solute and eventually decompose them. In combination with chemical methods (ozone treatment or enzyme treatment), the same degree of decomposition can be achievable with a lower dose. For further details, please see later chapters.

Potential obstacles to the application of radiation include (1) the cost of the accelerator and (2) public acceptance of radiation (which is sometimes low). Regarding the former, the price of accelerator as the initial investment is most influential to cost, the technology would not be disseminated. Communication with the public has been emphasized.

Exercises

1. Consult a book on radiation: how many water molecules are rearranged around a hydrated electron?
2. What kind of method is used to evaluate reaction rate constants between solutes and hydrated electron on OH radicals in aqueous solution?

Further Reading

1. Buxton GV (2004) Chapter 17: Radiation chemistry of water. In: Charged particles and photons interaction with matter. Mercel Dekker Inc, New York

Chapter 7
Nuclear Engineering and Effects of Radiation

Yosuke Katsumura and Hisaaki Kudo

Nuclear engineering consists of two fields: power (electricity) generation using nuclear energy from nuclear reactions and applications of radiation. The effects of radiation appear implicitly or explicitly and are strongly related to safety, so research on radiation chemical reactions becomes important. This chapter introduces three topics related to nuclear power generation: the water chemistry in nuclear reactors, reprocessing of spent nuclear fuel, and terrestrial disposal of high-level waste.

7.1 Water Chemistry of Nuclear Reactors

Most of the nuclear power plants currently in operation worldwide use water as a coolant and convert water to high-temperature steam, which spins the turbine to generate electricity. Nuclear power plants are classified into two types: boiling water reactors, which boil the cooling water, vaporize it, and apply the steam to the turbine, and pressurized water reactors, in which water is first pressurized and heated and then converted to steam in a secondary loop containing a steam generator. Light water (H_2O) is typically used as the coolant, though some reactors, such as the CANDU reactor in Canada, use D_2O. In any case, cooling water at 300 °C under 100 atm runs through the reactor core consisting of nuclear fuel within a few seconds. Inside the nuclear fuel, fast neutrons are generated by the nuclear fission reactor, and γ-rays are also generated simultaneously. The core is in a high-radiation field, and the dose rate is estimated to be on the order of dozens of kilograys per second. It is a thousand to ten thousand times higher than that of

Y. Katsumura · H. Kudo (✉)
The University of Tokyo, Tokai, Ibaraki, Japan
e-mail: hkudo@tokai.t.u-tokyo.ac.jp

H. Kudo (ed.), *Radiation Applications*, An Advanced Course in Nuclear Engineering 7, https://doi.org/10.1007/978-981-10-7350-2_7

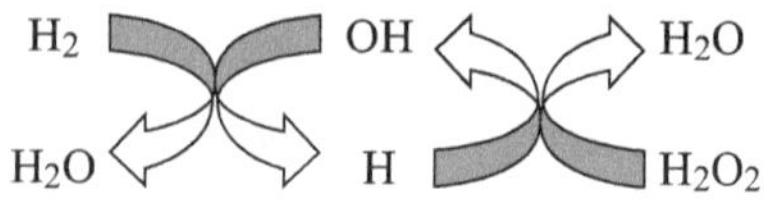

Fig. 7.1 Hydrogen induction to coolant of nuclear reactor

ordinary Co-60 γ-ray irradiation facilities, which is around 10 kGy/h. Moreover, more than half of the dose is due to fast neutrons. In the core, water is subjected to radiolysis; short-lived intermediates such as hydrated electrons, OH radicals, and hydrogen atom are generated, and radiolysis products such as H_2, H_2O_2, and O_2 are formed. Accordingly, radiolysis is one of the important factors that dominate the chemical environment of the cooling water. Indeed, substantial stress corrosion cracking begins to occur as the H_2O_2 concentration increases. As shown in Fig. 7.1, to maintain operation, the goal is to cause •OH radicals to react with H_2 and form hydrogen atoms and to change H_2O_2 into water, eventually reducing the H_2O_2 concentration.

As mentioned above, nuclear reactors operate under conditions of high temperature, pressure, and radiation, so it is difficult to observe radiolysis directly. Thus, numerical simulations, which were proposed in the 1960s, are essential. Simulations require two sets of data: the reaction rate constants between radiolysis products and the G values of high-temperature radiolysis by fast neutrons and γ-rays. For these reasons, for two decades since the early 1960s to early 1980s, research on high-temperature water radiolysis has been conducted intentionally in Denmark, Sweden, France, the UK, Canada, and Japan. To date, even though the data are still not perfect, practical data to 300 °C are available.

To describe the radiolysis of water, more than 40 reactions are necessary, as well as the ionic products of water, the pKa of molecules and radicals, and so on. Table 7.1 shows the temperature dependencies of the reaction rate constants at room temperature, 285 °C, and 300 °C. Further, Fig. 7.2 shows the temperature dependencies of the G values (primary field) of decomposition of water for linear energy transfer radiation. As shown in this figure, the G values of hydrated electrons, •OH radicals, H_2, and hydrogen atoms increase linearly from room temperature to 300 °C. The values for hydrated electrons and •OH radicals are 40% and 80% higher, respectively, at 300 °C than at room temperature. The reason is that •OH is less reactive than hydrated electrons at high temperatures. On the other hand, the value for hydrogen peroxide decreases with temperature. Further, hydrogen peroxide decomposes thermally, and the lifetime at 100, 150, and 200 °C is 50, 5–10, and 1 min, respectively.

Table 7.1 Reaction rate constants of radiolysis products of light water at 25, 285, and 300 °C

No.	Reaction	k_{25}	k_{285}	k_{300}
(1)	$e^-_{aq} + e^-_{aq} \rightarrow$	6.44E09*	2.93E11	3.29E11
(2)	$e^-_{aq} + H \rightarrow$	2.64E10	3.68E11	3.98E11
(3)	$e^-_{aq} + OH \rightarrow$	3.02E10	1.34E11	1.40E11
(4)	$e^-_{aq} + H_2O_2 \rightarrow$	1.41E10	2.53E11	2.76E11
(5)	$e^-_{aq} + O_2 \rightarrow$	1.79E10	2.57E11	2.78E11
(6)	$e^-_{aq} + O_2^- \rightarrow$	1.28E10	1.65E11	1.78E11
(7)	$e^-_{aq} + HO_2 \rightarrow$	1.28E10	1.65E11	1.78E11
(8)	$H + H \rightarrow$	5.43E09	9.20E10	1.00E11
(9)	$H + OH \rightarrow$	1.53E10	6.58E10	6.87E10
(10)	$H + H_2O_2 \rightarrow$	5.16E07	1.03E09	1.13E09
(11)	$H + O_2^- \rightarrow$	1.32E10	9.72E10	1.03E11
(12)	$H + HO_2 \rightarrow$	9.98E09	7.33E10	7.78E10
(13)	$H + O_2 \rightarrow$	9.98E09	7.33E10	7.78E10
(14)	$OH + OH \rightarrow$	4.74E09	2.00E10	2.09E10
(15)	$OH + H_2 \rightarrow$	4.15E07	1.26E09	1.40E09
(16)	$OH + H_2O_2 \rightarrow$	2.87E07	5.42E08	5.19E08
(17)	$OH + HO_2 \rightarrow$	1.08E10	3.10E10	3.20E10
(18)	$OH + O_2^- \rightarrow$	1.10E10	8.44E10	8.97E10
(19)	$HO_2 + HO_2 \rightarrow$	6.64E05	2.41E07	2.69E07
(20)	$HO_2 + O_2^- \rightarrow$	7.58E07	3.82E08	4.01E08
(21)	$O_2^- + O_2^- \rightarrow$	–	–	3.50E07
Equilibrium reaction				
(22)	$H_2O \rightarrow$	1.95E-05	7.57E-02	6.90E-02
(−22)	$H^+ + OH^- \rightarrow$	1.10E11	9.76E11	1.07E12
(23)	$H_2O_2 \rightarrow$	7.86E-02	4.40E01	4.15E01
(−23)	$H^+ + HO_2^- \rightarrow$	4.78E10	5.42E11	6.12E11
(24)	$H_2O_2 + OH^- \rightarrow$	1.27E10	1.37E11	1.47E11
(−24)	$HO_2^- + H_2O \rightarrow$	1.36E06	1.31E08	1.39E08
(25)	$H \rightarrow$	6.32E00	1.36E05	1.40E05
(−25)	$H^+ + e^-_{aq} \rightarrow$	2.25E10	4.31E11	5.15E11
(26)	$e^-_{aq} + H_2O \rightarrow$	1.57E01	8.28E03	9.94E03
(−26)	$e^-_{aq} + OH^- \rightarrow$	2.49E07	3.40E10	4.22E10
(27)	$OH \rightarrow$	7.86E-02	4.40E01	4.15E01
(−27)	$H^+ + O^- \rightarrow$	4.78E10	5.42E11	6.12E11
(28)	$OH + OH^- \rightarrow$	1.27E10	1.37E11	1.47E11
(−28)	$O^- + H_2O \rightarrow$	1.36E6	1.31E08	1.39E08
(29)	$HO_2 \rightarrow$	7.14E05	2.70E05	1.98E05
(−29)	$H^+ + O_2^- \rightarrow$	4.78E10	5.42E11	6.12E11
(30)	$HO_2 + OH^- \rightarrow$	1.27E10	1.37E11	1.47E11
(−30)	$O_2^- + H_2O \rightarrow$	1.36E06	1.31E08	1.39E08

(continued)

Table 7.1 (continued)

No.	Reaction	k_{25}	k_{285}	k_{300}
Reaction is alkaline solution				
(31)	$O^- + H_2 \rightarrow$	1.21E08	1.62E09	1.75E09
(32)	$O^- + H_2O_2 \rightarrow$	5.53E08	1.04E10	1.13E10
(33)	$OH + HO_2 \rightarrow$	8.29E09	1.56E11	1.70E11
(34)	$O^- + OH \rightarrow$	7.60E09	3.23E10	3.38E10
(35)	$e^-_{aq} + HO_2 \rightarrow$	3.50E09	6.33E10	6.90E10
(36)	$e^-_{aq} + O^- \rightarrow$	2.31E10	1.02E11	1.07E11
(37)	$O^- + O_2 \rightarrow$	3.70E09	3.04E10	3.24E10
(−37)	$O_3^- \rightarrow$	2.68E03	1.45E07	1.88E07

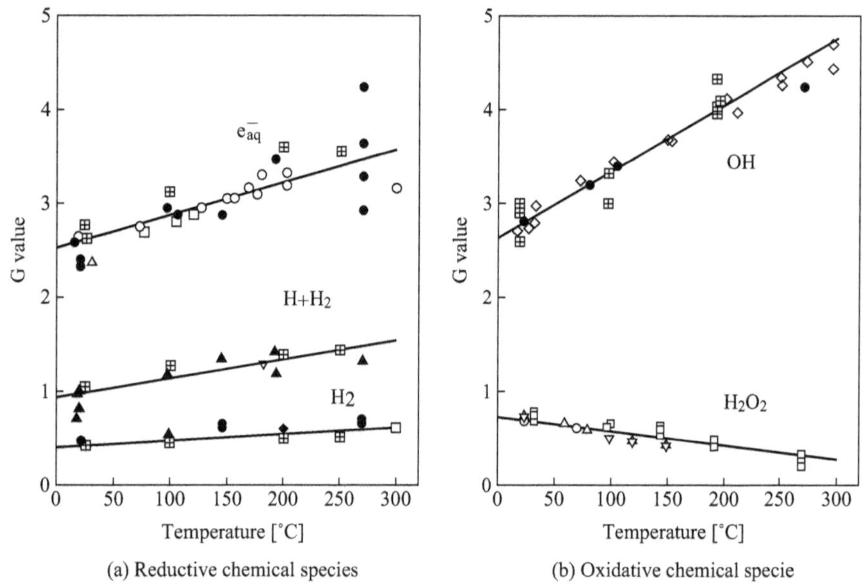

Fig. 7.2 Temperature dependence of G value (primary G value) of water radiolysis products

7.2 Effect of Radiation in Reprocessing of Spent Nuclear Fuel

7.2.1 Reprocessing and Radiolysis

Nuclear power generation uses a thermal energy of 200 MeV per fission of U-235. Simultaneously, fission products (FPs) are generated. A typical nuclear fuel is UO_2 in which the U-235 abundance is enriched from the natural value of 0.7% to several percent. Nuclear fuel is exchanged when one ton of UO_2 fuel is used to extract an energy of 30 GW•day^{-1}. One ton of spent fuel contains 30 kg of FPs as well as 950 kg of U, 9 kg of Pu, 140 g of Np, and 47 g of Am. Not all of the spent fuel is usable; it is reprocessed for reuse after the U and Pu are recovered from the other

FPs. The dominant U-238 is effectively used by changing it into fissile Pu. Reprocessing is expected to be able to (1) increase the available energy by converting the parent materials into fissile materials, (2) reduce the hazard and cost of handling high-level wastes, (3) reduce the cost of the nuclear fuel cycle, and (4) extract valuable materials from FPs.

Many reprocessing methods have been proposed to date, and the plutonium uranium refining by extraction (purex) method is widely used at present. This method uses tributyl phosphate [TBP, $(C_4H_9)_3PO_4)$] as an extractant; the TBP is dissolved in n-dodecane (n-$C_{12}H_{26}$, density 0.749) to a 30 vol% concentration, as TBP has a density of 0.973 g•cm^{-3} and is difficult to separate from water. First, the fuel is dissolved in nitric acid with a concentration greater than 10 mol•l^{-1}. Consequently, the uranium and plutonium are in the chemical forms of $UO_2{}^{2+}$ and Pu^{4+}, respectively. The solution is then diluted to 3 mol•l^{-1} nitric acid. This solution is mixed with the TBP-dodecane extractant, and after intense stirring, $UO_2{}^{2+}$ and Pu^{4+} are selectively extracted into the organic phase, leaving other FPs in the aqueous phase. This process is called the extraction process. Pu^{4+} in the organic phase is reduced to Pu^{3+} chemically or electrochemically. If the organic phase comes into contact with the dilute nitric acid aqueous solution, Pu^{4+} changes to the aqueous phase, leaving $UO_2{}^{2+}$ in the organic phase. Thus, uranium and plutonium can be separated.

The extractant must (1) have a high distribution ratio for uranium and plutonium, (2) have a low distribution ratio for FPs, (3) rapidly reach chemical equilibrium, (4) exhibit stability in water and acid, (5) be radiation resistant, (6) have a reasonable cost, (7) exhibit low inflammability, (8) exhibit a large difference in density from the aqueous phase, and (9) have an appropriate viscosity and surface tension. The TBP-dodecane system satisfies these requirements and has been widely used. Spent fuel contains many radioactive materials; therefore, the extraction process is conducted in a high-radiation environment. Radiolysis of the extraction solvent and nitric acid can cause (1) the loss of uranium and plutonium, (2) a decrease in the decontamination factor (ratio of radiation intensity) before and after treatment, that is, the degree of radiation removal, (3) precipitation, (4) a decrease in the decomposition yield, and (5) an increase in instability, with the possibility of criticality. Accordingly, understanding the degradation of the solvent is essential for successful reprocessing.

In this field, the intensity of radiation is usually expressed as the energy emitted from the solution in watt-hours per cubic decimeter. If all the energy is absorbed in the solution, 1 W•hr•dm^{-3} = 3600 J/kg = 3.6 kGy. An absorbed energy of 0.1 W•hr•dm^{-3} will not cause detrimental effects. At 1 W•hr•dm^{-3}, a slight effect is observed, and at 10 W•hr•dm^{-3}, a substantial decrease in the decontamination performance is observed.

To date, Japan has contracted with the UK and France to reprocess spent fuel. The first domestic commercial reprocessing facility is being tested for operation in Rokkasho-mura, Aomori Prefecture. It can treat 800 t•$year^{-1}$ with the purex method.

Table 7.2 G value of gamma radiolysis products of TBP

Products	Burr	Wilkinson	Wagner	Holland	Burger
H_2	1.73	1.11	1.59	2.02	–
CH_4	0.072	0.05	0.07	0.032	–
C_2	0.18	0.24	0.13	–	–
C_3 and C_4	0.66	0.45	0.75	–	–
DBP	2.44	1.52	2.25	3.71	1.9
MBP	0.14	0.12	0.39	–	0.3
Polymer	2.47	–	–	–	–

7.2.2 Radiolysis of TBP and n-Dodecane

Radiolysis of TBP has been reported since the 1950s; the main products are dibutyl phosphate (DBP), monobutyl phosphate, and hydrogen gas. C–O bonds tend to dissociate more readily than C–C bonds. TBP dimers are also observed. The yield of DBP is higher than that of hydrogen. The reports to date are summarized in Table 7.2. The G values differ between reports, possibly because of differences in the purity of the TBP used. Electron spin resonance measurements and pulse radiolysis studies of TBP suggest that solvated electrons and TBP react and produce butyl radicals and DBP.

$$e^-_{aq} + (C_4H_9O)_3P = O \rightarrow \bullet C_4H_9 + (C_4H_9O)_2P(O)O^-$$

Many reports have been made on the spectral characteristics of solvated electrons and ion pair recombination.

More reports are available for saturated hydrocarbons than for TBP. n-dodecane is a saturated hydrocarbon, and the G values of its products are summarized in Table 7.3. Hydrogen and dimers are the main products, and the decomposition of n-paraffin can be explained by the following reaction, where RH and R(−H) represent paraffin and the corresponding alkene, respectively.

$$RH \rightsquigarrow R\bullet + H\bullet$$
$$\rightsquigarrow R(\text{-}H) + H_2$$
$$R_1\bullet + R_2\bullet \rightarrow R_1(\text{-}H) + R_2H \rightarrow R_1R_2$$

7.2.3 Radiative Decomposition of Nitric Acid (Direct and Indirect Action)

The nitric ion is a scavenger of hydrated electrons; when it is concentrated, it is also known as a scavenger of precursors of hydrated electrons. Nitric ions have low

Table 7.3 G value of gamma radiolysis products of n-dodecane

Products	Report		
	Holland	Dewhurst	Gäuman
H_2			
CH_4	0.05	0.05	0.01
Dimer	–	–	1.4

reactivity with •OH radicals. In reprocessing, a wide range of nitric acid concentrations, from 0.1 to 10 M, is used. At high concentrations, the formation of the $NO_2^{\bullet}$ radical has been observed. The $NO_3^{\bullet}$ radical is similar to the sulfuric radical ($SO_4^{\bullet -}$), a strongly oxidative radical, and plays an important role, which makes this species of interest.

The formation mechanism of the $NO_3^{\bullet}$ radical was studied in a nanosecond-scale pulse radiolysis study of concentrated nitric acid. The $NO_3^{\bullet}$ radical, which has three characteristic absorption lines around a wavelength of 630 nm, can be deconvoluted to its components: a fast formation process just after the pulse and a slow formation process (on the order of several hundred nanoseconds). Adding different •OH scavengers to concentrated nitric acid reduces the slow process, leaving the fast process unchanged. For nitric Na and nitric Li solutions of the same concentration, only the fast process was observed, and the •OH scavenger was found to have less effect. This discussion of the change in the concentration of •OH scavengers in concentrated nitric acid suggests that the slow formation of $NO_3^{\bullet}$ radicals can be explained by the reaction between •OH radicals and molecular HNO_3 ($k = 1.4 \times 10^8$ l•mol^{-1}•s^{-1}). On the other hand, the rapid formation was found to be the result of direct action of radiation on the nitric acid.

Thus, for concentrated nitric acid, the $NO_3^{\bullet}$ radical forms through indirect and direct action. To examine whether similar phenomena would be observed for other acids, a pulse radiolysis study was conducted. It showed that there are fast and slow processes for $SO_4^{\bullet -}$ formation, the former by direct action and the latter by H abstraction from HSO_4^- or H_2SO_4 by •OH. Further, similar experiments were conducted for phosphoric acid, HClO, and $HClO_4$, and it was found that these phenomena are commonly observed for concentrated acids.

In radiolysis of nitric acid, hydrogen gas is known to evolve and to decrease with the concentration, while O_2 becomes dominant above the order of mol•l^{-1}. Simultaneously, increased formation of nitrous acid has been reported. To account for these results, direct action was taken into account, and a radiolysis model of a concentrated nitric acid solution was established.

7.2.4 TBP, n-Dodecane, and Nitric Acid Mixture System

So far, the radiolysis solvent used in reprocessing has been described. In a real process, solvents are used in mixtures, which makes the radiation effect more complicated. Further, the process occurs under an air atmosphere, so oxygen is

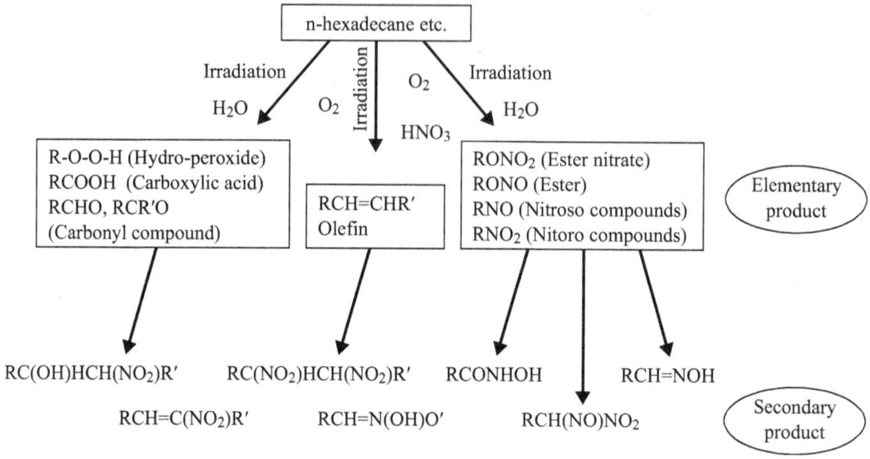

Fig. 7.3 Radiolysis products of organic solvent

involved in the reaction. Peroxy radicals would be formed, and oxidation would proceed, forming ketones, alcohols, and carboxylic acids. Consequently, miscellaneous products are found. The reactions are summarized in Fig. 7.3. DBP and other products form complexes with FP, and the separation efficiency is decreased. Accordingly, to reduce the radiolysis, in a real process, a separation column having a short contact time is used, and the solvent is washed frequently. Therefore, various actions to avoid accumulation of the products are performed. Further, a radiation-resistant extractant has been proposed, but further discussion from many viewpoints is necessary before it can be applied to real processes.

7.2.5 Radiation Chemistry of Actinide Ion

Transuranium elements generated during the fission of uranium are called actinides; they include uranium, neptunium, plutonium, curium, and americium. They have many oxidative states. For example, uranium has four oxidative states: U^{3+}, U^{4+}, UO_2^{+}, and UO_2^{2+}. Elucidation of the mechanism of radiation oxidation-reduction has attracted interest. Many actinides are alpha emitters, and radiation chemistry by α-rays is also of interest. These topics are assumed to represent important fundamental engineering data, and many studies have been performed in the former Soviet Union.

7.3 Once-Through, Reprocessing, and Disposal of High-Level Radioactive Waste

The uranium and plutonium separated in reprocessing are used again as fuel. The simultaneously separated FPs are dissolved into fused glass and then solidified. The glass is then sealed in a metal container called a canister made of thick copper, titanium, or carbon steel. The canister will be disposed of deep in the ground and surrounded by a silty soil called bentonite, through which water does not permeate. The idea is to ensure long-term safety by setting multiple barriers of glass, solidification, the canister, bentonite, and the deep underground environment. Space disposal to the outer Solar System or disposal in the ocean or arctic zones has been discussed, but it was concluded that terrestrial disposal is reasonable for long-term isolation from our biosphere, decay of the radiation, and avoidance of the effects of radioactive materials. However, not all countries have adopted a reprocessing policy; some countries, for example, the USA, Canada, and Sweden, do not reprocess spent fuel. They selected a single use, or once-through, approach; spent fuel is sealed into a canister without reprocessing and subjected to terrestrial disposal.

Long-term isolation of high-level radioactive waste from the biosphere is necessary, but underground water exists deep in the ground. A disposal site is designed to prevent groundwater from entering it, contacting the canisters, and corroding the metal, and to prevent glass or radioactive materials from dissolving into groundwater, diffusing, and reaching the biosphere. In the design process, it is necessary to evaluate in advance the effect of radiation on the corrosion of metal, dissolution of radioactive material into groundwater, and diffusion of radioactive material. For this purpose, radiolysis of water has been investigated. Groundwater originating in rainfall contains carbonic acid (up to 50 m mol•l^{-1}) and other trace solutes. Moreover, some groundwater is affected by seawater and contains NaOH. Radiolysis of carbonic acid aqueous solution has been conducted. Some countries have discussed the burial of spent fuel in the rock salt layer deep in the ground. This layer has not been subjected to groundwater for a long time, and a similar condition is expected to obtain for a long time into the future as well. When this type of disposal is considered, upon entering of ground water radiation effect on concentrated saltwater as high as 5 mol•l^{-1} is of concern, and a fundamental study has been conducted. Experiments on the dissolution of glass into an aqueous solution upon long-term exposure have also been conducted, and they confirmed that under the expected environment, the effect is not significant. For once-through, the radiation effect was discussed in terms of the behavior of groundwater and the dissolution of spent UO_2 into groundwater when groundwater enters, corrodes the canister, and contacts the fuel, and the cladding is damaged. Dissolution of UO_2 as the matrix causes dissolution of the FPs and actinides in the fuel. When UO_2 is oxidized to UO_2^{2+}, it becomes water soluble. It is important to understand the process of oxidization and the effect of radiation on the process.

As described in this chapter, the effect of radiation on groundwater has been studied fundamentally and experimentally as part of the evaluation of the safety of terrestrial disposal of spent fuel. The recommended reference is the volume on the engineering of radioactive waste in this series.

Exercises

1. Water contains oxygen at concentration of 2.5×10^{-4} mol/l under equilibrium with a vitrogen (80% and oxygen 20%). Is it appropriate to use as coolant for nuclear reactor? If it is not appropriate, consider how it is improved.
2. How is radiation chemical scheme different between dilute aqueous solution, concentrate aqueous solution, and organic solvent?

Further Readings

1. Katsumura Y (2004) Radiation application in nuclear technology. In: Charged particles and photons interaction with matter. Chapter 23. Mercel Dekker Inc., New York
2. Kent MC, Sims HE (1992) The yield of γ-radiolysis products from water at temperatures up to 270°C. AEA-RS-2302, PWR/CTG/ (92) 084
3. Kent MC, Sims HE (1992) The yield of γ-radiolysis products from water at temperatures up to 300°C. In: Proceedings of the 6th international conference of water chemistry of nuclear reactor systems. British Nuclear Energy Society, London, p 153
4. Elliot AJ, Chenier MP, Ouellette DC (1993) Temperature dependence of g values for H O and D O irradiated with low linear energy transfer radiation. J Chem Soc Faraday Trans 89:1193
5. Sunaryo GR, Katsumura Y, Hiroishi D, Ishigure K (1995) Radiolysis of water at elevated temperatures—II. Irradiation with γ-rays and fast neutrons up to 250°C. Radiat Phys Chem 45:131
6. Katsumura Y, Sunaryo GR, Hiroishi D, Ishigure K (1998) Fast neutron radiolysis of water at elevated temperatures relevant to water chemistry. Prog Nucl Energy 32:113
7. Stefanic I, LaVerne JA (2002) Temperature dependence of the hydrogen peroxide production in the γ-radiolysis of water. J Phys Chem A 106:447
8. Burr JG (1958) The radiolysis of tributyl phosphate. Radiat Res 8:214
9. Willkinson R, Williams TFJ Chem Soc 2098
10. Wagner RM, Kinderman EM, Towle LH (1959) Radiation stability of organophosphorus compounds. Ind Eng Chem 51:45
11. Holland JP, Merklin JF, Razvi J (1978) The radiolysis of dodecane-tributylphosphate solutions. Nucl Inst Methods 153:589
12. Berger LL, McClanahan ED (1957) Abstracts of ACS meeting at Miami
13. Dewhurst HA (1957) Radiation chemistry of organic compounds. I. n-Alkane liquids. J Phys Chem 61:1466
14. Rappoport S, Gaeuman T (1973) Radiolyse des hydrocarbures. 20e communication. n-Alcanes en phase liquide de l'heptane au dodécane. Helv Chim Acta 56:531
15. Rappoport S, Gaeuman T (1973) Radiolyse des hydrocarbures. 21e communication. Mécanismes radicalaires et non-radicalaires dans les nalcanes en phase liquide. Helv Chim Acta 56:2861

16. Huggard AJ, Warner BF (1963) Investigations to determine the extent of degradation of TBP/ Odorless Kerosene solvent in the new separation plant, Windscale. Nucl Sci Eng 17:638
17. Nagasaki S, Nakayama S (kyohen) (2011) Genshiryoku kyoukasyo Housyasei haikibutsu no kogaku, Omu sya, Tokyo (in Japanese); Nagasaki S, Nakayama S (eds) (2011) Engineering of radio-active wastes, in Nuclear Engineering Textbooks series, Ohmsha, Ltd., Tokyo (in Japanese)

Chapter 8
Radiation Processing of Polymers and Its Applications

Masao Tamada

8.1 Polymers and Processing Technologies

Plastics (synthesized resins) are man-made polymeric materials, and the name originates from the Greek word "plastikos" (made from soils by mixing and kneading). Plastic was initially a semi-synthesized mode of natural polymers. Natural rubber contained in the sap of the rubber tree was crosslinked and cured by sulfur-adding vulcanization; this was developed as ebonite in 1851. Because of its high mechanical strength, it was used for the body of pens and bowling balls. In 1868, celluloid (cellulose nitrate) was developed by chemical treatment of cellulose to improve the disadvantage that it burns easily. Celluloid is currently essential for ping-pong balls. In 1907, phenol resin (Bakelite) was developed through the reaction of phenol and formaldehyde at high temperature and pressure, where the phenol is a byproduct of coal tar. This was the first plastic made from small molecules of raw materials other than plant-originated materials.

As plastic became used as a material, its structure was not well recognized. It was assumed to be merely an aggregation of many small molecules. In 1930, H. Staudinger proposed that it was a compound in which many monomers were bonded. Later, structural analysis with X-rays and viscosity measurements of polymer solutions revealed that the plastics were composed of covalently bonded monomers.

Once the repeating structure of monomers resulting from polymerization of monomers in sequence was clarified, W. Carothers of DuPont in the USA synthesized polyamide resin for the first time. It was commercialized as nylon 6-6 and used instead of silk to make stockings. An advertisement described it as a "fiber made from coal, water and air, thinner than a spider's web, and stronger than steel."

M. Tamada (✉)
National Institutes for Quantum and Radiological Science and Technology, Gunma, Japan
e-mail: tamada.masao@qst.go.jp

H. Kudo (ed.), *Radiation Applications*, An Advanced Course in Nuclear Engineering 7, https://doi.org/10.1007/978-981-10-7350-2_8

Later, in 1956, polycarbonate and polyacetal were developed, and in 1965, polyimide was developed as an engineering plastic to replace metal because it was lightweight and had high mechanical properties. These polymers have excellent thermal endurance and durability, and as plastics they have low expansion coefficients; thus, they were used in mechanical engineering and are currently used for insulation in electronic circuits. They are also used as structural materials for automobiles, airplanes, and ships after they are reinforced with inorganic fiber for higher strength.

Polymer materials are not limited to engineering plastics; after further high functionalization, they have been widely used in various fields such as electrically conducting materials, liquid crystal materials, biomaterials, molecule-sensing materials, separation materials, optical materials, photofunctional materials, and magnetic materials. For functionalization, a specific chemical structure related to the intended purpose should be introduced, and monomers having miscellaneous functional groups have been synthesized. However, as further varieties were developed, the requirements for polymers moved beyond those that can be achieved with single monomers. Thus, various methods of blending, copolymerization, and graft polymerization have been researched as diverse methods of adding functions to polymers. Figure 8.1 shows the structures obtained by each method. Blending, which mixes polymers, is a simple process, but phase separation can occur depending on the compatibility of the polymers in the mixture, resulting in nonhomogeneous products. In the copolymerization, reactivity between monomers must be considered to control the polymer composition for necessary functionalization. Thus, attention was given to radiation graft polymerization and the radiation crosslinking reaction as the techniques of polymer modification. Graft polymerization has the advantage that combinations of hydrophilic and hydrophobic polymers or polar and nonpolar polymers can be used regardless of their compatibilities and relativities. Crosslinking reactions can improve solubility and heat resistance by introducing new covalent bonds between polymer chains.

Figure 8.2 shows a typical technique of radiation processing for polymers. Upon irradiation by an electron beam or γ-rays that induces ionization and excitation in polymers, radicals are formed after dissociation and act as the initial reactive species.

In graft polymerization, polymers having radicals are in contact with monomers, leading to grafting. From trunk polymers, different polymer chains evolve as branches. Well-known methods of forming radicals are radiation, ultraviolet light, plasma, and chemical initiators. In particular, when radiation is used to form radicals for grafting, the radiation energy is so high that there is no limitation on the types and shapes of trunk polymers and grafted monomers, and various combinations are available. In addition, it is possible for radiation-induced grafting to introduce functions to the deep inner region of the substructures.

Crosslinking is recombination between the formed radicals. When radiation is used, the solubility and the heat resistance can be improved without chemical crosslinking agent. Heating is not necessary, and the thermal degradation is

Fig. 8.1 Techniques to functionalize polymer materials and structures attained

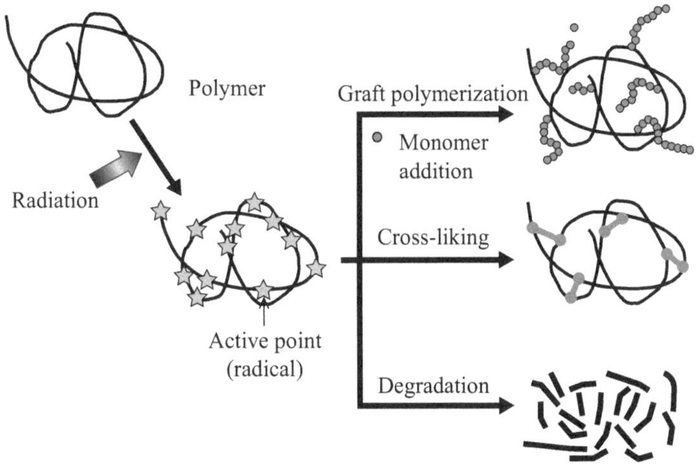

Fig. 8.2 Typical techniques to process polymer with radiation

negligible. Even if the crosslinking density is low, the reactions are between polymer chains, and the properties change substantially.

Degradation occurs when the formed radicals cause scission of the polymer chains. Scission and crosslinking can happen simultaneously and in general, which one dominates is determined by the chemical structure. Polyethylene (PE, -CH_2-CH_2-) is classified as a crosslinking type. Polyvinyl chloride [-(CH_2-CHR)-], where R-Cl, is classified as a crosslinking type, whereas polypropylene ($R=CH_3$) and polystyrene ($R=C_6H_5$) are classified as degradation types. Polymethyl methacrylate [-(CH_2-CR_1R_2)-], where $R_1=CH_3$ and $R_2=COOCH_3$, is classified as a degradation type. The yields of crosslinking and scission can be evaluated using the Charlesby–Pinner equation. The atmosphere during irradiation can affect the results of irradiation; for example, PE, which is a crosslinking type, can change to a degradation type under an oxygen atmosphere. Polytetrafluoroethylene (PTFE) is known as a degradation type, but the polymer can be crosslinked by irradiation just above its melting temperature.

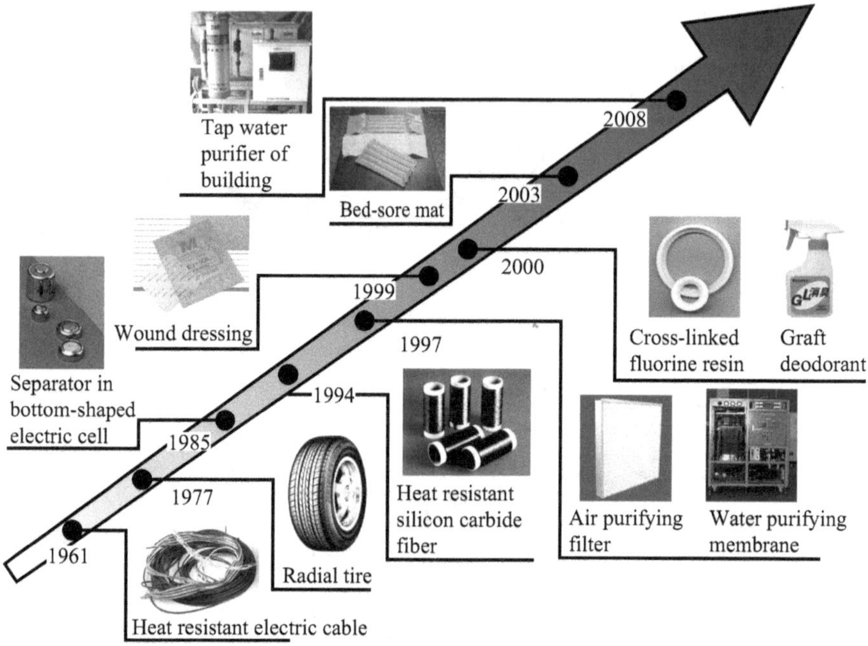

Fig. 8.3 Examples of commercialized technique of polymer processing by radiation

As mentioned above, by applying radiation-induced graft polymerization and crosslinking reactions, it is possible to endow polymers with new functions and improvement of their properties that are impossible to achieve by other methods of conventional chemical synthesis, copolymerization, and blending. These reactions, induced by high-energy electron beams and γ-rays, occur at low temperature. A chemical initiator is not necessary, and scaling is easy.

By applying radiation processing techniques to polymers, many products have been commercialized. Figure 8.3 shows the examples of commercialized products. In the 1960s, by crosslinking of PE, heat-resistant cables, heat-shrinkable tubes, and foamed mats were commercialized. The heat-resistant cables which have been used for electrical wiring around car engine establish the safety of cable sheath at high temperature. The heat-shrinkable tube is widely used to prevent leakage from drain pipes. Foamed mats are widely used for, e.g., the interior of automobiles, bath mats, and swimming beat mats. In the 1970s, radiation crosslinking began to be used to improve the processing performance in the fabrication of radial tires. More than 90% of tires currently undergo radiation crosslinking on the production line. The entire tire is finished by vulcanization by adding sulfur. In the 1980s, radiation graft polymerization was used to produce a commercial separator membrane for high-performance button-shaped battery cells. Lower self-discharge was realized by endowing a PE film with the function of electrical conductivity. In the 1990s, heat-resistant silicon carbide fiber was developed by crosslinking to improve the shape stability during sintering. Another important development in crosslinking is wound dressing by a crosslinked polyvinyl alcohol (PVA) gel. Graft polymerization enabled the commercialization of an air purification filter for ammonia in semiconductor factories and a self-reusable electrolysis membrane. The 2000s saw commercialization of the following: crosslinking of fluororesin, a deodorant obtained by graft polymerization, a bedsore prevention mat made of a biodegradable gel, and purification of circulation water for air conditioning using a metal adsorbent obtained by graft polymerization. In the coming years, the commercialization of a solidifier for stock-breeding waste using a biodegradable dry gel, a heat-resistant polylactic acid obtained by crosslinking, and a high-performance fuel cell obtained by graft polymerization is expected.

8.2 Graft Polymerization

Radiation graft polymerization can be classified into the simultaneous method and the pre-irradiation method, depending on the irradiation procedures applied to the trunk polymer. In the simultaneous method, the trunk polymer is irradiated in a monomer solution for graft polymerization. The dose rate should be low enough that the monomer can permeate the trunk polymer and be graft polymerized. Although the technique is simple, a substantial amount of homopolymer (polymers made of monomers not used for graft polymerization) is formed and must be removed after the reaction. On the other hand, in the pre-irradiation method, the trunk polymer is first irradiated and then placed in contact with a monomer and graft polymerized. In the pre-irradiation method, the reaction and graft reaction steps are separated from each other; thus, graft polymerization is easy to control, and this method is suitable for commercialization (i.e., large-scale production).

An outline of graft polymerization by the pre-irradiation method using polyethylene (PE) as the trunk polymer is described below. PE consists of crystalline regions and amorphous (noncrystalline) regions. When an electron beam irradiates PE to 200 kGy, radicals (as many as 10^{18} g^{-1}) are formed in the crystalline and amorphous regions. After irradiation, when the substrate is exposed to air, oxygen permeates the amorphous region and deactivates the radicals. However, it is difficult for oxygen to permeate the crystal region, so it deactivates only the radicals that diffuse to the surface of the crystalline region. Next, by contact with the monomer in the absence of oxygen, graft chains evolve from the surface of the crystalline region. Accordingly, even 20 days after electron beam irradiation, graft polymerization is possible with little trouble if the irradiated PE substrate is stored at the temperature of dry ice (−78 °C), with little deactivation of radicals. During storage at 7 °C for three days after irradiation, 50% of the radicals survive. Thus, it is possible to hire an irradiation service corporation to irradiate the trunk polymer and then keep the irradiated polymer at low temperature and perform graft polymerization at distant facilities.

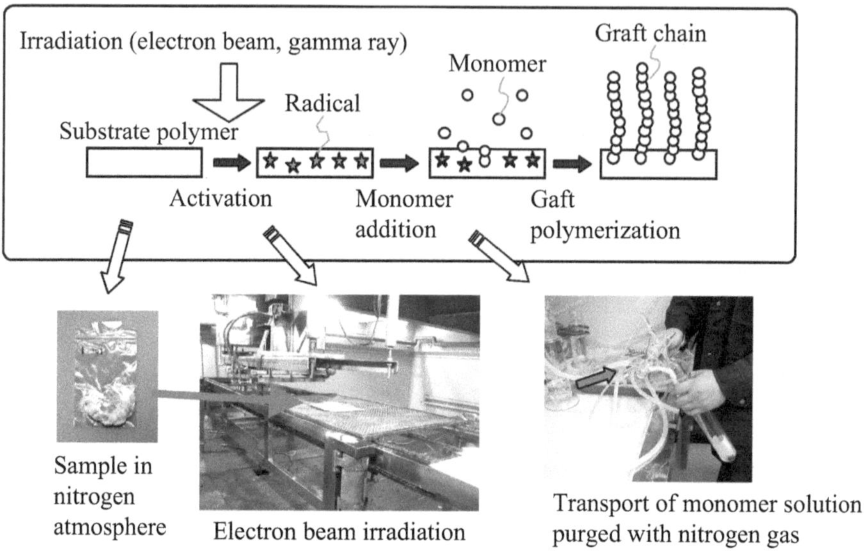

Fig. 8.4 Experimental procedure for graft polymerization by pre-irradiation method

In graft polymerization by the pre-irradiation method, the radical concentration and degree of grafting can be controlled by varying the dose and reaction time, respectively. The grafting rate increases with increasing monomer concentration or increasing reaction temperature. If the trunk polymer is swollen by the solvent used, the permeation forms grafts to the deep interior. If the trunk polymer precipitates the solvent, graft polymerization is enhanced by the gel effect, which reduces the termination reaction by increasing the viscosity.

Figure 8.4 shows the graft polymerization process for the pre-irradiation method. The trunk polymer is placed in a zippered plastic bag, the air inside of which is replaced by nitrogen gas. It is then irradiated with electron beam to the prescribed dose (up to ~200 kGy). Upon irradiation, radicals are formed as reactive sites in the trunk polymers. The trunk polymer is usually moved in an evacuated glass reaction vessel. The monomer solution, which is bubbled (and saturated) with nitrogen in advance, is transported, and then graft polymerization is performed. The temperature is controlled to 40–60 °C. The degree of grafting (D_g) is the weight increase after grafting as a percentage of the weight before grafting. D_g is given by the formula:

$$D_g\,[\%] = 100(W - W_0)/W_0$$

where W and W_0 are the weight of the trunk polymer after and before grafting, respectively.

When a nonwoven PE is selected as the substrate, irradiated to 200 kGy by an electron beam, and placed in contact with a solution of dimethyl sulfur oxide solvent containing 10% glycidyl methacrylate at 40 °C, the degree of grafting reaches 170% after 2 h. A degree of grafting of 170% means that a graft chain weighing as much as 1.7 times the weight of the trunk polymer is produced.

Even for a water-insoluble monomer such as glycidyl methacrylate, it was found that an emulsion system prepared by dispersing the monomer in water with a surfactant increased the yield of graft polymerization by as much as ten times compared to that obtained using an organic solvent. This can reduce the dose required for grafting, which is beneficial for commercialization.

8.2.1 Application of Graft Polymerization

When graft polymerization is adopted, it is possible to introduce a chemical structure of interest into the existing trunk polymer. Thus, by introducing a chelate group having high affinity to metal ions, a high-performance metal adsorbent is obtained. The shape of the trunk polymer can be selected with little limitation, and it is possible to make a fiber-like metal adsorbent. Adsorbents, forms spherical particles, 0.3–1.2 mm in diameter, currently on the market are produced by copolymerization of styrene derivatives and divinylbenzene. The shape stability of the resin adsorbent is improved by making a crosslinking structure and increasing the divinylbenzene component. However, the styrene derivative component used to introduce functional groups decreases, and diffusion into the interior of the resin is reduced, resulting in decreasing adsorption rate. In graft polymerization, functional groups are provided to the existing trunk polymer, which has fewer restrictions on synthesis. Graft polymerization is an excellent procedure for synthesizing metal adsorbents.

Research and development activities on metal adsorbents began in 1845, when natural zeolite was found to have ion-exchange functionality. In 1935, ion-exchange resins such as sulfonic acid or amine groups were synthesized. Since the 1950s, when environmental pollution became serious because of emerging economic development, research and development of chelate resins having high selectivity to metals were conducted. Research on the production of chelate adsorbents by graft polymerization began in the 1980s, and nonwoven fabric was developed as a substrate in the late 1990s. Fiber-shaped adsorbents having a much faster adsorption rate than crosslinked resin collecting materials have been developed.

For example, in the methods for handling graft-polymerized fabric adsorbents and crosslinked resin adsorbents, the initial step is the same, as both materials are soaked in solutions containing metal ions. However, after the metal is adsorbed, the crosslinked resin materials must be removed by percolation, whereas the grafted fabric can be removed by tweezers. Therefore, the two materials differ greatly in their ease of handling.

Figure 8.5 shows the procedure for synthesizing a metal adsorption fabric by graft polymerization. A substrate of PE nonwoven fabric is irradiated by an electron beam. After irradiation, monomers having a phosphate group or amino group are

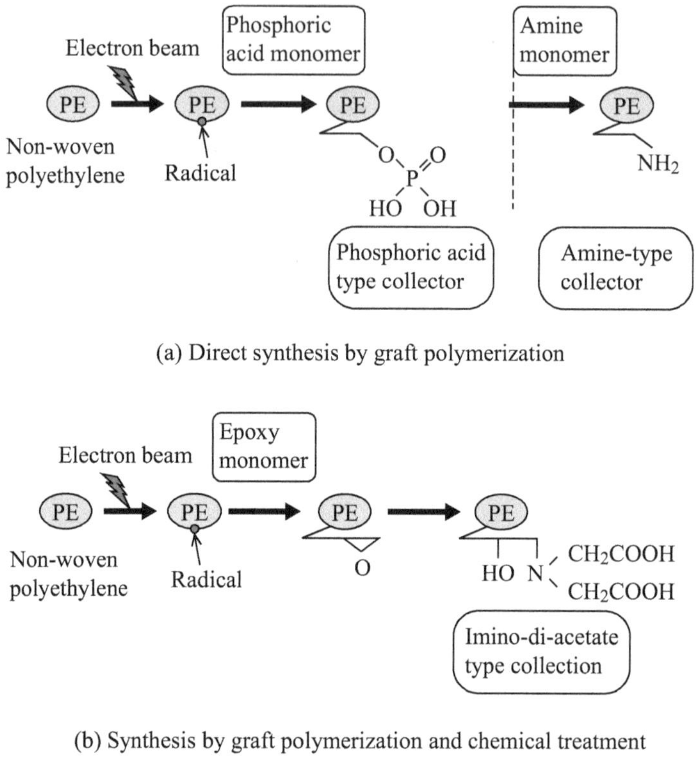

Fig. 8.5 Synthesizing procedure of metal collector by graft polymerization

directly grafted, and metal adsorbents having the functional group of interest can be synthesized in one step. If glycidyl methacrylate having an epoxy group is graft polymerized, the introduced epoxy group can be endowed with various functional groups by chemical treatments. In Fig. 8.5, after graft polymerization, sodium iminodiacetate is reacted, and metal adsorption materials of the iminodiacetate type are obtained. In this case, the precursors of metal adsorbents are synthesized by graft polymerization. Subsequently, functional groups having high affinity to metals are introduced by chemical treatment in a total of two stages. When acrylonitrile is grafted and subsequently treated with hydroxylamine, adsorbents incorporating an amidoxime group are obtained. Similarly, when glycidyl methacrylate is grafted and subsequently treated with N-methyl glucamine, glucamine-type adsorbents are obtained. When zirconium oxide is loaded onto phosphate-type adsorbents, the adsorbents can absorb arsenic, which exists as a negative oxy-ion. Figure 8.6 compares the adsorption capacities of crosslinked resin and graft-polymerized fabric. Adsorbents were packed in a column, 7 mm in inner diameter, and an arsenic (V) solution with a concentration of 5 mM was applied at different space velocity (SV = flow/adsorbent volume). For the crosslinked resin at $SV = 10\ \mathrm{h}^{-1}$, the breakthrough capacity for arsenic was 0.12 mol per kilogram of

Fig. 8.6 Comparison in absorption capacity between cloth-shaped collector (graft polymerization) and crosslinked resin

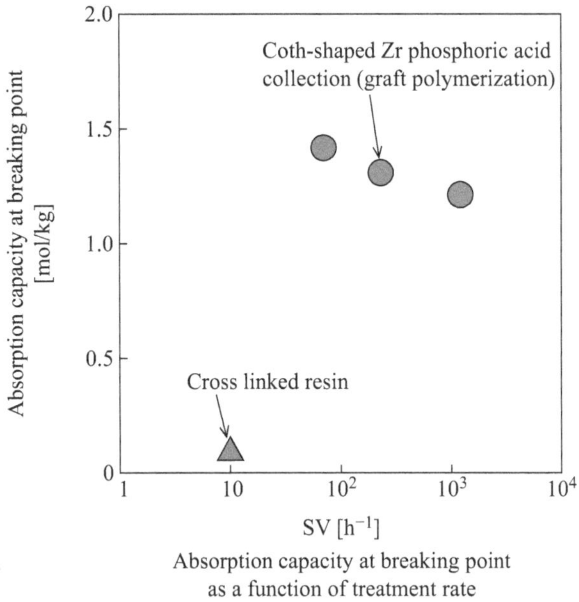

resin, where breakthrough was set as the concentration at which the concentration at the outlet becomes 5% of that at the inlet. Above $SV = 10\ h^{-1}$, arsenic was not absorbable. In contrast, for the grafted fabric, even at $SV = 1000\ h^{-1}$, the breakthrough adsorption capacity was still seven times higher than that of the crosslinked resin. This example demonstrates the excellent adsorption properties of grafted adsorbent fabric.

8.2.2 *Application of Grafted Metal Adsorbents*

1. Removal of Cadmium from Processing Residue of Scallops

Scallops are a specialty product of Aomori Prefecture, Japan, and their production amounts to hundreds of thousands of tons per year. Around a thousand tons of scallop glands have been disposed of as waste. Although the glands contain protein, fat, docosahexaenoic acid, and eicosapentaenoic acid, they also contain the harmful cadmium at about 10–20 ppm and so must be combusted as industrial waste without utilization.

Boiling chemically changes the protein in the gland, and cadmium ions become tightly bonded. To remove the cadmium using adsorption by metal adsorbents, it is necessary to dissolve the cadmium ion from the raw gland. Organic acids such as maleic acid, tartaric acid, malonic acid, and malic acid are effective for extraction, and cadmium can be dissolved effectively similarly to the cases of inorganic acids such as hydrochloric acid and sulfuric acid. Figure 8.7 shows the cadmium

Fig. 8.7 Adsorption property to Cd ion extracted from gland of scallop with malic acid

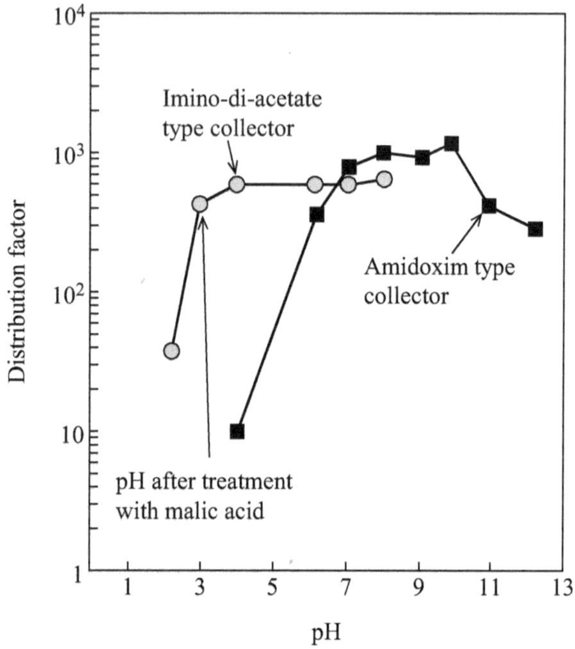

adsorption properties with respect to the distribution coefficient for cadmium adsorbents of the iminodiacetic acid and amidoxime types synthesized by radiation grafting. The distribution coefficient is the ratio of the metal concentration in the adsorbent to that in solution under equilibrium. The pH of malic acid after cadmium extraction was around 3.3, and the iminodiacetic acid-type adsorbent was more appropriate for cadmium removal than the amidoxime-type adsorbent [2]. This was demonstrated using an apparatus made to treat 30 kg of glands. A malic acid solution was circulated in a column filled with the iminodiacetic acid-type adsorbent; after 24 h, 95% of the cadmium was removed, and the concentration was decreased to 1.3 $mg \cdot kg^{-1}$ (1.3 ppm). This concentration satisfies the limit specified by law regulations for feed and fertilizer safety (2.5 ppm and 5 ppm, respectively), and it was found that the treated glands were applicable as animal feed or fertilizer.

Collection of Uranium from Seawater

The uranium concentration of seawater is as low as 3.3 $mg \cdot t^{-1}$. However, if the entire amount of seawater is considered, the amount of uranium is 4.5 billion tons, which are 1000 times the buried resources on land. Around the Japanese islands, the tidal wave called "Groshio" carries 5.2 million tons of uranium annually, and only 0.2% of it would provide the amount necessary for nuclear power generation.

Figure 8.8 shows the ocean test for amidoxime-type adsorbents prepared by graft polymerization. The adsorbents were hung from floating frame at 7 km offshore of

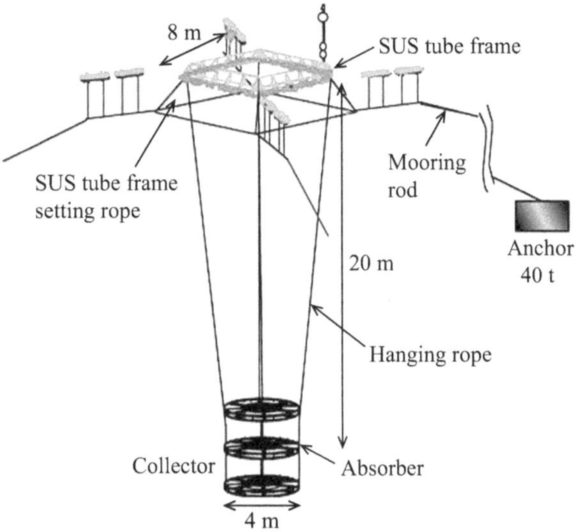

Fig. 8.8 Mooring system of amidoxime-type collector used in performance test ocean

Aomori Prefecture, Sekine-hama, Japan. To obtain as much as 6000 m^2 of adsorbents, rolled nonwoven fabric, 1.5 m in width and 200 m in length, was γ-irradiated and treated with grafting for 50 h at 40 °C. Then it was chemically treated with hydroxylamine. The obtained adsorbent fabric was cut into pieces, 29 cm × 16 cm in size, stacked in a cassette cases, and then set in a 4 m × 4 m absorption beds. The adsorption beds were hung from the floating frame at a depth of 20 m below the sea surface. The tests were conducted 12 times over 3 years, and a total of 1 kg of yellow cake was recovered from seawater.

This floating frame method is affected by bad weather or affects ship traffic, so another method using standing mooring on the seabed was considered. This new method does not require a structure above the sea surface, which reduces the recovery cost. A braid-type adsorbent that can stand up from the seabed was prepared using PE fiber as the trunk polymer, synthesizing amidoxime-type adsorbents and braiding the fibers. In the ocean around the Okinawa Islands in Japan, a performance test was conducted, as shown in Fig. 8.9, where the braid adsorbent stood at a depth of 100 m. In this test, braid uranium adsorbents having the ability of uranium adsorption were set between 15-m-long braid PE and placed in the sea. The braid adsorbents were connected to anchors using a detach equipment; when the braid adsorbents were released from the anchors by sending a signal to the detach equipment, they floated free and were recovered.

Using data from the collection tests in the sea near Okinawa with braid-type adsorbents, the cost for collecting uranium in seawater was analyzed. The amount of uranium collected annually was assumed to be 1200 t (the amount required for six nuclear power plant units). It was also assumed that the collecting performance was 2 grams of uranium per kilogram of adsorbent for 60 days of immersion in seawater, that the adsorbent can be used repeatedly six times, and that the cost of the

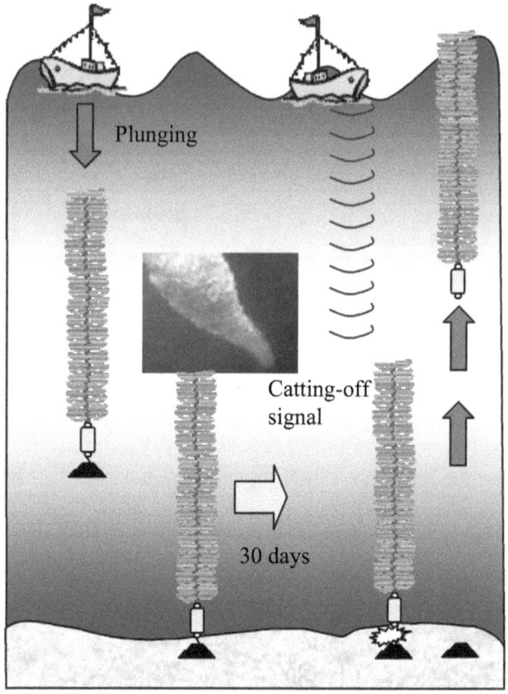

Fig. 8.9 Ocean test procedure of mooring-shaped collector carried out in Okinawa marine ocean

production of adsorbent, recovery, elution, and purification of uranium was 105 billion Japanese yen per year. This cost was divided by the amount of uranium collected annually (1200 t), yielding a cost of 88 thousand Japanese yen per kilogram of uranium. If research and development to improve the properties of the adsorbents are continued so that the collecting performance increases to 6 kilograms of uranium per kilogram of adsorbent, and the adsorbent becomes reusable for up to 18 times, the present plausible cost is 25 thousand yen, which is equivalent to 96 USD per pound of U_3O_8 and 1.5 times the spot price in August 2008. In this case, the initial investment is 107.7 billion Japanese yen, which is about one-third of the construction cost of a 1-million-KW-class nuclear power plant. In Japan, nuclear power generation accounted for 34% of the total electricity generated, and all the uranium fuel was imported. If uranium in seawater can be a feasible resource, energy security will become robust, and this will provide the time margin until an ideal nuclear fuel cycle system is established.

8.3 Crosslinking

Radiation crosslinking has been conducted only since 1952, when A. Charlesby found that it improved the heat resistance and the solubility of polyethylene. Crosslinking in polymers forms a three-dimensioned network structure through new bonds between chains due to combination of radiation-induced radicals. The radicals migrate within and between molecules through hydrogen abstraction and addition. When chains containing radicals have a high mobility and the radicals are located close to each other, crosslinking occurs readily. Thus, in crystals or below the glass transition temperature, the crosslinking efficiency is not high because the motion of chains is limited.

The degree of crosslinking can be evaluated after the soluble (not crosslinked) fraction is extracted by a solvent and the crosslinked fraction is separated as follows:

$$\text{Gel fraction}\,(\%) = 100 \quad W/W_0,$$

where W is the weight of the residual gel and W_0 is the weight of the sample before extraction.

The atmosphere during irradiation greatly affects the crosslinking. In the presence of oxygen, the irradiation-induced radicals react with oxygen and do not form crosslinking. However, crosslinking is preferred if the sample is thick or the dose rate is high since oxygen is supplied through diffusion inside the polymer. An antiaging reagent or antioxidant can scavenge radicals and inhibit crosslinking.

If crosslinking can be introduced by irradiation, practically important properties such as the mechanical, thermal, electrical, and chemical properties can be improved; thus, the crosslinking technique is used in many industrial fields. After crosslinking, the tensile strength generally increases once, reaches a maximum, and then decreases. The elongation decreases, and the friction resistance increases. In the thermal properties, the melting point and crystallinity are reduced, and elasticity is observed above the melting point because of scission of polymers in the crystalline region. The electrical properties change little, but the dielectric dissipation factor increases upon irradiation in the presence of air. As the chemical properties, the antisolvent stability improves.

8.3.1 Applications of Crosslinking

The crosslinking reaction is widely used, mainly to improve the heat resistance of sheaths for electrical wires and cables, plastic foam, and radial tires.

Electrical wires and cables are covered with polyethylene or polyvinyl chloride to insulate the metallic conductor from the exterior. The demand is increasing for cables having high heat resistance, shape stability, and mechanical strength as automobiles, computers, and cellular phone are miniaturized or made lightweight. In Japan, 20 companies are currently performing crosslinking of electrical cables, processing as much as 1.2 million km per year.

Plastic foam is made by increasing the viscosity in the molten state by crosslinking. The procedure is as follows: polyethylene and a foaming agent are mixed and produce many gases upon thermal degradation; the mixture is shaped in sheets, irradiated by an electron beam for crosslinking, and heated to the decomposition temperature of the foaming agent, finally yielding the foam material.

For radial tires, electron beam crosslinking has been conducted as a pretreatment for the body-reinforced layer. Finally, vulcanization using sulfur is applied to fabricate the tires. Radiation crosslinking increases the mechanical strength of the raw rubber and improves the processibility so that the amount of rubber needed can be reduced. The product quality becomes stable with air leakage.

Machinery and materials having heat resistance and high strength can be made by radiation crosslinking of composites of silicon carbide and metal or ceramics. Polycarbosilane fiber, an organic silicon polymer, cannot be used at high temperature (1200 °C) because of reactions with residual oxygen. However, by electron beam crosslinking under a helium atmosphere, silicon carbide fibers having less oxygen can be synthesized while maintaining the original shape, and their heat resistance can be improved so that they withstand 1700 °C. This lightweight fiber, which has high heat resistance and high mechanical strength, is expected to find applications in space exploration and fusion reactors.

Water-absorbable hydrogel can be obtained through electron beam crosslinking by irradiating a concentrated (ca. 10%) aqueous solution of water-soluble polyvinyl alcohol. This hydrogel, which retains moisture and is transparent, was commercialized as a wound dressing to heal scrapes and burns under wet conditions.

Polytetrafluoroethylene (PTFE, also known by the commercial name Teflon™) is used, for example, to line the inside of frying pans, and can be crosslinked when it is irradiated at high temperature (just above the melting point). Crosslinking improves the elongation at the break by more than ten times compared to non-crosslinked PTFE. The crosslinked PTFE can be fabricated as a transparent film. The friction resistance is also improved more than 1000 times, and it has been commercialized for use in bearing seals or O-rings.

8.3.2 Improvement of Biodegradable Polymers by Crosslinking

Polymers made of cellulose from plants and polymers made by microorganisms are called biopolymers; they can be decomposed by microorganisms and are environmentally friendly. Most of these polymers undergo scission upon irradiation and are classified as the degradation type. However, it was recently found that such polymers could be crosslinked by adding crosslinkers or by irradiating them under conditions where polymer radicals are located close to each other.

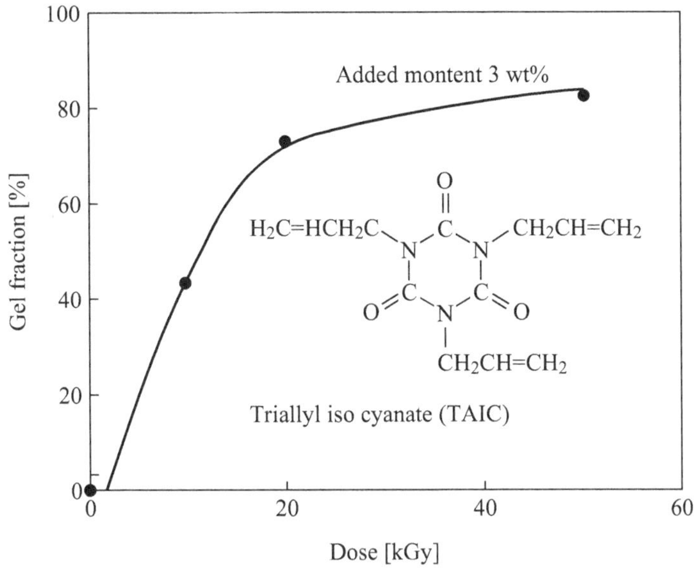

Fig. 8.10 Crosslinking property of polylactic acid with crosslinker

Crosslinking of Polylactic Acid

Polylactic acid (PLA) is synthesized by condensation polymerization of lactic acid, which is produced from fermentation of starch. Because it is transparent and biocompatible, this material is expected to be used in the bodies of electronic devices and regenerative biomedical materials. However, PLA is deformed above 60 °C, so it cannot be used as a biomedical material because medical devices are sterilized at 120 °C. The radiation crosslinking properties were investigated after PLA incorporated with various crosslinkers at the melting point of PLA (180 °C). Polyfunctional triallyl isocyanurate (TAIC) was found to be very effective, as shown in Fig. 8.10. Upon electron beam irradiation, TAIC reacts with PLA to form a crosslinked structure. For 1% TAIC concentration, crosslinking does not proceed. In the presence of 3% TAIC concentration, crosslinking becomes dominant, as shown in Fig. 8.10, and a gel fraction of ca. 80% is obtained at 20 kGy. Crosslinked PLA has excellent heat resistance, showing a deformation of less than 10% above 100 °C. Crosslinking provides a memory effect. For example, tubular PLA with addition of TAIC is irradiated for crosslinking. Then, the crosslinked PLA is heated to 200 °C and expanded twice, followed by cooling to room temperature while maintaining the expanded shape. This expanded tube, upon heating to 160 °C, recovers its original shape before expansion. Its mechanical strength at 80 °C is three times that of a heat-shrinkable polyethylene tube. It also has the advantage of being visually transparent, which enables visual inspection of the interior of the tube (e.g., a cable or wire) to detect shorting at the connection.

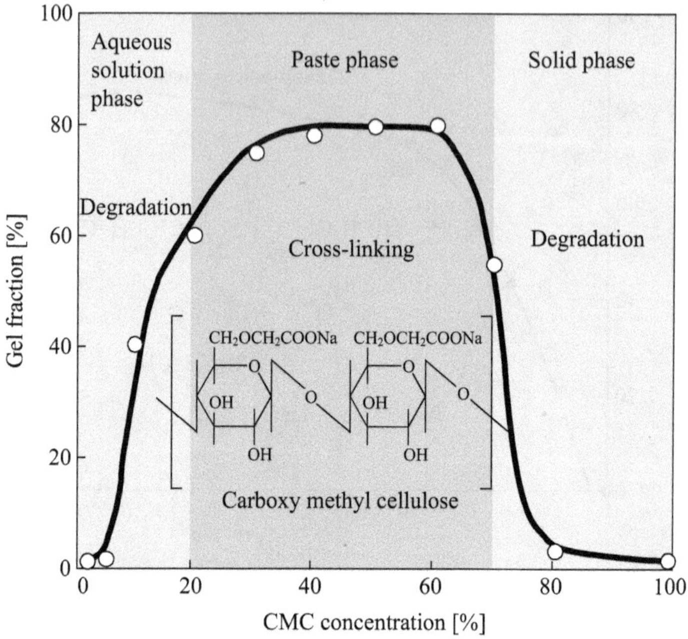

Fig. 8.11 State of carboxymethyl cellulose and crosslinking

Crosslinking of Natural Polysaccharides

Starch, cellulose, and chitin/chitosan are typical natural polysaccharides. These polymers and their derivatives do not undergo crosslinking and have been classified as the degradation type. However, in 2000 it was found that water-soluble polysaccharides can undergo crosslinking in a high-concentration aqueous solution. Carboxymethyl cellulose (CMC), shown in Fig. 8.11, is widely used as an additive to adjust the viscosity of ice cream. When more than 5% CMC is dissolved in water, the viscosity of the solution increases; therefore, studies have been limited to those with a dilute aqueous solution or solid. However, when more than 5% of CMC is mixed with water, the solution (mixture) becomes paste-like. Crosslinking was observed upon irradiation in this state. The crosslinking reaction depends on the CMC concentration, degree of substitution (DS) of the carboxymethyl group, and molecular weight. Figure 8.11 shows the gel fraction of CMC having a DS of 2.2 after irradiation to 10 kGy; a wide range of CMC concentrations between 30% and 60% yields a high gel fraction. For a DS of 1.3, CMC concentrations of 20–30% are most effective. The crosslinking points are the carboxymethyl groups, and only the paste-like state (which has little fluidity) can undergo crosslinking.Crosslinked CMC gel absorbs much amount of water. Its absorbance capacity is about 1.7 times higher than that of poly(sodium acrylate) gel, but the swelling rate is slow. Crosslinked CMC gel has a slightly higher biodegradability because irradiation occurs simultaneously, and the molecular weight is reduced.

Fig. 8.12 Biodegradable gel used for treatment of excrement of livestock

Carboxymethyl starch, carboxymethyl chitin, and carboxymethyl chitosan can be crosslinked only when they are irradiated in the paste-like state. In addition, it has been confirmed that methyl cellulose and hydroxypropyl cellulose can undergo crosslinking only when they are irradiated in the paste-like state.

Crosslinked CMC gel is decomposed by microorganisms in the soil, so it is applied to treat livestock excrement. Livestock excrement must be composted; stockpiling it in the field is prohibited so that it does not leak nitric nitrogen or harmful viruses or bacteria into rivers or underground. As shown as Fig. 8.12, livestock excrement has a water content of 80–90%, which makes it difficult to ferment and compost. Thus, mixing it with sawdust and reducing the water content to 65–70% enhance composting. Instead of sawdust, dried crosslinked hydrogel can be used, and adding 0.2% hydrogel could reduce the amount of sawdust waste by one-sixth. Sawdust is becoming difficult to obtain, so mixing with CMC hydrogel is expected to become more common.

By filling a mat with CMC paste and irradiating it, the mat can be given an excellent ability to keep warm and disperse stress to the human body. A trial was made using a surgical operation mat after it was warmed to human body temperature. It was demonstrated that no patient suffered bedsores in 178 trials. Therefore, this mat was commercialized as a bedsore prevention mat.Gel obtained by irradiation of paste-like carboxymethyl chitin and carboxymethyl chitosan can adsorb noble metal ions such as gold, and application and a biodegradable adsorbent are expected.

4% aqueous solution was irradiated and added to 20 ppm

Fig. 8.13 Plant growth promotion effect by radiation degradation of alginic acid

8.4 Degradation and Its Applications

PTFE (Teflon™) is highly resistant to acids and bases, but upon irradiation, it readily degrades. After its molecular weight is lowered by degradation, it still retains excellent nonadhesive properties, flame retardancy, and low friction resistance. By processing degraded PTFE into a powder, a high-performance lubricant and remover have been commercialized.

Polysaccharides such as alginate and chitosan, which occur in seaweed on kelp and the shells of shrimp and crab, respectively, degrade upon irradiation. The degraded polysaccharides show new activities. Degraded alginate and chitosan promote plant growth. Figure 8.13 demonstrates that rice shows evidently well-grown roots using degraded alginate. Degraded chitosan is found to lengthen the storage life of fruits.

Exercises

1. What are three methods of radiation processing of polymers?
2. Give examples of applications of radiation in industry.
3. Describe the problems and the countermeasures to develop radiation processing techniques to polymer materials for radiation application in industrial sector.

Chapter 9
Radiation Resistance of Semiconductors

Takeshi Ohshima and Shinobu Onodaa

9.1 Radiation Environment and Semiconductor Devices

9.1.1 Radiation Environment

There is a good side and a bad side to the relationship between radiation and semiconductors. The good side is that radiation can be actively utilized to improve material properties of semiconductors, and the bad side is that radiation can degrade the characteristics of semiconductor devices. Examples of active utilization of radiation include (1) conductive control such as neutron transmutation doping using thermal neutrons and the introduction of impurities through ion implantation and (2) carrier lifetime control through electron beam irradiation. With regard to the bad side, using semiconductor devices in a radiation environment seems to be a very special case that is irrelevant to our life. However, in the present day in which the domain of human activities has dramatically expanded to the extent where it is not limited to the Earth, using semiconductor devices in a radiation environment is no longer unrealistic. For example, it is known that there is a lot of radiation in the space environment, and failure of semiconductor devices due to radiation effects is a very serious problem for space satellites.

There is a strong radiation field around the Earth which consists of the solar flare, the galactic cosmic ray from other galaxies, and the radiation trapped by the magnetic field of the Earth. The solar flare is released as a result of solar flares, which are sudden explosions on the surface of the Sun. The solar flare consists mainly of protons and heavy particles. It is known that the amount and energy of the solar flare vary over time in a complex manner because of the fact that the solar flare is a result of sudden explosions. The galactic cosmic rays are high-energy radiation

T. Ohshima (✉) · S. Onodaa
National Institutes for Quantum and Radiological Science and Technology, Gunma, Japan
e-mail: ohshima.takeshi@qst.go.jp

H. Kudo (ed.), *Radiation Applications*, An Advanced Course in Nuclear Engineering 7, https://doi.org/10.1007/978-981-10-7350-2_9

that comes from outside the solar system. It is said that the galactic cosmic rays are generated by explosions of supernovas. The galactic cosmic rays are typically composed of 85% proton, 14% alpha particles, and 1% (or less) heavy ions. The heavy ions mainly contain boron (B), carbon (C), oxygen (O), silicon (Si), and iron (Fe). The galactic cosmic rays have high levels of energy to the order of several GeV. The radiation trapped by the magnetic field of the Earth forms a doughnut-shaped radiation belt around the Earth in which high-energy electrons and protons are permanently present. The part of the trapped radiation belt which is located below the altitude (as measured from the surface of the Earth) of 15,000 km is called the "inner belt," and the part of the trapped radiation belt which is located above the altitude of 15,000 km is called the "outer belt." The high-energy electrons are widely distributed in the outer part of the inner belt and in the outer belt, and there are high electron density areas in the outer belt. The high-energy protons are mostly present in the inner belt. These were discovered by the Geiger-Muller detector installed on board the Explorer launched in 1959. The captured radiation belt is also called the "geomagnetically captured radiation belt" or the "Van Allen radiation belt" (after the physicist who discovered the belt).

9.1.2 Types of Radiation Effects on Semiconductors

The radiation effects on semiconductor devices are the Total Ionizing Dose (TID) Effect, the Displacement Damage Dose (DDD) Effect, and the Single Event Effect (SEE). Each of these effects will be explained in detail later in this book. In general, these effects shorten the lifetime of space satellites and can cause temporary malfunctions and failures of electronic equipment. Table 9.1 summarizes the types of radiation, the main effects of the individual types of radiation, the interactions caused by the individual types of radiation, and the parts of semiconductors which can be affected by the individual types of radiation.

As shown in Table 9.1, gamma rays are utilized mainly to investigate the Total Ionizing Dose Effect. Gamma rays are also utilized to investigate the Displacement Damage Dose Effect, although such use of gamma rays is very infrequent. The parts of semiconductor elements which are most susceptible to gamma rays are silicon dioxide (SiO_2) of such parts as gates of MOS (metal oxide semiconductor) transistors and capacitors and interfaces between silicon dioxide and silicon. Electron beams are utilized mainly to investigate the Displacement Damage Dose Effect. Electron beams are advantageous in that, although the Displacement Damage Dose Effect is caused by gamma rays as well, electron beams allow irradiation tests to be conducted more efficiently (which means that irradiation tests can be conducted in short periods of time even in the case of high-dose irradiation) because the displacement cross section caused by electron beams is larger. There is also a practical advantage to the use of electron beams. That is, electron beams allow levels of displacement damage in the space environment to be confirmed directly because there are a lot of electrons in the space environment as mentioned above.

Table 9.1 Types of radiation and their effects

	Gamma rays	Electron beams	Heavy particle rays
Main effect	Total Ionizing Dose Effect (Displacement Damage Dose Effect)	Displacement Damage Dose Effect	Single Event Effect
Interaction	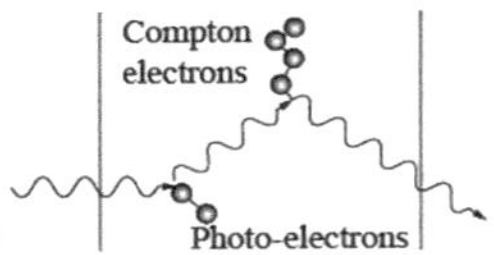	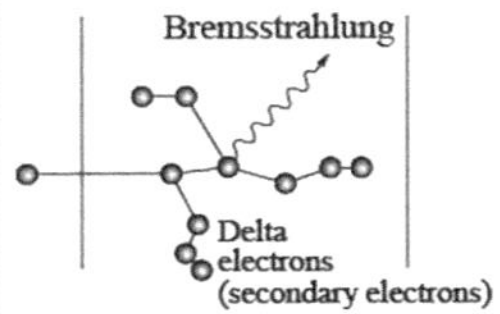	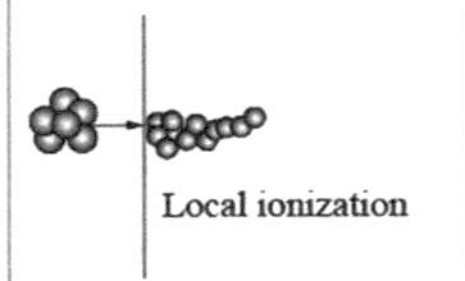
Affected parts of semiconductors	Insulating films of MOS transistors, etc.	Semiconductor crystals of solar cells, diodes, etc.	Semiconductor crystals of diodes, MOS transistors, etc.

For the same reason, proton beam-based irradiation tests are also being conducted frequently as are electron beam-based irradiation tests. The part of semiconductor elements which can be affected by electron beams and proton beams are not silicon dioxide formed on surfaces of semiconductor, but are semiconductor crystals themselves. Alpha rays and heavy particle rays are utilized mainly to investigate the Single Event Effect. One characteristic of alpha rays and heavy particle rays which distinguishes them from gamma rays, electron beams, and proton beams is that alpha rays and heavy particle rays deposit energy to extremely smaller areas and cause high-density ionization. This is the main reason for the fact that the incidence of even a single heavy ion can cause a significant effect on semiconductor devices such as soft errors and hard errors. The part of semiconductor devices that can be affected by alpha rays and heavy ions is semiconductor crystals themselves, as in the case of the Displacement Damage Dose Effect.

9.1.3 History of the Radiation Effects on Semiconductors

The Total Ionizing Dose Effect and the Displacement Damage Dose Effect started to be viewed as problems at about the same points of time. It is said that Telstar-I, which was launched in 1962, is the first satellite that stopped functioning as a result of the Total Ionizing Dose Effect. On the day before the launch of Telstar-I, the United States conducted a high-altitude nuclear experiment (the Starfish Prime). Beta rays (electron beams) generated by the high-altitude nuclear experiment were trapped by the Earth's magnetism and dramatically increased the dose in the radiation belt. This caused failures of semiconductor devices installed on board Telstar-I, including diodes and narrow-base transistors, due to the Total Ionizing Dose Effect, and, 1 year later (in 1963), Telstar-I completely stopped functioning. Subsequent studies showed that the Total Ionizing Dose Effect is a phenomenon that occurs when silicon dioxide films of MOS transistors and MOS capacitors and their interfaces are affected. The Total Ionizing Dose Effect on MOS capacitors will be explained in Sect. 9.2.

The Displacement Damage Dose Effect was reported in relation to solar cells installed on board such satellites as Transit 4B, Traac, Ariel, and OSO-1, which were launched at about the same points of time as Telstar-I. In Japan, the Displacement Damage Dose Effect was reported in relation to silicon solar cells installed on board Kiku 6 (ETC-VI), a satellite launched in 1994. Kiku 6 was not put into the originally intended geostationary orbit because of problems in the orbit change jet equipment called the "apogee engine" and was operated on an orbit that crossed the high-dose Van Allen belt. During the operation of Kiku 6 on this orbit, it was discovered that the power generation capability deteriorated sharply whenever the satellite entered a domain with a radiation dose above a certain threshold. Because of the discovery of the Displacement Damage Dose Effect in relation to solar cells and because of the fact that deterioration of the power generation capability of solar cells affects the operating lifetime of solar cells significantly and the fact that shielding is not possible, many studies have been conducted on the Displacement Damage Dose Effect on solar cells, and abundant information has been obtained both in the area of materials research and the area of space applications. The Displacement Damage Dose Effect on solar cells will be explained in Sect. 9.3.

With regard to the Single Event Effect, the possibility that the effect can occur was pointed out by J. T. Wallmark and S. M. Marcus in 1962. In 1975, D. Binder et al. discovered that the Single Event Effect can cause electrical errors in flip-flop circuits installed on board satellites. Since then, many studies have been conducted on the Single Event Upset (SEU), which is a phenomenon wherein high-energy heavy ions induce errors in highly integrated memory cells. When the Single Event Upset (which is also called the "soft error") occurs, the data stored as 0 and 1 will be reversed to 1 and 0, respectively. The reversed data can be restored to its original state by rewriting the data. In the 1980s, studies on the Single Event Upset caused by protons started to be conducted as well. Recently, a wide range of new Single Event Effects have been discovered which cannot be explained with the traditional theory of the previously discovered Single Event Upset. These new Single Event Effects include the multiple bit upset (MBU), single event latch-up (SEL), single event functional interrupt (SEFI), single event transient (SET), single event burnout (SEB), single event gate rupture (SEGR), and single event hard error (SEHE). The Single Event Effect is observed not only in the space environment in which a lot of radiation exists but also on the ground, because the Single Event Effect is caused by the incidence of a single ion. In the latter half of the 1970s, it was reported that Single Event Effect was caused by alpha rays emitted by radioisotopes contained in packaging materials and solder such as uranium and thorium. This problem has already been solved by the introduction of the technique to remove the radioisotopes contained in materials and raw materials. Aside from the problem of packaging materials and solder, J. F. Ziegler et al. reported in 1979 that Single Event Effect was caused by ground neutrons. Ground neutrons are neutrons that are generated by the collision of space radiation and molecules present in the atmosphere and fall onto the ground with various levels of energy. Incidence of ground neutrons onto semiconductor elements can cause a nuclear reaction, although the probability that it will cause a nuclear reaction is fairly low. If a nuclear reaction is caused, a nuclear product (heavy ions) is generated. This nuclear product causes Single Event Effect. The most important feature characterizing neutron-induced Single Event Effect is the fact that they occur not in special radiation environments such as space and nuclear facilities but on the ground on which we live.

Single Event Effect that occurs on the ground is not different from Single Event Effect that occurs in space in the sense that they are caused by an abnormal transient current in a circuit generated by the incidence of a single heavy ion into a semiconductor device. The process of generation of abnormal transient currents which trigger Single Event Effect will be explained in Sect. 9.4.

9.2 Total Ionizing Dose Effect

9.2.1 Interactions Between Gamma Rays and Semiconductors

As described in Sect. 9.1.2, gamma rays are utilized to investigate the Total Ionizing Dose Effect. Because evaluation tests of semiconductor devices sometimes span a long period of time, radioisotopes (especially 1.17 MeV and 1.33 MeV gamma rays emitted by cobalt 60) are used in many cases rather than electron or heavy ion accelerators. In general, incidence of a gamma ray onto a material causes such interactions as the Photoelectric Effect, the Compton Effect, and pair production. Photonuclear reactions are also caused. However, the photonuclear reactions can be ignored in the case where gamma rays are used in Total Ionizing Dose Effect evaluation tests, because the cross section of photonuclear reaction is very small. As the gamma rays repeat the interactions and lose energy, the orbital electrons will be ionized and excited. The secondary electrons generated by the ionization (which are called "delta rays") will repeat elastic scattering and inelastic scattering (ionization loss and bremsstrahlung) and lose energy (for details, refer to Sect. 9.3). If an atom is excited, the ground state is reached through transformation into heat energy and release of the characteristic X-ray and energy will be lost.

The likelihood of occurrence of the Photoelectric Effect, the Compton Effect, and pair production depends on the energy of the incident photons. In the case of silicon, there is a strong likelihood that the Photoelectric Effect, the Compton Effect, and pair production occur in the low-energy domain (up to several hundred keV), middle-energy domain (several hundred keV to several MeV), and high-energy domain (higher than several MeV), respectively. Therefore, in the case of cobalt 60 gamma rays used in Total Ionizing Dose Effect evaluation tests, the main interaction is the Compton Effect. Because the energy of the gamma rays scattered by the Compton Effect can be as high as about 1 MeV, it is likely that the Photoelectric Effect will occur following the Compton Effect. Electron pair generation will be negligible. Outlines of the Photoelectric Effect, the Compton Effect, and pair production follow:

Photoelectric Effect

The Photoelectric Effect is a phenomenon wherein the energy of electromagnetic radiation is totally absorbed through transformation into the ionization energy of orbital electrons and the kinetic energy of photoelectrons. The likelihood of occurrence of the Photoelectric Effect is strong for the inner orbital electrons of the K shell, L shell, etc. In particular, the Photoelectric Effect on K shell electrons accounts for 80% of the total. As the gamma ray energy increases, the photoelectron generation cross section becomes smaller. The photoelectron generation cross section becomes larger as the substance's atomic number increases. For this reason, this is not negligible for substances with a large atomic number such as metals used for electrodes of integrated circuits.

Compton Effect

The Compton Effect is a phenomenon wherein, as electromagnetic radiation passes through the vicinity of atoms, photons collide against orbital electrons, impart part of their energy to the orbital electrons, and displace them. The photons lose their energy by the amount imparted to the electrons, and the photons are scattered in different directions. Strictly speaking, the Compton Effect is a phenomenon that can only occur in relation to free electrons. However, it can occur in relation to outer orbital electrons as well, because the bounding energy for the outer orbital electrons is small. That is, the likelihood of occurrence of the Compton Effect is strong in relation to outer orbital electrons, whereas the likelihood of occurrence of the Photoelectric Effect is strong in relation to core electrons. In the case of silicon, the reaction cross section of the Compton Effect becomes larger than the reaction cross section of the Photoelectric Effect at an energy level of about 200 keV or higher.

Pair production

Pair production is a phenomenon wherein, as high-energy electromagnetic radiation passes through the vicinity of atoms, photons vanish in the Coulomb electric fields of the atomic nuclei and "electron-positron" pairs are created. This phenomenon can only occur in the Coulomb electric fields of atomic nuclei. It does not occur in free space. This phenomenon can occur only when the photon energy is 1.02 MeV (which is the sum of the electron rest mass and the positron rest mass) or larger. The likelihood of occurrence of pair production becomes stronger in proportion to the square of the atomic number of the material, and the cross section becomes larger as the incident photon energy increases.

9.2.2 Total Ionizing Dose Effect on MOS Capacitors

Ionization interaction causes generation of electron-hole pairs in semiconductor devices. The generated electrons and holes can act as carriers in the semiconductor devices. However, the generation of such carriers in a semiconductor crystal does not mean that transient current will be generated immediately, because the electrons and holes will recombine after a certain period of time. As the electrons and holes generated by the ionization have electric charges of opposite polarities (negative and positive), the electrons and holes will move in the direction of the other polarity (at a speed proportional to their respective mobility) if an electric field exists in the semiconductor crystal, in which case the electrons and holes can be taken out as transient current. Schottky junction and p-n junction type semiconductor detectors are examples of application of this principle. Ionization also causes generation of electron-hole pairs in silicon dioxide films (SiO_2 films) of such parts as gates of MOS transistors and MOS capacitors basically in the same manner as in the case of semiconductor crystals, although the density of electron-hole pairs generated is different. If an electric field has been applied in the oxide film, the electrons in the oxide film will be swept out of the oxide film to the electrode (where they will vanish) as soon as they are generated, because their mobility is 5–6 orders of magnitude greater than the mobility of the holes. On the other hand, the holes in the oxide film will be slowly swept out of the oxide film. During this hole sweeping, holes will be trapped in defects present in the vicinity of the interface between Si and SiO_2, and fixed positive charges will be generated as a result of the trapping. It has been reported that these defects in the vicinity of the interface are made by oxygen deficiency states in the oxide film (these are called "E' centers"). The fixed charges generated are positive charges because the oxygen deficiency traps holes. Aside from these fixed positive charges, charge states called "interface states" will appear in the band gap at the interface between the oxide film and silicon. This is because of the uncombined hands left as a result of the absence of covalent bonding of silicon atoms and oxygen atoms at the interface (these are also called "dangling bonds" or "Pb centers"). Figure 9.1 shows schematic diagrams of the presence of fixed positive charges and interface states at the gates of a MOS transistor and a MOS capacitor. As the amount of gamma ray irradiation increases, the fixed positive charges and interface states increase, and electrical characteristics changes are observed.

Charge state changes of interface states can be explained on the basis of donor-type (electron provider-type) levels and acceptor-type (electron receptor-type) levels. It is said that, in the band gap of a silicon semiconductor, acceptor-type levels are present in the energy domain whose energy is larger than the intrinsic Fermi energy E_i and donor-type levels are present in the energy domain whose energy is smaller than E_i. The density distribution is normally a U-shaped distribution. A donor-type level is an interface state that is neutral when electrons exist at that level and becomes positively charged when it releases electrons. This means that donor-type levels are neutral when they are lower than the Fermi energy E_f and are positively charged when they are higher than E_f. An acceptor-type level is an interface state that is neutral when no electron exists and becomes negatively charged when it accepts electrons. This means that acceptor-type levels are neutral when they are higher than E_f and are negatively charged when they are lower than E_f. On the basis of these characteristics of donor-type and acceptor-type levels, the charge states of donor-type and acceptor-type levels when the charge state of the interface state is the flat band state, mid-gap state, and threshold state will be explained below taking an n-type semiconductor as an example and using band diagrams.

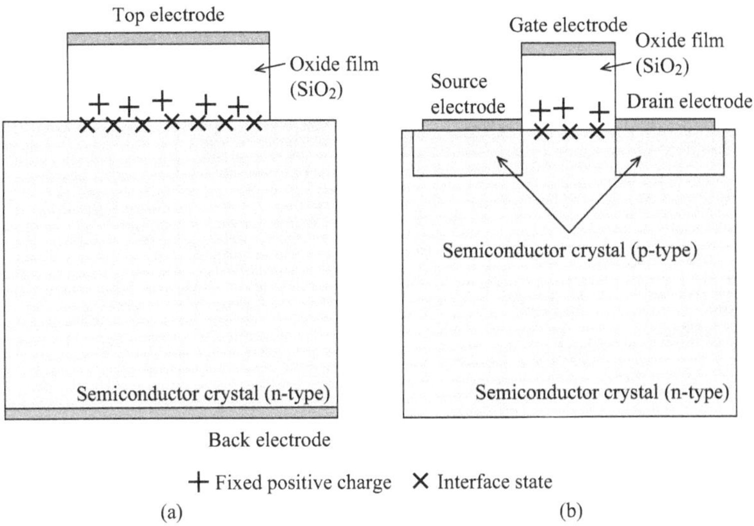

Fig. 9.1 Cross sections of an nMOS capacitor and a p-channel MOS transistor. (**a**) Cross section of a MOS capacitor. (**b**) Cross section of a MOS transistor

Flat band state

In the flat band state, the energy levels of all donor-type levels are lower than E_f, and the donor-type levels are neutral, as shown in Fig. 9.2a. The energy levels of some of the acceptor-type levels are lower than E_f, and these acceptor-type levels are negatively charged, but all of the acceptor-type levels whose energy levels are higher than E_f are neutral. Therefore, the interface state as a whole is negatively charged in the flat band state. This means that the applied voltage in the flat band state is shifted to the positive voltage side relative to the ideal state.

Mid-gap state

In the mid-gap state, all interface states are neutral, as shown in Fig. 9.2b. Therefore, the applied voltage in the mid-gap state (V_{MG}) is not affected by interface states.

Threshold state

In the threshold state, all acceptor-type levels are neutral, as shown in Fig. 9.2c. However, some of the donor type levels are positively charged, and the interface state as a whole is positively charged. This means that the applied voltage in the threshold state is shifted to the negative voltage side relative to the ideal state.

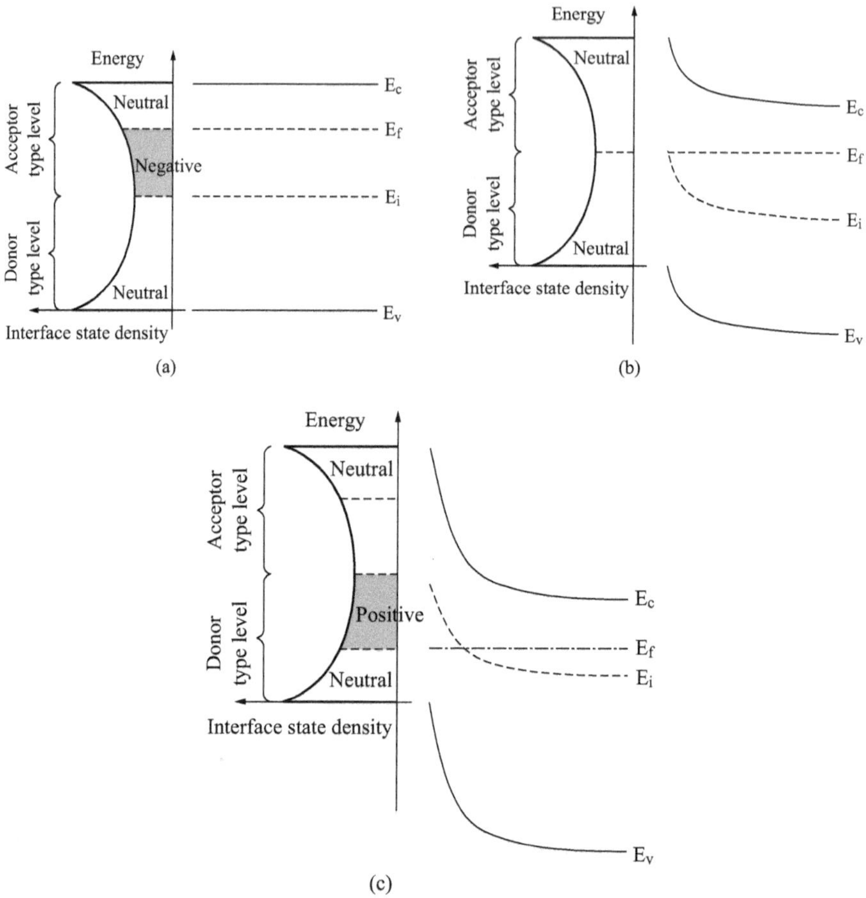

Fig. 9.2 Band diagrams for a MOS capacitor. (**a**) Flat band state. (**b**) Mid-gap state. (**c**) Threshold state

In summary, an interface state changes its charge state depending on the relationship with the Fermi level, whereas a fixed positive charge does not.

Figure 9.3 shows the effects of fixed positive charges and interface states on the capacitance-voltage (*CV*) characteristic of a MOS capacitor. Figure 9.3a is the *CV* curve representing the case where only fixed positive charges are formed in the oxide film. When fixed positive charges are introduced, the *CV* curve will shift in the negative voltage direction. That is, the behavior will look as if positive voltage was applied by fixed positive charges. Figure 9.3b is the *CV* curve representing the case where only interface states are formed at the oxide film interface. The interface states are negatively charged when the voltage is higher than the mid-gap voltage (flat band state) and positively charged when the voltage is lower than the mid-gap voltage (threshold state). Therefore, the *CV* curve will shift in the positive voltage direction when the voltage becomes higher than the mid-gap voltage and in the

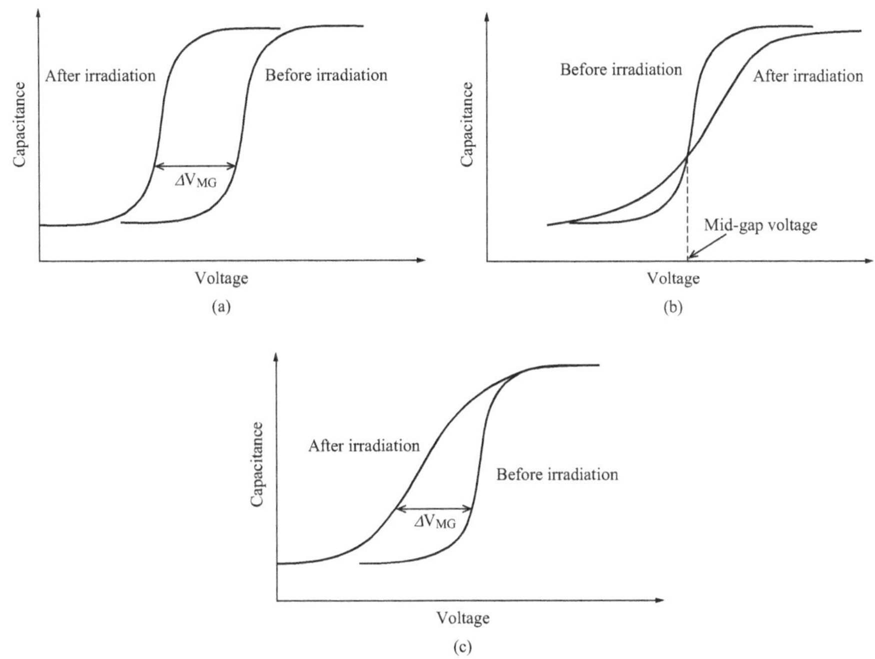

Fig. 9.3 Capacitance-voltage curve of an nMOS capacitor before and after irradiation. (**a**) Fixed electric charge. (**b**) Interface state. (**c**) Fixed positive electric charge and interface state

negative voltage direction when the voltage becomes lower than the mid-gap voltage. When the voltage is equal to the mid-gap voltage, the amount of fixed positive charges in the silicon oxide film can be estimated by determining the amount of the change in the middle-gap voltage, because there is no effect of interface states. Figure 9.3c is the *CV* curve representing the case where both fixed positive charges and interface states are formed.

Figure 9.4 shows the *CV* curves before and after gamma ray irradiation of a MOS capacitor made on an n-type silicon substrate. The gamma ray irradiation dose in terms of absorbed dose is 6.6 kGy (SiO_2). The oxide film thickness of the MOS capacitor used in the gamma ray irradiation experiment is 100 nm, the diameter of the circular aluminum electrode of the capacitor is 100 μm, and the impurity concentration in the silicon substrate of the capacitor is 3 to 5 × 10^{15} cm^{-3}. As can be seen from Fig. 9.4, the *CV* curve after gamma ray irradiation was shifted to the negative voltage side by as much as approx. 12 V by the effect of fixed positive charges, and the gradient of the curve was made slightly gentler by the effect of interface states.

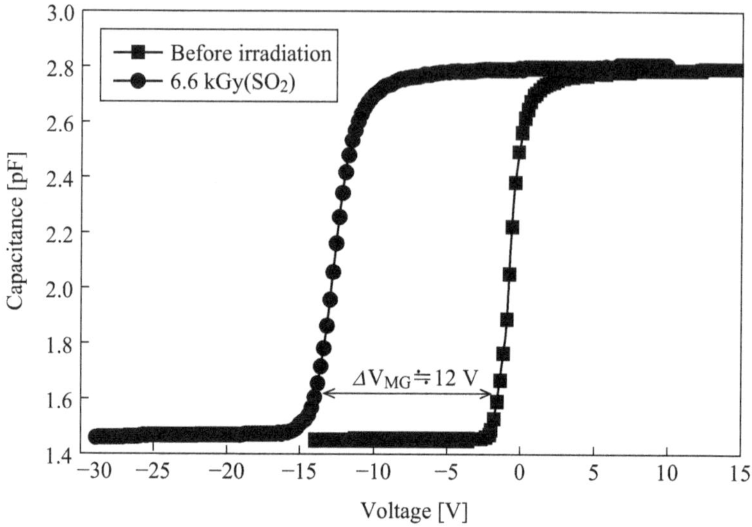

Fig. 9.4 CV curve of an nMOS capacitor before and after 6.6 kGy (SiO_2) gamma ray irradiation

9.2.3 Absorbed Dose

Normally, “absorbed dose” is used as the evaluation index for the Total Ionizing Dose Effect. Absorbed dose (D_e) is the radiation energy absorbed (dE (J)) by unit weight (dm (kg)) of the irradiated substance and is defined by the following formula:

$$D_e = \frac{dE}{dm} \quad [\mathrm{J/kg}] \tag{9.1}$$

The unit for absorbed dose is Gy, and the absorption of an energy of 1 J per 1 kg of the irradiated substance is defined as 1 Gy (1 J/kg). Absorbed dose per unit time is called “absorbed dose rate” and is expressed using such units as Gy/h and Gy/s. Absorbed dose values are usually expressed with the name of the irradiated substance indicated in parentheses after the unit (e.g., 1 Gy (SiO_2)), because absorbed dose is dependent on the type and density of the irradiated material. For electromagnetic radiation such as gamma rays, Formula (9.1) can be rewritten as follows:

$$D_e = \Psi\frac{\mu}{\rho} \tag{9.2}$$

where Ψ, ρ, and μ/ρ represent the energy fluence (J/m^2), the density of the irradiated material (g/m^3), and the mass energy absorption coefficient (cm^2/g), respectively. The energy fluence is the radiation energy that passes through unit area of the irradiated material. The mass energy absorption coefficient varies depending on the irradiated material. In the case of a gamma ray of 1.25 MeV (the average of the 1.17 and 1.33 MeV gamma rays emitted by cobalt 60), the mass energy absorption coefficient for water (H_2O) is 0.0296 cm^2/g, the mass energy absorption coefficient for dry air is 0.0266 cm^2/g, the mass energy absorption coefficient for silicon (Si) is 0.0265 cm^2/g, and the mass energy absorption coefficient for silicon oxide film (SiO_2) is 0.0266 cm^2/g. Absorbed dose for charged particles such as electrons and heavy ions can be determined by the following formula:

$$D_e = \Phi \frac{(dE/dx)_{\text{ion}}}{\rho} \tag{9.3}$$

where Φ represents the number of radioactive rays that pass through unit area of the irradiated material (m^{-2}), which is called "particle fluence." $(dE/dx)_{\text{ion}}$ represents the radiation energy lost through ionization during unit distance travel of the radiation (eV/m), which is called "electric stopping power." According to H. Bethe, the electric stopping power of electrons whose energy is E_e can be expressed with the following formula:

$$\left(\frac{dE_i}{dx}\right)_{\text{ion}} = \frac{2\pi e^4 z_i^2}{m_0 v_i^2} nZ \left[\ln \frac{m_0 v_e^2 E_e}{2I^2(1-\beta)} - \left(2\sqrt{1-\beta} - 1 + \beta^2\right) \ln 2 + (1-\beta)^2 + \frac{1}{8}\left(1 - \sqrt{1-\beta^2}\right)^2 \right] \tag{9.4}$$

where x represents the distance (m) along the track, E_e represents the energy of the electrons (eV) ($e^2 = 1.439809 \times 10^{-13}$ MeV $\cdot$ cm), m_0 represents the rest mass (kg) of the electrons, v_e represents the velocity (m/s) of the electrons, Z represents the atomic number of the irradiated material, n represents the density of the atoms (cm^{-3}), and I represents the average excitation energy (keV). For alpha particles and heavy particles, the electric stopping power is expressed by the following formula:

$$\left(\frac{dE_i}{dx}\right)_{\text{ion}} = \frac{4\pi e^4 z_i^2}{m_0 v_i^2} nZ \ln \frac{2m_0 v_i^2}{I} \tag{9.5}$$

where E_i, z_i, and v_i represent the energy of the heavy particles (eV), the atomic number of the heavy particles, and the velocity (m/s) of the heavy particles, respectively. For the average excitation energy, an excellent empirical formula has been proposed by Berger and Seltzer:

$$I = \left(9.76Z + 58.5/Z^{0.19}\right) \quad [\text{eV}] \tag{9.6}$$

For silicon, the average excitation energy calculated with this formula is 304.37 eV.

The electric stopping power for the incidence of a 1 MeV electron beam onto silicon calculated with Formula (9.4) is approx. 3.5 MeV/cm. In the case of heavy particles, $(dE/dx)_{ion}/\rho$ is called "linear energy transfer (LET)." LET serves as an important evaluation index for the Single Event Effect described in Sect. 9.4.

9.2.4 Dosimeters that Utilize the Total Ionizing Dose Effect

As shown in Figs. 9.3 and 9.4, the *CV* curve of the nMOS capacitor shown in Fig. 9.1a changes when affected by fixed positive charges and interface states. The drain current-gate voltage ($I_d - V_g$) curve of the p-channel MOS transistor shown in Fig. 9.1b also changes when affected by fixed positive charges and interface states. Specifically, the gate voltage at which the drain current starts flowing (threshold voltage) will be shifted. Semiconductor devices have been developed that measure absorbed dose using the relationship between the amount of threshold voltage change and the absorbed dose. These are called "radiation-sensitive field-effect transistors" (RADFETs). RADFETs are lightweight semiconductor devices with a simple structure and low power consumption and are being used for absorbed dose measurement in the space environment.

9.3 Displacement Damage Dose Effect

9.3.1 Effects of Electron Beams on Semiconductors

As described in Sect. 9.1.2, electron and proton beams are mainly used to investigate the Displacement Damage Dose Effect in ground tests. In evaluations of the Displacement Damage Dose Effect on solar cells, it is often the case that 1 MeV energy is used for electrons and 10 MeV energy is used for protons.

Incidence of high-energy electrons onto a material causes the electrons to repeat elastic and inelastic scattering (ionization loss and bremsstrahlung) and lose energy. Most of the electron energy loss is caused by the inelastic scattering. The breakdown of the inelastic scattering, that is, the ratio of the ionization loss $(dE_e/dx)_{ion}$ to the radiation loss $(dE_e/dx)_{rad}$, is expressed by the following formula:

$$\frac{(dE_e/dx)_{\text{ion}}}{(dE_e/dx)_{\text{rad}}} = \frac{800}{E_e Z}{}_e \tag{9.7}$$

where Z represents the atomic number of the irradiated substance and E_e represents the energy of the electrons in MeV. The electron energy at which the two losses become equal is called "critical energy." In the case of silicon ($Z = 14$), the critical energy is 57 MeV. Therefore, in the case of 1 MeV electron beams used in evaluation tests for the Displacement Damage Dose Effect, most of the energy is consumed as ionization loss.

For the elastic scattering, whose probability of occurrence is much lower than the inelastic scattering, the law of conservation of momentum and the law of conservation of kinetic energy stand before and after the collision. If high-energy electrons pass through the vicinity of atomic nuclei, Coulomb's force acts between them. Because the mass of the electrons is much smaller than the mass of the atomic nuclei, the electrons will be scattered. The probability of occurrence of this scattering is very high and follows Rutherford's scattering equation that takes the relativity into consideration. On the other hand, the loss in the energy of the electrons is very small, because the mass of the electrons is much smaller than the mass of the atomic nuclei. That is, there is little energy transfer to the atomic nuclei. There are rare occasions where large amounts of energy are transferred to atomic nuclei, in which case crystal atoms are displaced from their positions. Therefore, the probability that crystal atoms are displaced is very low. The atoms displaced from their positions are called "primary knock-on atoms (PKAs)" and form point defects (interstitial atoms) of the crystal. The locations previously occupied by the atoms that have been displaced also form point defects (atomic vacancies) of the crystal. The numbers of these defects increase with increasing irradiation dose. In 1 MeV electron beam irradiation tests on silicon solar cells, irradiation of up to 10^{16} to 10^{17} cm^{-2} in terms of particle fluence is made. To allow the absorbed dose effect of gamma rays to be discussed using the same index as for the Displacement Damage Dose Effect of electron beams, it is necessary to convert particle fluence into absorbed dose. A particle fluence of 10^{16} to 10^{17} cm^{-2} can be converted into an absorbed dose of 2.4 to 24 MGy (Si) by using Formulas (9.3) and (9.4) in combination. Figure 9.5 shows the relationship between the absorbed dose and the effect of the Total Ionizing Dose Effect on MOS transistors and the effect of the Displacement Damage Dose Effect on solar cells. As shown in Fig. 9.1b, a MOS transistor consists of a silicon dioxide film made on a semiconductor crystal. A solar cell consists of a semiconductor crystal only and does not have an oxide film. In the several ten to several hundred kGy range in which MOS transistors fail to operate properly as a result of the Total Ionizing Dose Effect, the Displacement Damage

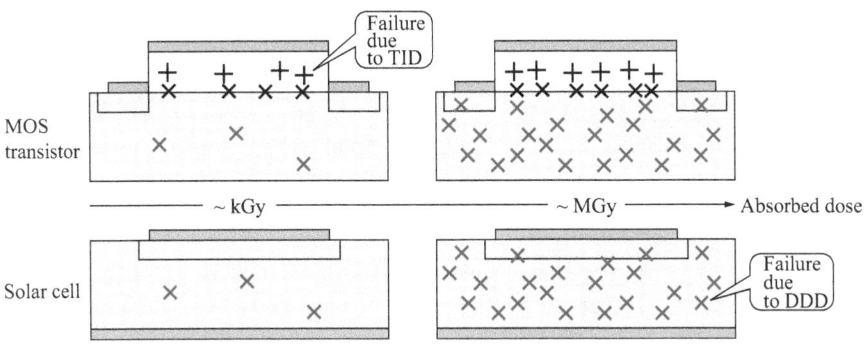

Fig. 9.5 The effect of the Total Ionizing Dose Effect on MOS transistors and the effect of the Displacement Damage Dose Effect on solar cells

Dose Effect is negligible. If the absorbed dose increases beyond 1 MGy, the Displacement Damage Dose Effect will become no longer negligible, and the degree of deterioration of the power generation capability of solar cells increases accordingly, and eventually the solar cells fail. MOS transistors are also affected by the Displacement Damage Dose Effect, but they fail at absorbed doses lower than 1 MGy due to the Total Ionizing Dose Effect.

Point defects formed in the semiconductor crystal as a result of the Displacement Damage Dose Effect (interstitial atoms and atomic vacancies) couple with other point defects or trace impurities present in the crystal or vanish in voids. Ultimately, various kinds of complex defects will be formed. When such defects are present in the semiconductor crystal, energy levels called "generation and recombination centers" are formed in the band gap of the semiconductor. If generation and recombination centers are formed, the lifetime, mobility, etc. of the electrons and holes deteriorate. Because these parameters are main determining factors for the performance of semiconductor devices, deterioration in these parameters causes such problems as deterioration of electrical characteristics of solar cells, leakage current increases at p-n junctions, forward current reductions, and reductions in the amplification rates of bipolar transistors.

9.3.2 Displacement Damage Dose Effect on Solar Cells

Figure 9.6 shows the current-voltage characteristic of a space silicon solar cell before and after irradiation. In the irradiation test, 1 MeV electron beams were used. The current at an applied voltage of 0 V is called "short-circuit current" (I_{SC}) (unit: A) and the voltage at which the current becomes 0 A is called "open-circuit voltage" (V_{OC}) (unit: V). The electric power of a solar cell can be calculated by multiplying the voltage by the current. The maximum of the electric power is called "maximum power" (P_{max}) (unit: W). The gray rectangle in Fig. 9.6 shows the maximum power of a solar cell that has been subjected to irradiation of 10^{16} cm^{-2} in terms of particle fluence. It can be seen from the figure that the Displacement Damage Dose Effect significantly deteriorates the current-voltage characteristic. Figure 9.7 shows the relationship between the short-circuit current, open-circuit voltage, maximum power (as standardized at the initial values of a silicon solar cell), and the particle fluence. The particles used in the irradiation test are proton beams, and the energy is 10 MeV. The short-circuit current, open-circuit voltage, and maximum power constantly decrease as the amount of irradiation increases, and the solar cell ceases to output power when irradiated with a particle fluence of approx. 2×10^{14} cm^{-2}.

Next, we will consider the mechanism of deterioration of the power generation capability of solar cells caused by the Displacement Damage Dose Effect such as that shown in Fig. 9.7. To understand this mechanism, it is first necessary to understand the principle of the power generation by solar cells. If energy that is higher than the band gap is deposited to a semiconductor, "electron-hole" pairs are

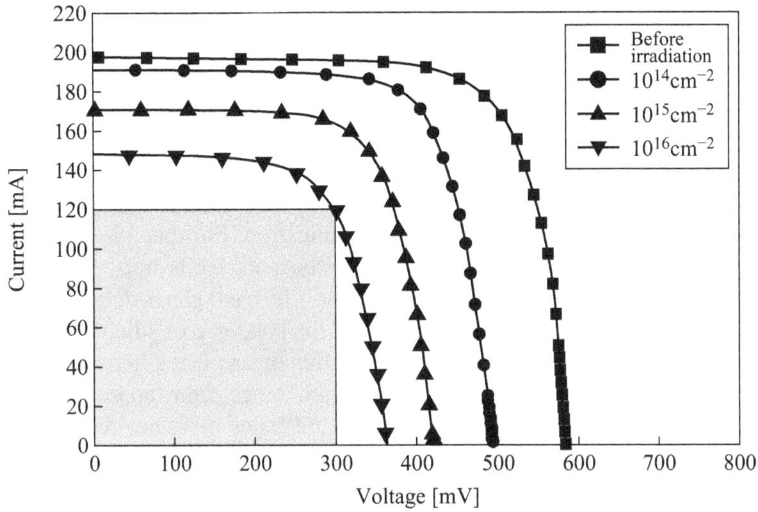

Fig. 9.6 Current-voltage characteristic of a silicon solar cell before and after 1 MeV electron irradiation

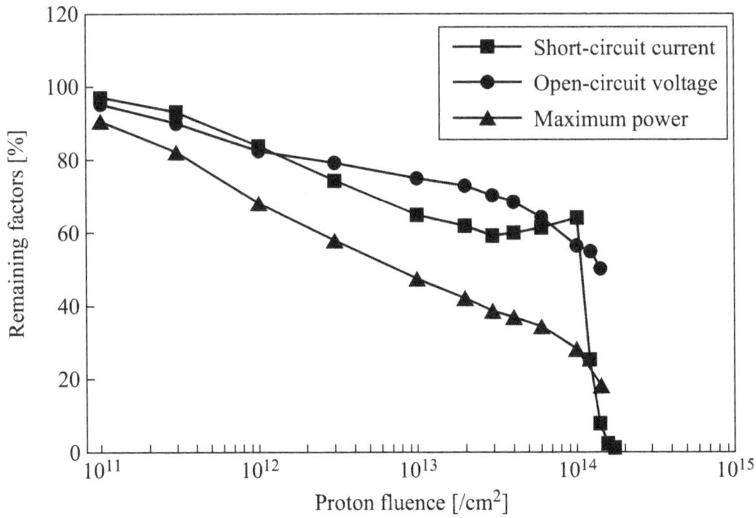

Fig. 9.7 Relationship between 10 MeV proton fluence and the remaining factor of short-circuit current, open-circuit voltage, and maximum power

generated through ionization. Solar cells generate power by taking out the generated electrons and holes in an appropriate manner. Here, taking them out "in an appropriate manner" is important, because it is not possible to extract electricity simply by subjecting a semiconductor to light. This is because, although the generated electrons and holes move around in the semiconductor through diffusion, they disappear by recombination after a certain period of time. For this reason, it is necessary to make an electric field (potential difference) in the semiconductor to extract the electrons and holes to take out the power. Solar cells are structurally p-n junction diodes and are designed to take out electrons and holes making use of the potential difference created by the p-n junction. That is, solar cells generate power by using electric fields formed in them to take out the electrons and holes generated by light before they disappear through recombination. For this reason, solar cells must satisfy the following requirements. Firstly, solar cells must be capable of generating electron-hole pairs utilizing sunlight. The band gap is determined by the properties of the semiconductor. For example, the band gap of silicon, which is the most popularly used material, is 1.12 eV, and this means that silicon has sufficient sensitivity to sunlight, which has a wavelength range from approx. 250 nm to 2600 nm. Secondly, the generated electrons and holes must not recombine until they reach the p-n junction (the electric field region). This means that the carrier (the generated electrons and holes) lifetime (τ) must be long. Because the relationship defined by Formula (9.8) stands between the carrier lifetime and the diffusion length (L) and diffusion coefficient (D), the second requirement means that the diffusion length must be long.

$$L = (D \cdot \tau)^{1/2} \tag{9.8}$$

Thirdly, high-quality solar cells must have a potential difference in them. That is, a p-n junction must have been formed properly.

Incidence of energetic particles such as electrons and protons onto a solar cell causes crystal defects to be generated through the Displacement Damage Dose Effect. Because some of the generated crystal defects act as recombination centers, the lifetime of the carriers (electrons and holes) generated gradually shortens (in other words, the diffusion lengths shorten). This means that the carriers which have been generated at locations farther from the p-n junction (i.e., locations near the back side of the solar cell) but would have been able to reach the p-n junction through diffusion if there had been no irradiation gradually become unable to reach the p-n junction. It is known that the relationship defined by the following formula stands between the particle fluence (ϕ) and diffusion length reductions:

$$\left(\frac{1}{L_\phi}\right)^2 = \left(\frac{1}{L_0}\right)^2 + K_L \cdot \phi \tag{9.9}$$

where L_ϕ and L_0 represent the diffusion length after (ϕ) and before irradiation, respectively. K_L is called "diffusion length damage coefficient" and is used as the index of the degree of deterioration. The deterioration of the power generation capability seen in the 10^{11}–10^{13} (cm^{-2}) particle fluence region in Fig. 9.7 is definitely the result of the diffusion length reductions caused by the generation of defects due to the Displacement Damage Dose Effect.

As can be seen from the figure, in the higher particle fluence region (higher than 5×10^{13} (cm^{-2})), the short-circuit current, which monotonously decreased in the lower particle fluence region, increases temporarily and then decreases sharply. Some of the defects generated by the Displacement Damage Dose Effect have characteristics that make them capable of capturing majority carriers (i.e., carriers that are inherently present in the semiconductor). If the density of such defects increases to the extent where it is close to the density of majority carriers in the semiconductor, a carrier depletion state develops. For carrier depletion, it is known that the relationship defined by the following formula usually stands:

$$p_\phi = p_0 - R_c \cdot {}^c\phi \text{ or } p_\phi = p_0 \exp\left(-R \cdot \frac{\phi}{p_0}\right) \tag{9.10}$$

where p_0 and p_ϕ represent the carrier density without irradiation and the carrier density after irradiation with radiation, respectively. R_c is called "carrier removal rate" and represents the magnitude of the carrier density reduction effect. The potential difference in a p-n junction is related to the hole density in the "p" area and the electron density in the "n" area, and solar cells are optimally designed by appropriately controlling the concentrations of acceptor impurities and donor impurities that release holes and electrons. However, if the carrier density decreases and the carrier depletion state is created, the "p" and "n" areas of the p-n junction will become high-resistance insulators, and the p-n junction will become incapable of serving as a p-n junction. As a result, the short-circuit current will decrease sharply. The temporary increase in the short-circuit current before the sharp decrease is due to the temporary electric field region expansion before the complete carrier depletion which made it temporarily possible to collect the generated carriers that had become unable to reach the junction. This reflects the fact that, in p-n junctions, the electric field region becomes longer as the carrier density in the substrate decreases.

9.3.3 Displacement Damage Dose

Absorbed dose cannot be used as an index for evaluating the displacement damage, because it is based on the energy deposited through ionization. That is, the Displacement Damage Dose Effect is generated not by the cumulative dose generated by ionization but by the displacement of elements that comprise the crystal. For this reason, an evaluation index called "displacement damage dose (D_d)" was proposed in the latter half of the 1980s at the US Naval Research Laboratory. Displacement damage dose (D_d) is expressed by the following formula:

$$D_d = \Phi \cdot NIEL \quad [\text{MeV/g}] \tag{9.11}$$

NIEL ($\text{MeVcm}^2\text{/g}$) is called "non-ionizing energy loss" and is dependent on the energy of the incident radiation. *NIEL* is the radiation energy lost as a result of the displacement during unit distance travel of the radiation divided by the density, which means that the unit for displacement damage dose is (MeV/g). On the other hand, the unit for absorbed dose is (J/kg), as shown in Formulas (9.1), (9.2), and (9.3). Both units are the same in the sense that they represent energy per unit weight. The only difference between the two units is whether the energy loss is caused by ionization or displacement. *NIEL* is defined by the following formula:

$$NIEL(E_{\mathrm{in}}) = \frac{1}{M}\int_{E_d}^{E_{p(\max)}} \frac{d\sigma_p}{dE_p} \cdot E_p \cdot L \cdot dE_p \; \left[\mathrm{eVcm^2/g}\right] \tag{9.12}$$

where E_{in} (eV), M (u), $d\sigma_p/dE_p$ (cm^2/eV), E_{p} (eV), and L (nondimensional) represent the energy of the incident radiation, the mass of the irradiated sample, the differential displacement cross section, the displacement energy, and the Lindhard partition function, respectively. E_d (eV) represents the minimum energy that causes displacement (threshold energy), and $E_{p(\max)}$ (eV) represents the maximum of the displacement energy. Now, let us consider the case where a solid that consists of atoms of mass M is irradiated with electrons of mass m_0. The collision between incident particles and target atoms imparts displacement energy E_p to the target atoms hit by incident particles. If E_p is larger than a certain threshold energy (E_d) ($E_p > E_d$), the target atoms hit by incident particles are displaced from their positions in the lattice. The displacement energy is expressed by the following formula that uses scattering angle (θ) as a parameter:

$$E_p = E_{p(\max)} \sin^2\left(\frac{\theta}{2}\right) = \frac{2m_0}{M}\frac{1}{m_0 e^2}\left(E_e + 2m_0c^2\right)E_e \sin^2\left(\frac{\theta}{2}\right) \quad [\mathrm{eV}] \tag{9.13}$$

where E_e (eV) represents the incident electron energy and c (m/s) represents the speed of light. The displacement energy becomes highest when $\theta = \pi$. In the case of substances with a small atomic number (Z), the differential displacement cross section can be expressed with the following formula (the Mckinley–Feshbach equation that approximates the Mott solution):

$$\begin{aligned} d\sigma_p = {} & \frac{4\pi a_0 Z^2 E_R^2}{m_0^2 c^4}\frac{1-\beta_2}{\beta^4}\left[1 - -\beta_2 \frac{E_p}{E_{p(\max)}}\right. \\ & \left.\pi\alpha\beta Z\left\{\left(\frac{E_p}{E_{p(\max)}}\right)^{1/2} - \frac{E_p}{E_{p(\max)}}\right\}\right]\frac{E_{p(\max)}}{E_p^2} dE_p \; \left[\mathrm{cm^2}\right] \end{aligned} \tag{9.14}$$

where a_0 (m), E_R (eV), β (= v/c), and α (= 1/137.036) represent the Bohr radius, the Rydberg energy, the ratio of the electron speed to the speed of light in vacuum, and the fine-structure constant, respectively. The displacement cross section can be calculated by integrating Formula (9.14) from E_d to $E_{p(\max)}$. The Lindhard partition function (L) for an irradiated substance whose atomic number is A is calculated with the following formulas in the case where the displacement energy is high:

$$L = (1 + kg)^{-1} \tag{9.15}$$

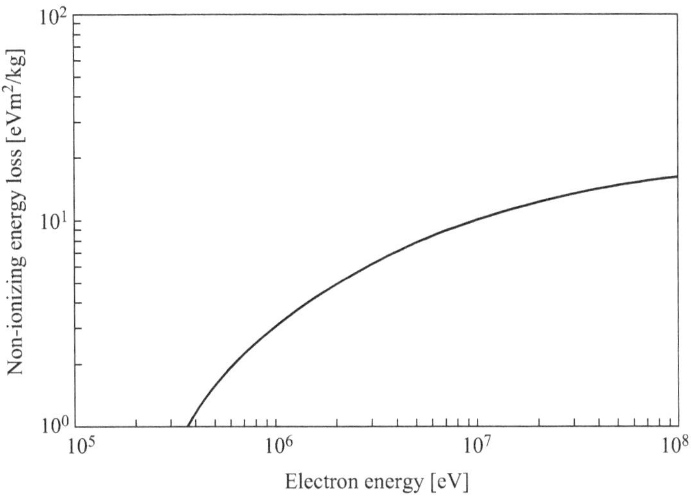

Fig. 9.8 Relationship between the electron energy and the non-ionizing energy loss (NIEL)

$$k = 0.1337 \cdot Z^{1/6}\left(\frac{Z}{A}\right)^{1/2} \tag{9.16}$$

$$g = 3.4008{\varepsilon_p}^{1/6} + 0.40244{\varepsilon_p}^{3/4} + \varepsilon_p \tag{9.17}$$

$$\varepsilon_p = \frac{E_p}{86931 \cdot Z^{7/3}} \tag{9.18}$$

In the case where the displacement energy is low, L is calculated with the following formula:

$$L = 1 - E_p\left(2E_p/AZ\right)^{1/2} \tag{9.19}$$

E_p in Formulas (9.18) and (9.19) is the displacement energy expressed in keV and is nondimensional. Figure 9.8 shows the relationship between the *NIEL* for silicon calculated with Formula (9.12) and the electron energy. The *NIEL* of 1 MeV electron in silicon is 3.1×10^{-5} MeVcm^2/g. The *NIEL* of proton beams can also be calculated in the same manner from the differential displacement cross section, displacement energy, and Lindhard partition function. The NIEL of 10 MeV proton beams in silicon is 9.4×10^{-3} MeVcm^2/g.

9.3.4 Monte Carlo Code-Based Calculation of the NIEL of Heavy Particles

The *NIEL* of heavy particles can also be calculated using Formula (9.12) in the same manner as in the case of electrons and protons. However, the *NIEL* can be calculated easily by using the SRIM (Stopping and Range of Ions in Matter) code, which calculates the loss in the energy of the charged particles and other parameter values using the Monte Carlo method. As will be described in Sect. 9.4.2, the SRIM code is frequently used for the calculation of the electric stopping power. In the calculation of the *NIEL*, VACANCY.TXT is used in the SRIM code output file. VACANCY.TXT consists of the distance in the direction of entry of the incident particles along the track of the incident particles (x), the rate of generation of point defects by the incident particles (I_v (x)), and the rate of generation of point defects by the recoil particles (R_v (x)). The relationship defined by the following formula stands between the *NIEL* and the distance:

$$NIEL(x) = 10^5 M_{NIEL}(I_V(x) + R_V(x))/\rho \tag{9.20}$$

where

$$M_{NIEL} = \frac{(E_p/0.4 + 2)}{1000} \tag{9.21}$$

9.3.5 Displacement Damage Dose-Based Analyses

This section presents examples of evaluations that utilize the displacement damage coefficient. Figure 9.9 shows how the remaining factor of maximum power of a

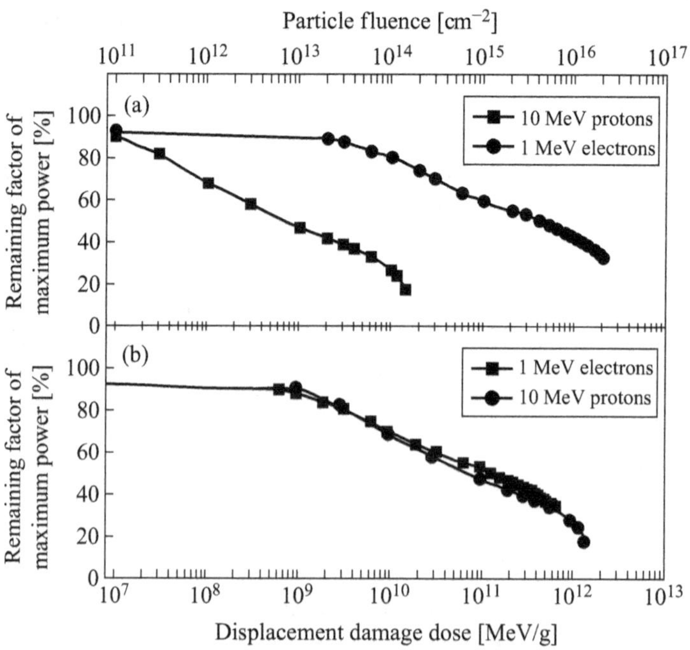

Fig. 9.9 Relationship between displacement damage dose and remaining factor of maximum power

space silicon solar cell changes with particle fluence and displacement damage dose as a result of 1 MeV electron irradiation and as a result of 10 MeV proton irradiation. As can be seen from Fig. 9.9a, which shows the remaining factor of the maximum power changes with particle fluence caused by irradiation of 1 MeV electrons and by irradiation of 10 MeV protons, the deterioration in the remaining factor of the maximum power caused by 10 MeV proton irradiation is about 3 orders of magnitude greater than the deterioration caused by 1 MeV electron irradiation. On the other hand, in Fig. 9.9b, which shows the remaining factor of the maximum power changes with displacement damage dose caused by 1 MeV electron irradiation and by 10 MeV proton irradiation, the remaining factor curve for the 1 MeV electrons and the remaining factor curve for the 10 MeV protons match well. This means that *NIEL* is an important concept in obtaining a rough understanding of the Displacement Damage Dose Effect and the use of displacement damage dose, which is a physical quantity based on *NIEL*, allows degrees of deterioration of solar cells in power generation capability to be compared on the basis of the amount of damage irrespective of differences in the type of radiation or energy. That is, displacement damage dose can be used as a very convenient index when making estimations as to what degree of crystal damage will cause what degrees of deterioration of characteristics of solar cells from the standpoint of engineering in solar cell development. In the development of solar cells for space applications, various materials and structures are evaluated, and it is important to achieve solar cells with even better radiation resistance.

Because electron and proton beams with various levels of energy exist in outer space, it is necessary in the development of solar cells to confirm how solar cells deteriorate in power generation capability as a result of their irradiation with electrons and protons with various levels of energy and estimate their lifetime in outer space.

Since the deterioration of solar cells in power generation capability can be expressed and evaluated in a unified manner by using displacement damage dose as a parameter, it can be said that displacement damage dose can be an effective means of breaking away from the traditional style of development in which irradiation tests are conducted using electron and proton beams with various levels of energy to clarify characteristic deterioration curves of solar cells, thereby reducing the costs and time for irradiation tests and solar cell development. Lastly, let me remind you that the characteristics of point defects generated by the incidence of radiation into semiconductors and subsequently formed complex defects naturally vary widely depending on the type of semiconductor, and looking at radiation damage microscopically from the point of view of materials science and truly understanding the origins, structures, and electronic properties of defects requires detailed examination of individual defects.

9.4 Single Event Effect

9.4.1 Initial Process of the Single Event Effect

As described in Sect. 9.1.2, alpha rays and heavy ions are used to investigate the Single Event Effect. This can be divided into the case where alpha rays emitted by radioisotopes are used and the case where heavy ions accelerated by an accelerator are used. The most commonly used radioisotope is Americium 241, which emits alpha rays of approx. 5.5 MeV. In accelerator-based experiments, semiconductor devices are irradiated with diverse heavy particles ranging in energy from several MeV to several GeV.

High-energy heavy particles have very strong electric stopping power, and incidence of a single heavy particle onto a p-n junction that forms part of an integrated circuit (IC) causes electron-hole pairs to be generated at a high-density along the track. A distribution of high-density electrons and holes is called an "ion track." Generally speaking, gamma rays, electron beams, and proton beams do not cause the Single Event Effect because their electric stopping power is much lower than that of heavy particles. However, there have been reports in recent years showing that protons, which have electric stopping power that is weaker than that of heavy particles, can cause the Single Event Effect. This is attributable to the technological advances that have made it possible to make integrated circuits more minute and reduce the driving voltage of integrated circuits.

Carriers such as electrons and holes that have been generated by the ionization interaction of heavy particles can be drifted by the electric field present in the p-n junction or drawn to the electrode through diffusion. If this happens, transient currents flow in the external circuit. Figure 9.10a shows a schematic diagram of the initial process of the generation of transient current in the p-n junction. The broken line shows the area that has an electric field, which is called the "depletion layer." The broken line is curved in a way that it envelopes the track of the incident particle. This is called the funneling effect. Figure 9.10b shows the changes with time in the transient current that flows in the external circuit and the electric charge (Q). The electric charge can be calculated by integrating the transient current waveform.

The time required for the transient current to converge (i.e., the time required for the electric charge to saturate) is usually very short (approx. several ten picoseconds to several nanoseconds).

The generated transient current flows in the external circuit, alters the operation state of the IC, and causes various Single Event Effects including malfunctions. Here, I will explain the relationship between the transient current and the effects generated, taking as an example the Single Event Upset in dynamic random access memories (DRAMs) and static random access memories (SRAMs), whose history of study is longest among the Single Event Effects.

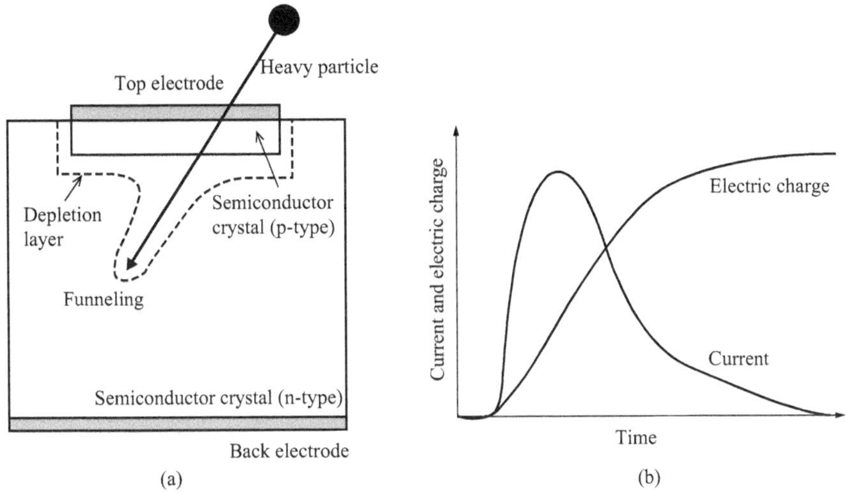

Fig. 9.10 Cross section of a p-n junction diode and the change with time in the current and electric charge. (**a**) Cross section of a p-n junction diode (**b**) Changes with time in the current and electric charge

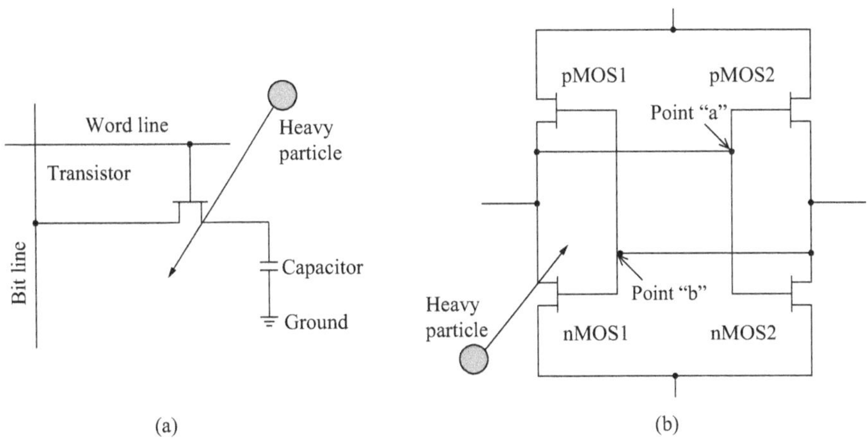

Fig. 9.11 Memory cell. (**a**) DRAM (**b**) SRAM

Single Event Upset in DRAMs

Figure 9.11a shows a memory cell of a DRAM. Each memory cell consists of MOS transistors and capacitors. The source, gate, and drain of the MOS transistor are connected to the bit line, word line, and capacitor, respectively. Information, that is, 0 or 1, is stored in the capacitor by letting the capacitor accumulate electric charge. The accumulated electric charge will be discharged over a certain period of time. Therefore, it is necessary to rewrite data into the memory cell at certain intervals of time. This is called the "refresh operation." The structure of the drain is the same as the p-n junction structure shown in Fig. 9.10a. The transient current induced by the incidence of heavy particles into the drain flows into the capacitor. If the transient current converges and electric charge is accumulated by the capacitor in a period of time, that is, much shorter than the refresh interval, the stored data will be reversed. That is, the information stored as "0" will be reversed to "1" and vice versa. The minimum electric charge required for the Single Event Upset to occur is called the "critical electric charge" (Q_c). The smaller the critical electric charge, the lower the Single Event Upset resistance is.

Single Event Upset in SRAMs

Figure 9.11b shows a CMOS (complementary metal oxide semiconductor)-type memory cell. Here, let us consider the case where incident ions get into the drain part of nMOS transistor 1 (nMOS1). The drain part is a p-n junction, and reverse voltage is applied to the drain. Therefore, the "electron-hole" pairs generated at the p-n junction will be accelerated by the electric field present in the depletion layer, which will cause transient current to flow. The generated transient current lowers the electric potential at point "a," which makes the electric continuity between the drain and source of pMOS transistor 2 (pMOS2) better than that before the irradiation. This means that the electric potential at point "b" is raised and the electric continuity of nMOS1 becomes better. When the electric continuity of nMOS1 becomes better, the electric potential at point "a" will be lowered further. If the electric potentials at points "a" and "b" become close to their threshold values through repetitions of these electrical potential changes, the stable state of the cell is reversed. That is, upsetting of the stored data will occur. As in the case of DRAMs, the smaller the critical electric charge, the lower the Single Event Upset resistance of SRAMs is.

As described above, the Single Event Upset occurs in a DRAM or SRAM if transient current that acts as the trigger occurs and the electric charge obtained by integrating the transient current exceeds the critical electric charge. The likelihood of occurrence of the Single Event Upset is becoming increasingly strong as a result of the fact that the critical electric charge is being made smaller year by year by recent advances in the integration technology. For this reason, semiconductor device type-specific measures to increase radiation resistance are being taken. The following are examples of such measures:

1. The critical electric charge of a DRAM is determined by the capacity of the capacitor. Therefore, the frequency of occurrence of the Single Event Upset in a DRAM can be lowered by increasing the capacity of the capacitor.
2. For SRAMs manufactured using the 120 nm technology, a technology has been developed that reduces the frequency of occurrence of the Single Event Upset to approx. the one hundredth of the normal frequency encountered with the traditional technology by adding capacitors at both ends of the memory cell.

3. The electric charge collected can be reduced by using an SOI (silicon-on-insulator) substrate, which has a triple-layer structure that is comprised of an oxide film layer sandwiched between two silicon layers, instead of using a traditional silicon substrate.

In addition to the abovementioned semiconductor device-level hardening, circuit-level and system-level hardening are being taken including error detection and error correction. An example of the error detection/correction is the ECC (error correction code). This is also called error detection and correction (EDAC) method-based relief. These functions can be used for a wide range of purposes because they can be implemented in software. Apart from the Single Event Upset in memory devices such as DRAMs and SRAMs, many problem phenomena in other types of semiconductor devices have been reported including the single event transient in logical large-scale integrated circuits (logic LSIs) and the single event burnout in power MOS transistors. The mechanisms of occurrence of these phenomena are different from the mechanism of occurrence of the Single Event Upset in memory devices such as DRAMs and SRAMs. However, there is a similarity between all Single Event Effect in that, in the initial process of their occurrence, electrons and holes are generated at a high density and collected.

9.4.2 Electric Charge Generation and Its Depth Distribution

If a heavy particle hits and gets into a substance, most of its energy will be consumed to ionize and excite orbital electrons of the crystal atoms. The secondary electrons generated by the ionization will cause further ionization and excitation that will be repeated in a cascading manner as a result of the interaction between electrons and the material. The radiation energy lost through ionization during unit distance travel of a radiation divided by the density is called "linear energy transfer (LET)" and is expressed by the following formula:

$$\mathrm{LET} = \frac{(dE/dx)_{\mathrm{ion}}}{\rho} \quad \left[\mathrm{MeVcm^2/g}\right] \tag{9.22}$$

The $(dE/dx)_{\mathrm{ion}}$ is the same as the electric stopping power in Formula (9.5). At the same time, crystal atom displacement due to elastic collision occurs at a frequency lower than that of the ionization and excitation. The radiation energy lost per unit distance through elastic scattering is called the "nuclear stopping power" in contradistinction to the electric stopping power mentioned above. In the case of high-energy heavy particles, the electric stopping power is normally several orders of magnitude greater than the nuclear stopping power. The total stopping power is expressed by the following formula:

$$(dE/dx) = (dE/dx)_{\text{ion}} + (dE/dx)_{\text{nucl}} \tag{9.23}$$

where $(dE/dx)_{\text{nucl}}$ represents the nuclear stopping power.

The total electric charge (Q (C)) of the electrons and holes generated by the ionization is expressed by the following formula:

$$Q = e\frac{E_i}{E_{\text{eh}}} \tag{9.24}$$

where E_{eh} (eV) represents the mean energy required for the generation of an electron-hole pair and e (C) represents the elementary charge. It is known that the mean energy required for the generation of an electron-hole pair is 3.6 eV for silicon. The distance a heavy particle travels after hitting and getting into a substance until the particle loses all of its energy through repetitions of elastic and inelastic scattering is called the "range." The higher the energy, the longer the range is, and the stronger the stopping power, the shorter the range is. If the range of the heavy particle is expressed as R, Formula (9.24) can be rewritten as follows:

$$Q = \frac{e}{E_{\text{eh}}}\int_0^R \frac{dE_i}{dx}dx \tag{9.25}$$

The energy of an alpha particle released by Americium 241 is approx. 5.5 MeV, and its range in silicon is approx. 30 μm. Heavy particles present in radiation environments in outer space have energy on orders of MeV to GeV or higher. Heavy particles may have a long range that allows them to penetrate through a semiconductor device. On the other hand, the sizes of those parts of semiconductor devices which are sensitive to the Single Event Effect (such as the drain of an MOS transistor) are very small. For example, the sizes of such parts are approx. tens of nanometers in the case of memory devices and approx. tens of micrometers in the case of power MOS transistors. This fact makes the electric charge per unit distance an important parameter in studies of effects of radiation on semiconductor devices. The electric charge per unit distance (dQ/dx) can be calculated from the electric stopping power and Formula (9.24) as follows:

$$\frac{dQ}{dx} = e \cdot \frac{(dE/dx)_{\text{ion}}}{E_{\text{eh}}} \quad [\text{pC}/\mu m] \tag{9.26}$$

A heavy particle with an LET value of 100 MeVcm^2/mg generates an electric charge of approx. 1 pC/μm in silicon.

In the field of the study of radiation effects on semiconductor devices, it is often the case that the ionization loss or linear energy transfer is calculated using a Monte Carlo simulation code. As mentioned in Sect. 9.3.4, the SRIM code proposed by Ziegler et al. is being widely used. In the SRIM code-based computation, the result of the electric stopping power calculation is output in IONIZ.TXT. Figure 9.12 shows the result of the SRIM code calculation of the relationship between the silicon penetration depth of a 5.5 MeV incident alpha particle (in the direction of the

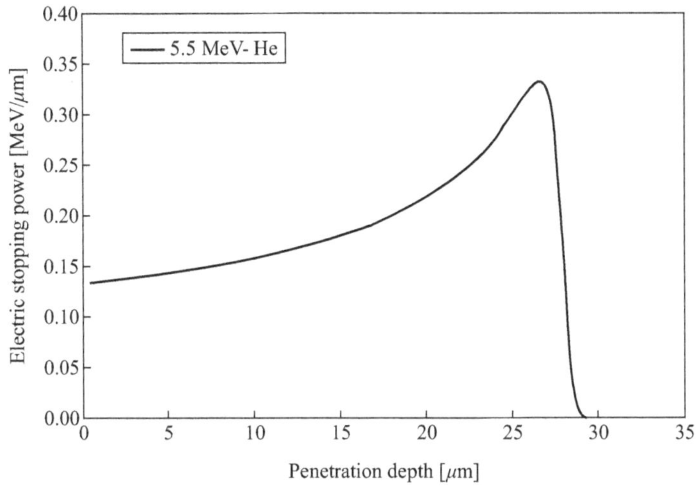

Fig. 9.12 Relationship between the electric stopping power and penetration depth calculated using the SRIM code

travel of the particle) and the electric stopping power. The electric stopping power peaks just before the heavy particle stops traveling in the substance and then drops sharply. This relationship is called the "Bragg curve" after the name of the scientist who discovered it (W. H. Bragg).

9.4.3 Generation of Charges and Its In-Plane Distribution

The integration technology for highly integrated semiconductor devices is now approaching the physical limit. As a result, the areas of individual cells have become very small. For example, in the case of an SRAM manufactured with a 45 nm process, the area of each cell is approx. 0.25 μm^2/cell or less, and a several hundred nm-square area contains multiple MOS transistors. On the other hand, the spatial distribution of the ion tracks, that is, the charges (electrons and holes (carriers)) generated by ionization, becomes wider as the energy of the heavy particle increases. In the case of incidence into silicon of a heavy particle with energy on the order of MeV to GeV which is present in the space environment, the extent of spread of the charges generated reaches the μm order. These facts mean that there are multiple memory cells among the ion tracks generated by the incidence of a single heavy particle. For this reason, studies have been conducted on effects of the special spread of ion tracks on the charge collection process.

In the field of radiation chemistry, the penumbra model proposed by A. Chatterjee et al. in 1973 is being widely used to determine the spatial distributions of charges generated by the incidence of heavy particles into material. This is a double-cylinder-type model that divides the area that is imparted energy by heavy particles into the core, which is the area that is imparted energy directly by the heavy particles, and the penumbra, which is the area that is imparted energy indirectly by secondary electrons. The model assumes that the core is the area within a radius of approx. 1 nm from the center of the track (which is called the "core radius"). The electric charge density within the core does not depend on distance from the center of the track. The charge density in the area located outside the core radius and within the penumbra radius varies in reverse proportion to the square of the distance from the center of the track. The penumbra model can be used to determine the typical charge density of ion tracks formed in water and gas.

In the field of the study of radiation effects on semiconductors for space applications, the model of ion tracks formed underwater proposed by R. Katz is being widely used. By using the electron density of silicon instead of the electron density of water and adding a correction term in the ion track model developed for water, the ion tracks in silicon can be calculated. The ion track model for silicon is as shown below. The carrier density distribution ($n_{\mathrm{eh}}(r)$) is expressed by the following formula:

$$n_{\mathrm{eh}}(r) = D(r)\frac{\rho}{E_{\mathrm{eh}}} \quad \left[\mathrm{cm}^{-3}\right] \tag{9.27}$$

where r represents the distance (radius) from the track of the heavy particle. $D(r)$ represents the energy deposition distribution and is expressed by the following formulas:

$$D(r) = D_1(r)(1 + k(r)) \quad [\mathrm{J/kg}] \tag{9.28}$$

$$D_1(r) = \frac{N_{\mathrm{Si}} e^4 Z^{*2}}{\alpha_\delta m_0 c^2 \beta^2 r}\left[\frac{(1 - (r + \theta_\delta)/(T_\delta + \theta_\delta))^{1/\propto\theta}}{(r + \theta_\delta)}\right] \tag{9.29}$$

where N_{Si} represents the density of electrons in silicon and is expressed by the following formula:

$$N_{\mathrm{Si}} = A_\nu \rho \frac{Z}{A} \tag{9.30}$$

where A_ν represents the Avogadro's number. Z^* represents the effective electric charge and, according to Barkas, is expressed by the following formula:

$$Z^* = Z\left\{1 - \exp\left(-125\beta Z^{-2/3}\right)\right\} \tag{9.31}$$

T_δ can be expressed with the following formula if the maximum energy (E_δ) of the secondary electrons (delta ray) is expressed in keV:

$$T_\delta = 6 \times 10^{-6} E_\delta^{\alpha\delta} \tag{9.32}$$

where E_δ is expressed by the following formula:

$$E_\delta = \frac{2m_0c^2\beta^2}{\sqrt{1-\beta^2}} \tag{9.33}$$

where

α_δ is 1.079 when E_δ is lower than 1 keV and 1.667 when E_δ is higher than 1 keV.
θ_δ can be expressed by the following formula:

$$\theta_\delta = 6 \times 10^{-6} I^\alpha \tag{9.34}$$

where I represents the ionization energy expressed in keV.

(1 + $k(r)$) in Formula (9.28) is the correction term. According to Fageeha et al., the correction term for silicon is expressed by the following formula:

$$k(r) = A\beta^{0.215}(r - 0.1)\exp\left(-\frac{r}{3.127 - 0.434\beta}\right) \tag{9.35}$$

where

A = "0" in the case where $\beta < 0.0081$
A = "$112\beta - 0.899$" in the case where $0.0081 < \beta < 0.091$
A = "$0.674\beta + 9.21$" in the case where $\beta > 0.091$

Figure 9.13 shows the ion track generated by the incidence of a 5.5 MeV alpha particle into silicon. Figure 9.13 also shows the ion track generated by the incidence into silicon of 150 MeV iron whose range is equivalent to that of a 5.5 MeV alpha particle. Figure 9.14 shows the horizontal spread of the electric charge density at the outermost silicon surface. It can be seen from the figure that carriers were generated at a high density along the track. In particular, the carrier density in the central part is several orders of magnitude higher than the impurity density of semiconductors and satisfies the high-injection condition. The ion track radius is 160 nm in the case of a 5.5 MeV alpha ray and 490 nm in the case of 150 MeV iron. In a state-of-the-art integrated circuit, there are multiple MOS transistors in a single ion track.

9.4.4 Charge Collection Process

The charges generated in a semiconductor device become transient current through the drifting and diffusion processes, and the transient current propagates in the circuit. Here, suppose that there is a p-n junction diode shown in Fig. 9.15. Let us assume that the heavy particles stop in the depletion layer and the ion track density satisfies the low-injection condition. The transient current is in proportion to the density of electrons (n_e) generated along the track of the heavy particles and the density of holes (n_h) generated along the track of the heavy particles. The electrons and holes generated are drifted by the electric field of the depletion layer of the p-n junction. The transient current is in proportion to the drift speed and continues

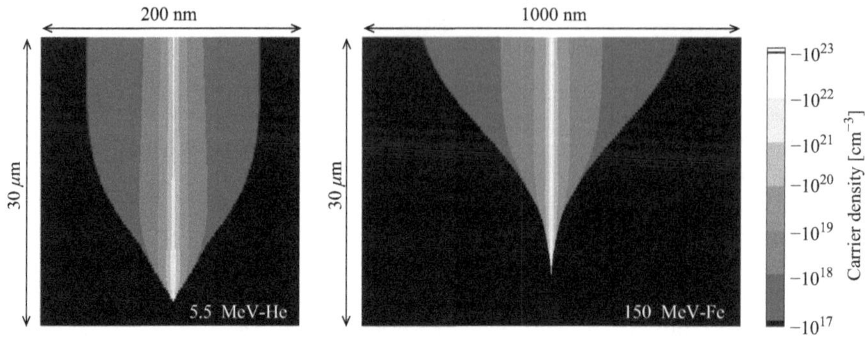

Fig. 9.13 Carrier density distribution (electrons and holes) in an ion track

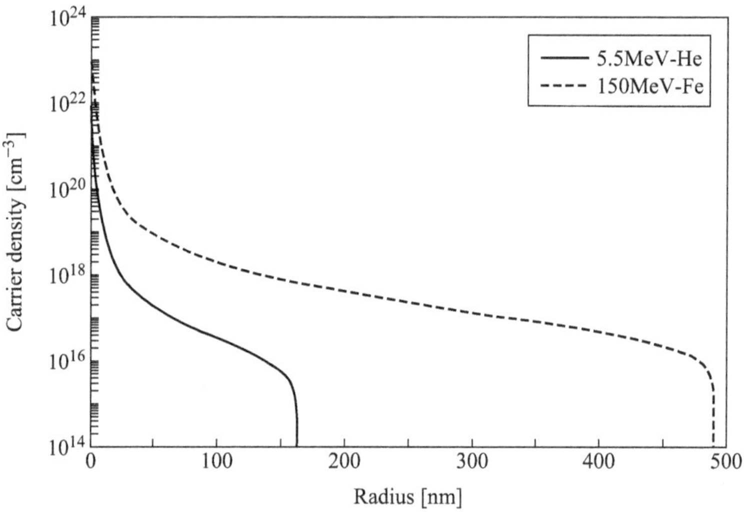

Fig. 9.14 Relationship between the carrier density (electrons and holes) in an ion track and the distance from the center of the track

flowing as long as the carriers keep drifting. If the drifting speed of the electrons is expressed as v_{de} and the drifting speed of the holes is expressed as v_{dh}, the minute electronic current element (di_e) induced by the electrons generated within a given position range between x_0 and $x_0 + dx_0$ during a given period of time can be expressed with the following formula:

$$di_e(t) = eSE_W(x)n_e(x)\nu_{\mathrm{de}}(x)dx_0 \tag{9.36}$$

where S represents the cross section of the ion track. E_w represents the electric field strength standardized in terms of potential and is defined by the following formula:

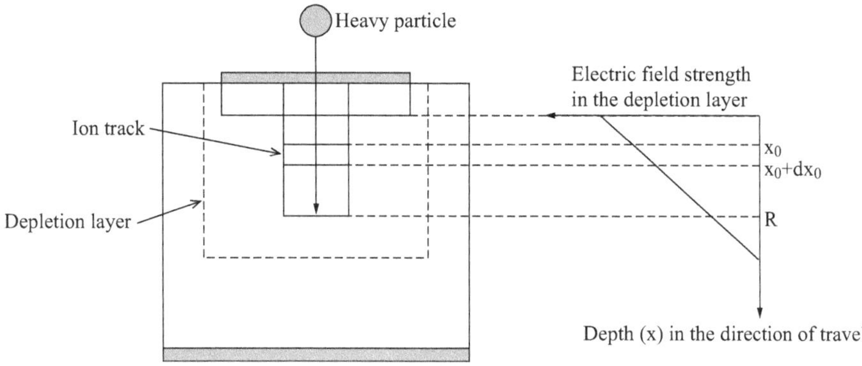

Fig. 9.15 p-n junction diode and its electric field strength distribution

$$E_W = \frac{dE}{d\psi} \tag{9.37}$$

where E represents the electric field and Ψ represents the potential difference. The electric field (E) is defined by the following Poisson's equation:

$$\frac{d^2\psi}{dx^2} = -\frac{dx}{dE} = -e\frac{N_D - N_A + n_h - n_e}{\varepsilon} \tag{9.38}$$

where ε represents the dielectric constant for the irradiated substance. N_A and N_D represent the impurity concentrations in the p and n layers, respectively. If one assumes that the p-n junction is an abrupt step junction-type p-n junction and solves Formula (9.38) on the premise that the depletion layer does not expand into the p layer, the electric field (E) and depletion layer width (W) can be expressed with the following formulas:

$$E = -E_{\max}\left(\frac{1-x}{W}\right) \tag{9.39}$$

$$E_{\max} = \frac{eN_D W}{\varepsilon} \tag{9.40}$$

$$W = \left(\frac{2\varepsilon(V_{\mathrm{bi}} - V)}{eN_D}\right)^{1/2} \tag{9.41}$$

where V_{bi} represents the built-in potential and V represents the voltage applied. The potential difference is expressed by "$\Psi = (V_{\mathrm{bi}} - V)$." If one differentiates E for Ψ and substitutes it in Formula (9.38), the following formula can be obtained:

$$E_W = \frac{1}{W} \tag{9.42}$$

The density of the carriers generated by the charged particles can be expressed by the following formula that uses the electric stopping power of the charged particles:

$$n(x) = \frac{(dE/dx)_{\text{ion}}}{SE_{\text{eh}}} \tag{9.43}$$

The carriers will be drifted by the electric field. The drifting speed (v_d) can be expressed as $v_d = \mu_0 E$ by using the carrier mobility (μ_0) in weak electric fields. When the electric field is strong, the drifting speed approaches the saturation value. The drifting speed in strong electric fields is empirically expressed with the Canali model:

$$v_d = \frac{\mu_0 E}{1 + \mu_0 E / v_s} \tag{9.44}$$

where t and v_s represent the time and saturation speed, respectively.

The time required for a carrier generated at a given location (x_0) to each Location x is calculated by substituting Formula (9.39) in Formula (9.44) and solving the differential equation. The boundary condition is "$x = x_0$ when $t = 0$." If the equation is solved using the electron mobility ($\mu_{0,\ e}$) and hole mobility ($\mu_{0,\ h}$), the following will be obtained for the electrons and holes, respectively:

$$t(x_0, x) = \frac{x - x_0}{v_s} + \tau_e \ln \frac{E_{\max}(1 - x_0/W)}{E_{\max}(1 - x/W)} \tag{9.45}$$

$$t(x_0, x) = \frac{x - x_0}{v_s} + \tau_h \ln \frac{E_{\max}(1 - x_0/W)}{E_{\max}(1 - x/W)} \tag{9.46}$$

where

$$\tau_e = \frac{\varepsilon}{e\mu_{0,e} N_D} \tag{9.47}$$

$$\tau_h = \frac{\varepsilon}{e\mu_{0,h} N_D} \tag{9.48}$$

By solving each of Formulas (9.45) and (9.46) for x, the $x = f(x_{0,}\ t)$ relationship can be determined. By substituting the obtained relationship in Formula (9.36), the minute current element induced by the electrons (di_e) can be determined. The minute current element induced by the holes (di_h) can also be derived in the same manner.

$$di_e(t) = eSE_W(t, x_0) n_e(t, x_0) v_{\text{de}}(t, x_0) dx_0 \tag{9.49}$$

where v_{de} represents the electron drifting speed. The current induced by the electrons in a given period of time ($i_e(t)$) can be expressed as follows by integrating the above formula for x_0:

$$i_e(t) = eS \int_0^W E_W(t, x_0) n_e(t, x_0) \nu_{\mathrm{de}}(t, x_0) dx_0 \tag{9.50}$$

The current induced by the holes in a given period of time ($i_h(t)$) can also be derived in the same manner. Therefore, the total induced current ($i(t)$) is expressed by the following formula:

$$\begin{aligned} i(t) = i_e(t) + i_h(t) \\ = eS \textstyle\int_0^W E_W(t, x_0) n_e(t, x_0) \nu_{\mathrm{de}}(t, x_0) dx_0 \\ +eS \textstyle\int_0^W E_W(t, x_0) n_h(t, x_0) \nu_{\mathrm{de}}(t, x_0) dx_0 \end{aligned} \tag{9.51}$$

The transient current generated when a heavy particle with a range shorter than the depletion layer width hits and gets into a p-n junction diode and the carrier density satisfies the low-injection condition can be calculated as described above. However, it is usually the case that the range of a heavy particle is longer than the depletion layer width and the carrier density in the ion track satisfies the high-injection condition. Therefore, the transient current is usually calculated by means of numerical analysis. In many cases, a commercially available device simulator (technology computer-aided design (TCAD)) is used in this analysis.

Figure 9.16 shows the changes with time of the electric field strength distribution obtained from a numerical analysis using a device simulator. A p-n junction diode was irradiated with a 5.5 MeV alpha particle. It was assumed that the impurity concentration in the p layer of the p-n junction diode, the impurity concentration in the n layer, and the reverse voltage were 10^{19} cm^{-3}, 10^{14} cm^{-3}, and 50 V, respectively. It can be seen from the figure that the electric field strength started changing after the incidence of the ion and a funneling was formed. The funneling peaked after the passing of several hundred picoseconds and disappeared after the passing of several nanoseconds.

This section discussed the generation of electric charges caused by the incidence of ions into p-n junctions, the process of the propagation of the generated charges, and the electric field changes caused by the propagation, in order to allow you to understand the basic process of Single Event Effects. However, this only accounts

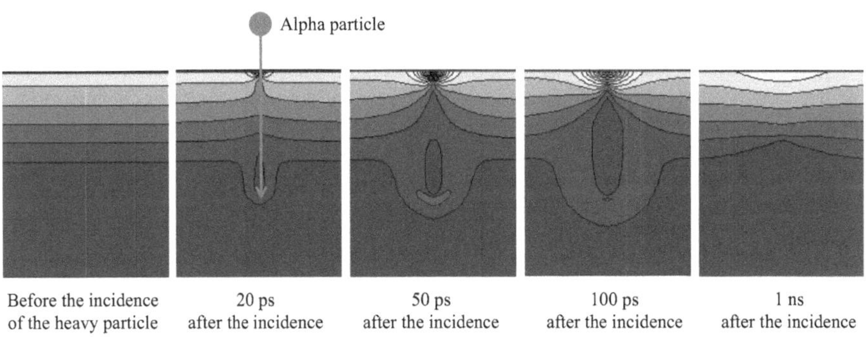

Fig. 9.16 Changes in the electric field strength

for the behavior of ion-induced charges in semiconductors that trigger Single Event Effects. What finally causes Single Event Effects that are actually observed in memories and logical circuits are secondary processes that follow the generation of the charges such as the process of the propagation of the charges in the circuit and the process of the generation of effects of the electric fields disturbed by the charges on the circuit. Therefore, truly understanding Single Event Effects requires total analysis including analysis of the circuit composition.

Exercises

1. State the three types of radiation effects on semiconductor devices and the main types of radiation that cause each of the three types of effects in real-world environments. In addition, consider and indicate examples of measures against each type of effect.
2. Cite examples of semiconductor devices used in strong radiation fields on the ground such as those in nuclear facilities and accelerator facilities and indicate the possible radiation effects occurring in each of the examples cited.

Further Readings

1. Ma TP, Dressendorfer PV (1989) Ionizing radiation effects in MOS devices and circuits. Wiley, New York
2. Holmes-Siedle A, Adams L (2002) Handbook of radiation effects, 2nd edn. Oxford University Press, Oxford
3. Velazco R (2007) Radiation effects on embedded systems. Springer, New York
4. Grove AS (1967) Physics and technology of semiconductor devices. Wiley, New York
5. Fleetwood DM, Dodd PE, Leray J, Marshall PW, Marshall CJ, Heidergott WF (1999) IEEE nuclear and space radiation effects conference short course, IEEE/NPSS Radiation Effects Committee.
6. Messenger SR, Burke EA, Summers GP, Xapsos MA, Walters RJ, Jackson EM, Weaver BD (1999) Nonionizing energy loss (NIEL) for heavy ions. IEEE Trans Nucl Sci 46(6):1595–1602
7. Fageeha O, Howard J, Block RC (1994) Distribution of radial energy deposition around the track of energetic charged particle in silicon. J Appl Phys 75:2317–2321
8. Li Z, Kraner HW (1993) Modeling and simulation of charge collection properties for neutron irradiated silicon detectors. Nucl Phys B Proc Suppl 32:398–409
9. Sze SM (1987) Handoutai debaisu – Kisoriron to purosesu gijutu -, Sangyo tosho (in Japanese); Sze SM (1987) Semiconductor devices – Basic theory and process technologies. Sangyo Tosho (in Japanese)
10. Ishino S (1979) Gensiryoku kougaku sirizu 8 syousha sonsyou. Toukyou daigaku syuppankai (in Japanese); Ishino S (1979) Nuclear Engineering Series 8: Irradiation damage. University of Tokyo Press, Tokyo (in Japanese)
11. Price WJ (1966) Houshasen keisoku. Korona sha (in Japanese); Price WJ (1966) Radiation measurement. Corona Publishing (in Japanese)
12. Otsuka N (1995) Q&A houshasen buturi. Kyouritu syuppan (in Japanese); Otsuka N (1995) Radiation Physics Q and A. Kyoritsu Shuppan (in Japanese)

Chapter 10
Utilization of Ionizing Radiation in Environmental Purification

Takuji Kojima

Sulfur oxides (SO_x, most of which is SO_2) and nitrogen oxides (NO_x) in the flue gas generated by the burning of coals and heavy oils at thermal power stations, etc. change into sulfuric acid, nitric acid, etc. as a result of the effect of sunlight in the atmosphere on them and get mixed into rain and fog to make that rain and fog acidic. The resultant acid rain and fog fall onto the ground surface and make soil, rivers, and lakes acidic, thereby causing such environmental damage as massive tree and fish death. On the other hand, volatile organic compounds (VOCs) used for paints for coatings and to wash off oils/greases get mixed into ventilation gas and released into the environment and cause adverse effects including health disorders such as cancers, photochemical smog, and ozone layer depletion. Photochemical reactions in the atmosphere that are associated with these have many paths and are very complicated.

Environmental pollutants (in particular, organic ones) that spread from wastewater and water, etc. seeped out from landfills and used sites of old factories affect ecosystems and our living environments, because the natural oxidation effect of the atmospheric oxygen and the purification effect of the decomposition by microorganisms, etc. are not sufficient enough to eliminate the effects of such environmental pollutants.

Environmental conservation and purification technologies to address such environmental problems, which are complementary with environmental monitoring technologies to grasp the statuses of pollutions, include ionizing radiation technologies [1]. These technologies aim at achieving environmental purification by artificially causing the above-mentioned natural oxidation reactions, etc. in controlled areas in short periods of time.

T. Kojima (✉)
National Institutes for Quantum and Radiological Science and Technology, Gunma, Japan
e-mail: kojima6@mvi.biglobe.ne.jp

H. Kudo (ed.), *Radiation Applications*, An Advanced Course in Nuclear Engineering 7, https://doi.org/10.1007/978-981-10-7350-2_10

10.1 Environmental Purification Technologies

As shown in Table 10.1, commonly used treatment technologies include the burning method, catalysis method, adsorption/collection method, biological method, and ultraviolet light, ionizing radiation, and plasma (discharging) methods. All of these methods have both advantages and disadvantages, and the most appropriate method is selected based on the type, concentration, and volume of the substance to be treated and the degree of CO_2 emission reduction that can be achieved, whether new pollutants such as external endocrine disruptors (which are often called hormone-disrupting chemicals) will be generated as a result of decomposition treatment and the simplicity, compactness, and economics of the process. The burning method, which is mainly used for thermal decomposition treatment of gaseous organic pollutants, has a large treatment capacity, but generates CO_2 and toxic decomposition gases. The catalysis method, which is a molecular weight reduction treatment method that uses catalysts to cause oxidation or reduction, requires the selection of catalysts suitable for the target pollutants and the heating of the treatment target gases, etc. to a temperature of about 300 °C. The adsorption/collection methods, which collect and remove pollutants using adsorbents such as activated carbon and zeolite or filters, are advantageous in that recovery and recycling is possible, but the amounts of pollutants that can be adsorbed are limited (which means that part of the pollutants may be released without being treated), and the decomposition of the pollutants requires additional treatment. The biological method, which utilizes decomposition by microorganisms, requires a long treatment time and the control of the nitrogen/phosphorus concentration or temperature/humidity to maintain the growth environment for the microorganisms.

The abovementioned methods are efficient when the concentration of the pollutant (in gas or water) to be treated is relatively high. However, it is required nowadays to further reduce the amounts of environmental hazardous substances released, which means that it is necessary to treat pollutants with relatively low concentrations and did not warrant treatment in the past. Methods to treat such pollutants, in the case where they are gases, include treatment methods that utilize chemical reactions caused by nonthermal equilibrium plasma. These plasma methods can be largely divided into electron beam methods [1] and corona discharge methods [2]. Electron beam methods generate plasma using a number of electrons whose kinetic energy has been accelerated to a level between several hundred keV and several MeV. Discharge methods generate plasma using electrons with energy between several eV and 10 eV generated by causing dielectric breakdown in part of atmospheric-pressure gas by giving a direct-current or alternate-current electric field with electric field strength on the order of kV/cm to the gas.

Both types of plasma methods for treating gases impart energy to the nitrogen, oxygen, water, and carbon dioxide gas molecules present in air, which is the main component of the gas, and cause ionization or excitation to generate ions, free radicals, and thermal electrons and also ion-molecule reactions to generate ions and free radicals. These resultant active species cause chemical reactions such as

Table 10.1 Comparison of methods to treat environmental pollutants contained in gases and water

	Temperature control requirement	Posttreatment for neutralization accompanies	Treatment speed	Treatment of low-concentration pollutant	Treatment volume	Notes
Burning method	Yes	Yes	Fast	○	Large	Generates CO_2, toxic gases, and/or airborne ash
Catalysis method	Yes	No	Medium	△	Medium	It is necessary to select the most suitable catalyst
Adsorption methods using activated carbon, etc.	No	Yes	Medium	△	Medium	Reuse is possible
Filter-based collection method	Yes	Yes	Medium	△	Medium	Reuse is possible
Biological method	Yes	No	Slow	○	Large	Control of nitrogen/phosphorus concentration, humidity, etc. is necessary
Ultraviolet (UV) methods	No	No	Fast	○	Medium	Efficiency may be reduced by soot particles, high turbidity, etc.
Plasma (discharging) methods	No	No	Fast	○	Large	Energy utilization efficiency is low (one tenth to one fiftieth of electron beam)
Ionizing radiation method	No	No	Fast	○	Large	Good public understanding of radiation and a large initial investment is necessary

oxidation of pollutants. Both plasma methods for treating gases are advantageous compared to other treatment methods in that (1) pollutants can be decomposed and removed in short periods of time even when their concentrations are very low, (2) it is possible to neutralize pollutants or reduce their toxicity through decomposition instead of moving them out of the surroundings (which means that secondary treatment after the treatment is not necessary), and (3) it is not necessary to control the temperature, etc. of the gas to be treated. In addition, electron beam methods are advantageous compared to discharge methods in that (1) large amounts of pollutants can be treated, (2) high treatment efficiency can be achieved even when the target pollutants coexist with other substances because the active species generation density is high, and (3) the energy utilization efficiency is high, although costs including the electron beam generation equipment are higher.

In the case where the treatment target is water, it is difficult to apply discharge methods, but electron beam methods can utilize chemical reactions caused by active species that have been generated in water basically in the same manner as in the case of the gas treatment mentioned above, although it is necessary to adapt the treatment process.

10.2 Physicochemical Reactions Caused by Ionizing Radiation [3]

Irradiation of gas or water with ionizing radiation causes such reactions as the ionization, excitation, and dissociation of gas or water molecules. These reactions are caused by the displacement of electrons of molecules and atoms (inner core and outermost core) by irradiation, their shifts to other orbits, and the resultant separation of atoms from each other. Such a state of coexistence of ions (most of which are positive ions) with electrons is called a "plasma state (or space)." Plasma can be divided into thermal equilibrium plasma, which is generated using heat much like in the case of nuclear fusion, and nonthermal equilibrium plasma, which is generated in a room temperature medium. Plasma can also be divided into strongly, intermediately, and weakly ionized plasma based on the ratio of the ionized species present to the number of atoms or molecules present in the medium. Plasma generated by irradiation, which is the main subject of this chapter, is weakly ionized plasma in most cases. Species that react with pollutants in a plasma space include ion species, exciting species, dissociating species, and secondary electrons. Ion species and exciting species usually react with gas components of water and oxygen in different ground states, to generate radical species, ozone, etc. It is also possible to utilize stable ion species left by these reactions in pollutant treatment. The following sections describe the individual generation reactions:

10.2.1 Free Radical Reactions

Ionizing radiation environmental purification technologies mainly use reactions caused by free radicals generated by irradiation. In the case where environmental pollutants are present in gases such as flue gas and exhaust gas, the energy of the ionized radiation is absorbed by the nitrogen, oxygen, water, and carbon dioxide gas molecules present in air, which is the main component. As a result, active species such as N, O, OH, and e^- are generated via such chemical species as N_2^+, O_2^+, and O_2^-. These active species first chemically react with the pollutants in their vicinity and then react with the oxygen contained in the gas, thereby causing oxidation decomposition. In particular, the OH radicals generated mainly by the following reactions play very important roles:

$$N_2 \rightsquigarrow N_2^+ + e \qquad (10.1)$$

$$N_2^+ + 2N_2 \rightarrow N_4^+ + N_2 \qquad (10.2)$$

$$N_2^+, N_4^+ + O_2 \rightarrow N_2, 2N_2 + O_2^+ \qquad (10.3)$$

$$N_2^+, N_4^+, O_2^+ + H_2O \rightarrow N_2, 2N_2, O_2 + H_2O^+ \qquad (10.4)$$

"Reactions caused by ionizing radiation" are often mistaken for "oxidation reactions." This misunderstanding stems from the common knowledge that the interested substance is oxidized. However, it should be noted that oxidation reactions are not the only reactions induced by irradiation. For example, in the case where the target pollutant is an organic matter, the organic matter is oxidized but oxygen is reduced.

The energy of ionizing radiation absorbed by a substance is expressed in terms of "absorbed dose (in gray (Gy))." One gray (Gy) equals 1 J/kg. The number of atoms/molecules generated per 100 eV radiation energy absorption is called the G value. For example, the G value for the generation of OH radicals in air is about 3.0. A G value of 1 corresponds to the generation of about 3.0 ppm of molecules/atoms (= $9.7 \times 10^{13}/cm^3$ molecules/atoms per cubic centimeter) at room temperature by irradiation of 1 kGy.

In the case where the environmental pollutant is contained in water (such as in the case of wastewater), most of the energy is also absorbed by the water, and the following reactive species are predominantly generated.

$$H_2O \rightsquigarrow e_{aq}^-, H, OH, H_2O_2, H_2 \qquad (10.5)$$

Among these reactive species, the ones that are mainly associated with decomposition are hydrated electrons (e^-_{aq}), hydrogen atoms (H), and OH radicals. For organic matters, hydrated electrons and hydrogen atoms act as reductants and OH radicals act as an oxidant. The generation G values of hydrated electrons, hydrogen atoms, and OH radicals are 2.7, 0.6, and 2.8, respectively. In the case where 1 kGy energy is imparted to water, the amount of reactive species generated at a G value of

1 in the water is approx. 0.1 mmol/L, and the water temperature increase is very small (approx. 0.2 °C). Since wastewater normally contains a trace amount of dissolved oxygen, the hydrogen atoms and hydrated electrons react with the oxygen at a fast rate to generate HO_2 radicals. However, their reactivity is far lower than that of the OH radicals. Therefore, it can be said that OH radicals play a major role in the decomposition of hazardous organic substances contained in wastewater as in the case of decomposition of hazardous organic substances present in the atmosphere.

10.2.2 Electron Attachment

In the case where the pollutant is a halide, it is sometimes the case that secondary electrons generated by the ionization attach to the pollutant, the halogen dissociates and the alkyl radicals generated as a result of the attachment and dissociation react with oxygen molecules to start an oxidation reaction. This reaction is called the "dissociative electron attachment reaction" and is observed, for example, in the form of the starting of decomposition of the carbon tetrachloride and part of the dioxin contained in a flue gas upon irradiation of the gas. However, in the case of air where oxygen is present and nitrogen oxides are generated as an irradiation product, the contribution of the secondary electrons to oxidation is smaller even if halides are present, because oxygen and nitrogen oxides have high electron affinity. On the other hand, there are also secondary electrons that attach to pollutants and pollutants negatively ionized with the halogen remaining without being desorbed.

10.2.3 Ozone Generation

In the case where a gas containing oxygen is irradiated with ionizing radiation, a reaction that generates ozone, which is an oxidizer, also occurs. This ozone generation reaction can be subdivided into a direct ozone generation reaction and an indirect ozone generation reaction. In the direct ozone generation reaction, plasma causes oxygen molecules to dissociate and the resultant oxygen atoms unite with other oxygen molecules, which results in the generation of ozone. In the indirect ozone generation reaction, nitrogen molecules, etc. that have been ionized and excited by plasma attack oxygen molecules to cause them to dissociate, which results in the generation of ozone. Because of the contribution of this indirect reaction, there is no proportional relationship between the concentration of oxygen contained in the gas and the amount of ozone generated. For example, the difference between pure oxygen gas (whose oxygen concentration is 100%) and air (whose oxygen concentration is 20%) in the concentration of ozone generated is only 20–30%. In the case of air, the concentration of ozone generated is about

10 ppm per 1 kGy which is possible to use in oxidation decomposition of pollutants. Use of ozone as it is has traditionally been considered difficult because its reactivity with organic substances is very low. Ozone is harmful to humans, and the option of decomposing ozone using a catalyst and utilizing the generated active oxygen to decompose organic matters has been considered in recent years.

10.2.4 Charging Phenomenon

In air that has been irradiated with ionizing radiation, many atoms and molecules with charges such as ions and electrons as well as many electrons are generated. These not only cause chemical reactions but also act as charged particles and cause such phenomena as transfers of charges to neutral substances in their vicinity and coagulation of polar substances in their vicinity. For this reason, in radiation-induced decomposition treatment in the atmosphere, intermediate products generated during the decomposition process can coagulate to form particulates with a diameter of submicron order and hinder the decomposition reaction. There are examples of development of technologies that actively control this charging phenomenon by applying an external electric field and separate and remove the hard-to-oxidize by-products of radiation-induced decomposition in their charged states using another external electric field.

In water, charging phenomena also occur such as the above-mentioned hydrated electrons. For these phenomena, technologies have been reported that make harmful metal ions contained in wastewater coagulate and precipitate by irradiation, thereby making it easier to remove them.

10.3 Radiation-Induced Decomposition of Organic Pollutants Contained in Flue Gas and Exhaust Gas

Technologies are being developed that utilize physicochemical reactions caused by ionizing radiation, such as the ones described above, to clean pollutant-containing flue gas and exhaust gas. The target pollutants are sulfur oxides and nitrogen oxides contained in flue gas from coal and thermal power stations, ultra-trace amounts of dioxins contained in flue gas from waste incineration plants, and photochemical smog-causing substances. Attempts are being made to apply these technologies to various highly volatile organic compounds contained in exhaust gas or ventilation gas from paint and chemical factories. The processes of chemical reactions caused by radiation are outlined below, taking as an example the case of oxidation decomposition-based removal of volatile pollutants (in particular, aromatic organic compounds) contained in exhaust gas from paint factories, etc.

In Japan, the organic compounds that are released into the atmosphere in the largest amounts include aromatic organic compounds such as toluene and xylene (which have benzene rings with their hydrogen atoms replaced by one or two methyl groups), which are mainly used as solvents in paints, methylene chloride (methane with two of its hydrogen atoms replaced by chlorine atoms) used in chemical synthesis processes and trichloroethylene (ethylene with three of its hydrogen atoms replaced with chlorine atoms), which is used to wash off oils/greases from equipment. Since the emission of organic compounds is being regulated to control photochemical smog, such organic compound gases must be cleaned by their oxidation decomposition to CO_2, which is not a photochemical smog-causing substance, or removed from the gas phase. It can be said that economical gas purification methods are ones that use radiation only in those parts of the purification treatment process in which the above-mentioned advantages of effects of radiation can be utilized to a sufficient degree and use other types of treatment (such as catalyst treatment) in combination according to the characteristics, form, etc. of the target pollutant in the remaining parts, rather than ones that use radiation in all parts of the purification treatment process.

The concentrations of aromatic organic compounds in polluted air are very low, and the radiation energy is absorbed by nitrogen, oxygen, moisture, etc., which are the main components. This results in the generation of ion species and radical species as mentioned in the previous section. Reactions associated with ion species are about 1000 times as fast as reactions associated with radicals. Ion species usually react with the main gas particles in the exhaust gas to contribute to the generation of radical species. Among these radical species, OH radicals are of importance. This is because the reaction rate constant of OH radicals with most organic compounds is large and OH radicals are harder to perish than N and O atoms.

In the initial process of reactions that occur at room temperature, the OH radicals generated by irradiation attach to one carbon atom of the benzene ring. This results in the generation of alkyl radical peroxides (R-OO•). However, in the case of exhaust gas at a temperature of several hundred degrees Celsius from a burning/incineration process, the OH radicals pull hydrogen atoms out of the benzene ring to generate benzene radicals and water molecules. In either case, alkyl radical peroxides such as benzene radicals and radicals generated as a result of addition of OH to the benzene ring subsequently react with other oxygen molecules or go through other processes to cause the benzene ring-opening reaction. The products of this benzene ring-opening reaction are straight chain type products that have aldehyde radicals or carboxyl groups attached to the carbon atoms located at both ends and have double bonds in molecules. These products are subsequently attacked by OH radicals and decomposed by oxidation into lower-class organic acids or inorganic substances. Unlike the case of the benzene ring, this oxidation decomposition reaction generates alkoxy radicals (RO•) through bimolecular reactions between alkyl radical peroxides or reactions with nitrogen oxides. These alkoxy radicals then go through intramolecular relaxation, etc. and cleaving of "carbon-oxygen" bonds or other processes to become aldehydes or organic acids with the carbon bonds retained. These reactions are expressed by the following formulas:

$$2R - OO\bullet \rightarrow 2RO\bullet + O_2 \tag{10.6}$$

$$R - OO\bullet + NO \rightarrow RO\bullet + NO_2 \tag{10.7}$$

There are also reports indicating that part of the NO_x reacts with alkoxy radicals to generate a nitrated organic substance.

This organic substance is less volatile, more soluble to water, and more resistant to oxidation decomposition by OH radicals than the original aromatic organic compound. Its molecules are polar molecules because of the presence of aldehyde radicals and carboxyl groups in the molecules, and this gives non-volatility and water solubility to the organic substance. When the organic substance reacts with OH, the carbon will be turned into alcohol, and the "carbon-carbon" bonds will not be cleaved because there are multiple double bonds in each molecule. Because of this, decomposition of this organic substance is not easy with radiation treatment alone.

Therefore, using other treatment methods suitable for the characteristics of the decomposition product in combination with radiation treatment is important to efficiently decompose the decomposition product through oxidation. Specifically, attempts are being made to use treatment methods that remove particulate decomposition products using a filter or static electricity and treatment methods that facilitate oxidation decomposition using a solid catalyst (a technique to cause catalytic activity using part of the radiation energy; an example is the active oxygen obtained by decomposing the generated ozone using a catalyst).

Examples of organic compounds to which other oxidation decomposition reactions than the reactions described above apply include organic chlorine compounds that have saturated bonds only or unsaturated bonds such as methylene chloride and trichloroethylene. In the case of methylene chloride, the secondary electrons generated by irradiation attach to the chlorine, and the alkyl radicals generated as a result of the removal of the chlorine react with oxygen molecules to cause oxidation decomposition of the methylene chloride. In the case of trichloroethylene, alkyl radicals are generated either by attachment of OH radicals or by attachment of secondary electrons and desorption of the chlorine. In either case, the alkyl radicals generated react with oxygen to cause oxidation decomposition. It should be noted that the reaction between the alkyl radicals generated by attachment of OH radicals and oxygen results in the generation of chlorine atoms, and these chlorine atoms attach to other trichloroethylene molecules to cause a chain reaction of successive similar oxidation decomposition (see Fig. 10.1). There are examples of development of technologies that reduce the concentrations of dioxins contained in flue gas from the incineration of municipal waste by decomposing them using radiation-induced physicochemical reactions in air such as the ones described above.

In addition, a technology has been developed which uses a treatment process that artificially causes the natural process of transformation of SO_x and NO_x contained in flue gas from thermal power stations, etc. into acid rain, etc. by irradiating the

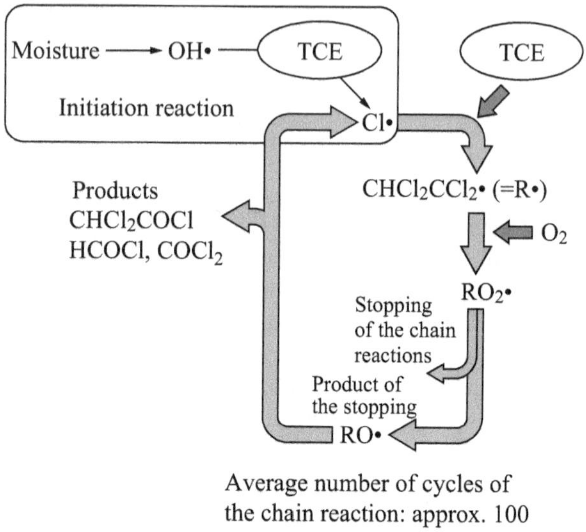

Fig. 10.1 Chain of the decomposition of chloroethylenes (Example of the case of trichloroethylene (TCE))

flue gas with electron beams in an environment in which ammonia is present, converts the generated sulfuric acid and nitric acid into powdery ammonium sulfate and ammonium nitrate, and collects and removes them using electric dust precipitators to clean the flue gas. This technology is based on a dry process that does not generate any wastewater and achieves desulfurization and denitrification at the same time and also has economic advantageous over other technologies such as the fact that the ammonium sulfate and ammonium nitrate generated as by-products can be used as fertilizers. Practical treatment plants based on this technology have been constructed in countries such as China and Poland (see Fig. 10.2).

10.4 Radiation-Induced Decomposition of Organic Pollutants Contained in Wastewater

Irradiation of wastewater with ionizing radiation decomposes pollutants contained in it and is therefore useful in sterilizing wastewater. Currently, ultraviolet-light-based ozone treatment is being popularly used.

10.4.1 Decomposition of Pollutants Contained in Wastewater

Pollutants present in rivers and soil can also be decomposed and removed by using OH radicals and hydrated electrons generated in the river/soil using ionizing

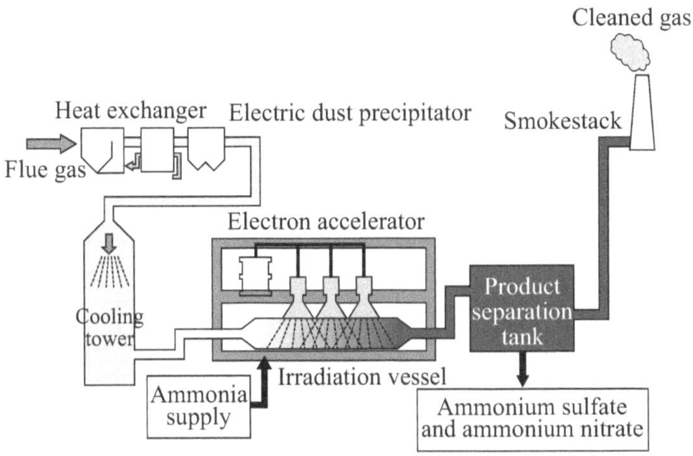

Fig. 10.2 Example of a flue gas treatment process of a thermal power station
Flue gas from the burning of fuel is cleaned by electron beam irradiation with ammonia supplied into the irradiation vessel to cause oxidation and neutralization reactions and removing the produced powdery ammonium sulfate and ammonium nitrate

radiation as in the case of pollutants in the gas phase, if the pollutants are harmful organic compounds described above. For volatile organic compounds such as trichloroethylene present in groundwater, etc., the abovementioned method of radiation-induced decomposition in air can also be used wherein pollutants are transferred into gas by aerating the air and treated.

A typical example of water purification is the application of a technique to decolor water containing ordinary acid and basic dyes by irradiation to wastewater from dyehouses, which is difficult to treat with the conventional activated sludge method (biological treatment). This decoloring reaction makes use of the alteration of organic compounds contained in dyes into colorless substances or substances that have no absorption region in a certain wavelength range by oxidation decomposition, and massive irradiation is required to decolor dye substances with irradiation alone. However, a treatment method is being considered in recent years wherein microorganism treatment is applied after irradiation with 1 kGy. This method makes use of the transformation of substances contained in dyes by irradiation into OH adducts and partially oxidized products which are more easily decomposed by microorganisms. It is known that this method makes the chemical oxygen demand (COD) and biochemical oxygen demand (BOD) 30–40% lower as compared to the case where only the microorganism treatment is used. Because of this economic advantage, a practical-scale wastewater treatment plant was constructed in South Korea (see Fig. 10.3).

Fig. 10.3 Electron beam irradiation equipment of the wastewater treatment facility with a practical-scale treatment capacity of 10,000 m^3/h for wastewater from dyeing (1MV, 400 mA, 3-head, 6-window)
Wastewater is injected from the nozzles located on the right and left sides in such a way that the wastewater jets form a thin layer of wastewater, and the thin layer of wastewater is irradiated with electron beams that are projected vertically from above. After this irradiation, microorganism treatment is applied

It is also possible to decompose humic acid, etc. contained in tap water, which reportedly induce generation of carcinogenic substances.

10.4.2 Sterilization of Wastewater Using Ionizing Radiation

Irradiation can be used to sterilize sewage and sewage sludge.

Most of the sludge from sewage treatment plants, etc. is currently being disposed of by incineration followed by landfill. However, incineration can cause air pollution or add to global warming. From this point of view, utilizing the sterilization effect of radiation is effective. This can be achieved with a lower absorbed dose than the dose required for the radiation-induced decomposition described above. Specifically, sewage sludge contaminated by bacteria is first sterilized by irradiation and then fermented using appropriate fermentative microorganisms, thereby making it possible to effectively utilize the sludge as animal feed, fertilizer, or soil improvement material. This radiation sterilization has an advantage over heat sterilization since no odor is generated and it is possible to kill pathogenic bacteria and parasite eggs. However, because the amounts of heavy metals and dioxins in wastewater have increased in recent years, removing them has become a task that must be addressed to put the radiation sterilization into practical use.

Exercises

1. From Table 10.1, select suitable treatment methods for exhaust gases whose maximum organic environmental pollutant concentrations in air are 10 ppm, 100 ppm, and 1000 ppm, respectively, taking into consideration the economics of the treatment methods.
2. "Micrograms/liter" is a unit for concentration used in regulations, standards, etc. that control releases of pollutants into the environment. Provide a rough estimate of the amount in grams of a pollutant whose concentration is 1 microgram/liter contained in a space whose volume is the same as that of Tokyo Dome (1,240,000 m^3).
3. It is not always the case that a technically superior exhaust gas treatment method is selected as a method to use in practical exhaust gas treatment. Give a reason for this citing an example.

References

1. Kojima T (2003) Denshi biimu o mochiita haien haisui shori gijutu to sono jiturei, Ouyou buturi, vol 72, pp 405–414 (in Japanese); Kojima T (2003) Electron beam-based flue gas and waste-water treatment technologies and examples of their use. Appl Phys 72:405–414 (in Japanese)
2. Chang JS (2000) Taiki osen bussitu no kankyou taisaku gijutu no genjou to kadai – Teion purazuma gijutu o cyushin toshite -, Ouyou buturi, vol 69, pp 268–277 (in Japanese); Chang JS (2000) Current status of and challenges for technologies to protect the environment from air pollutants – an analysis with a focus on low-temperature plasma technologies. Appl Phys 69:268–277 (in Japanese)
3. Nihon houshasen kagaku kai (hen) (2006) Houshasen kagaku no susume, Gakkai syuppan senta (in Japanese); Japanese Society of Radiation Chemistry (ed.) (2006) Introduction to radiation chemistry. Gakkai Shuppan Center (in Japanese)

Chapter 11
Production of Inorganic Materials

Masaki Sugimoto

11.1 Ceramics and Silicon Carbide

According to dictionaries, the term "ceramics" refers to porcelain in general. However, recently, this term has started to be used as a term that refers to inorganic material moldings that have been sintered at a high temperature above 1000 °C. Generally speaking, ceramics are chemically inactive, heat-resistant materials that have high hardness and strength but are brittle and lack toughness. When ceramics are used in industrial applications, they are sometimes molded together with other materials into composite materials to make up for these weaknesses. Because the present day is a time in which combating global warming is a task of utmost importance, development of advanced technologies aimed at reducing emissions of carbon dioxide, which is a greenhouse effect-causing gas, is being actively pursued, including the development of technologies to improve the efficiency of power generation gas turbines used in thermal power stations and the development of nuclear fusion reactors, which constitute an energy source that can be used without worrying about depleting resources. Because the environments in which these are used are aerobic atmospheres at temperatures much higher than 1000 °C, it is said that no materials can be used other than ceramics-based composite materials. This is one of the reasons for the fact that research and development is being actively pursued in various areas to produce ceramics with high strength and good toughness which have high heat and corrosion resistance suitable for use in aerobic atmospheres.

Ceramics are usually produced by compression-molding fine powder of an inorganic compound that contains silicon, carbon, nitrogen, boron, oxygen, or

M. Sugimoto (✉)
Department of Advanced Functional Materials Research, Takasaki Advanced Radiation Research Institute, Quantum Beam Science Research Directorate, National Institutes for Quantum and Radiological Science and Technology, 1233 Watanuki-machi, Takasaki, Gunma 370-1292, Japan
e-mail: sugimoto.masaki@qst.go.jp

H. Kudo (ed.), *Radiation Applications*, An Advanced Course in Nuclear Engineering 7, https://doi.org/10.1007/978-981-10-7350-2_11

metal oxides and sintering the resultant molding at a high temperature to solidify it (this is called the "sintering solidification method"). This method is advantageous in that it allows ceramics to be easily molded together with other materials for making up for the brittleness because of the small rate of density change during the molding and the small rate of volume reduction after the sintering, although tasks to be addressed remain in relation to the fine powder production and to sintering technologies. This method is not suitable for the production of fibers, which requires the production of minute moldings on the order of micrometers and smoothness, but can produce relatively large structural materials that are heat and catalyst resistant. Selection of the fine inorganic compound powder to use as the raw material for ceramics is made taking into consideration the application, heat resistance, corrosion resistance, strength, and/or hardness. One raw material for ceramics that has been attracting attention in recent years as a raw material that can be used for advanced materials is silicon carbide (SiC). SiC is a compound which does not occur naturally and was artificially created in the late nineteenth century by mixing coke, silicon, chaff, etc. and burning it in the field. At the time, it was regarded as a mix of carbon and corundum, which is a crystal of alumina, and was called "carborundum." Because it was chemically very stable, had high heat and corrosion resistance, and was second only to diamond in hardness, it was being sold as polishing powder. In the latter half of the 1970s, studies started to be conducted actively on technologies to produce monocrystal of SiC by sublimating fine powder of SiC in an inert atmosphere at a temperature of 1500 °C or higher. As a result, the range of application of SiC expanded, and SiC started attracting attention also as a power control semiconductor material which has high electron mobility on par with Si and high saturated electron drift speed and is capable of operating at high temperatures. Nowadays, SiC is also being utilized in the form of ceramic structural materials for use in high-temperature and corrosive environments, which are produced by compression-molding high-purity SiC powder and solidifying the resultant moldings by high-temperature sintering, and as coating materials for material surfaces that are exposed to high-temperature and corrosive environments.

Around 1975, a new technique to synthesize ceramics that contain SiC as the base material (SiC ceramics) using a silicon-containing polymeric material as the starting material (the precursor method) was invented. In this method, the silicon-containing polymeric material is molded into the desired shape, curing is applied to the resultant molding and then the molding is fired at a high temperature above 1100 °C to convert it into SiC ceramics. The term "curing" refers to the series of treatments to rid the silicon-containing polymeric material of its melting point and prevent changes in shape due to melt during the subsequent high-temperature firing. The curing is an indispensable process for precursor method-based production of SiC ceramics and is often achieved through cross-linking of the main chain of the silicon-containing polymeric material accompanied by heat treatment. The precursor method is advantageous in that it allows SiC ceramics of any shape with high toughness to be produced through a series of processes that consist of molding, curing, and firing of a silicon-containing polymeric material, which means that the method allows "flexible" SiC ceramic fibers to be produced which cannot be produced with the firing solidification method. However, the precursor method is not suitable for the production of gears, etc. which requires precise control of dimensions or the production of structural materials of complex shapes, because a large amount of gas is released and the volume decreases by about 50–60% during the firing for converting into SiC ceramics.

SiC microtubes described in this chapter are discontinuous hollow yarns of SiC ceramics developed out of a study aimed at utilizing the selective oxidation phenomenon that occurs when a substance is irradiated with electron beams, wherein only the surface layer is oxidized, in the curing process of fibers made from silicon-containing polymeric materials. This chapter describes methods for producing polycarbosilane, which is a silicon-containing polymeric material, and polycarbosilane fibers, the effects of irradiation observed when electron beam oxidation curing is applied and electron beam irradiation-based hollow SiC ceramic yarn production technologies.

11.2 SiC Microtube Production Method

11.2.1 Silicon Polymetric Fibers

One silicon-containing polymeric material that is being widely used as a precursor for SiC ceramics is polycarbosilane (PCS). PCS is a translucent solid as shown in Fig. 11.1a that is blackish brown at room temperature. The molecular weight and melting point of PCS are approx. 500–10,000 and 200–350 °C, respectively.

The industrial PCS production method is as follows: first, metallic sodium is added to dimethyldichlorosilane ($(CH_3)_2SiCl_2$) to cause a dechlorination polycondensation reaction to synthesize polydimethylsilane (PDMS) ($\{-(CH_3)_2Si)\}n$). Then, the PDMS is heated to a temperature of 400 °C or higher in a pressure vessel or a nitrogen gas atmosphere to cause a thermal decomposition reaction that decomposes the PDMS to synthesize the PCS. Many synthesis methods for different polycarbosilane applications have been developed, and it is known that the molecular weight, structure, and composition slightly differ between synthesis methods. The PCS the authors used is a product purchased from Nippon Carbon Co., Ltd. with an average molecular weight of approx. 2600 and a melting point of approx. 240 °C. Another well-known silicon-containing polymeric materials that are being used as precursors for SiC ceramics are polyvinylsilane (PVS), polysilane, hydropolycarbosilane (with a Si to C content ratio of 1:1), and polymethacarbosilane. Polyvinylsilane is a viscous liquid that is colorless and transparent at room temperature as shown in Fig. 11.1b with a molecular weight of approx. 1000 which is sometimes used to make a polyvinylsilane-polycarbosilane mixture used to improve melt-spinning efficiency.

A method that is being used to produce continuous fiber from polymeric materials is the "melt-spinning" method, in which a polymeric material is heated and melted, the molten polymeric material is continuously extruded from a

Molecular weight: approx. 2,000 to 5,000
Melting point: 200 to 250°C

(a) polycarbosilane(PCS)

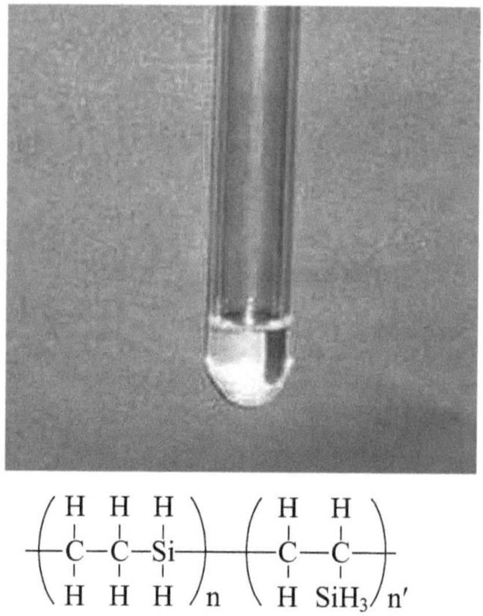

Molecular weight: approx. 500 to 1000
A liquid that is viscous at room temperature

(b) polyvinylsilane(PVS)

Fig. 11.1 Main properties of polycarbosilane and polyvinylsilane

spinning tube that has a small extrusion opening with the polymeric material maintained at an appropriate temperature and then the extruded polymeric fiber is spun using a wheel rotating at a constant speed. The material of the spinning tube, the shape of the extrusion opening, and the atmosphere into which the polymeric fiber is extruded vary, but this is a very important process that determines the performance of the fiber. Figure 11.2 shows the spinning equipment and the mono-hole-type spinning tube the authors fabricated out of glass tubes the authors had been using in the laboratory. The outer diameter of the fiber spun can be adjusted by mechanical processing such as increasing the reeling speed above the speed of extrusion of the fiber from the spinning tube and mechanically stretching (pulling) the fiber to make it thinner. However, it is difficult to apply such mechanical processing to hardened PCS fiber because it does not have sufficient toughness. Producing very thin fiber will require the use of advanced techniques in such processes as the fabrication of the extrusion opening of the spinning tube. Hollow yarns used to produce ultrapure water, etc. are made hollow as they are spun using an extrusion opening of a special shape, but it is said that continuous hollow yarns cannot be produced with this method when the molecular weight of the PCS is less than 10,000 because it has been difficult to achieve the proper level of viscosity in the melting process.

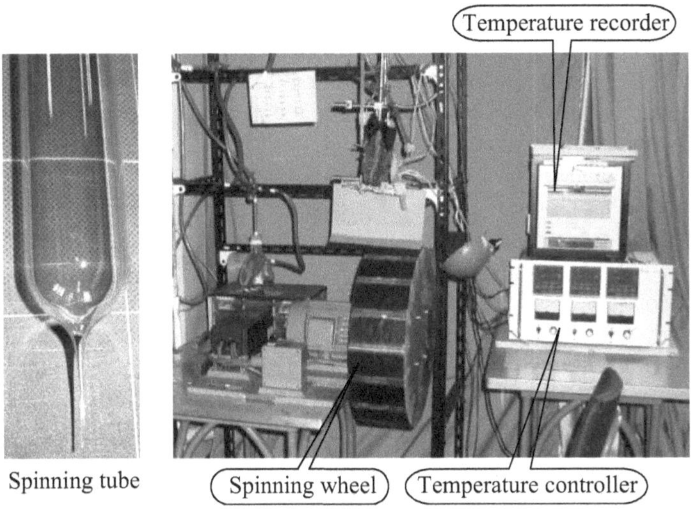

Fig. 11.2 The spinning equipment and the spinning tube

11.2.2 Electron Beam Oxidation Curing

For the curing of PCS fibers produced through melt spinning, the "thermal oxidation curing" method is usually used, in which the PCS fiber is thermally oxidized in an oxygen-containing atmosphere at 160–200 °C to achieve cross-linking. Although this is a method that can be used easily, use of this method to cure a large amount of PCS fiber for industrial purposes requires advanced temperature control technologies because heat is released in such curing. In addition, it is known that, because oxygen in excess of 10 mass% is taken in as a result of the thermal oxidation, if an SiC ceramic fiber produced through conversion by firing is subjected to a high temperature above 1200 °C, the oxygen used for the cross-linking will react with the carbon and silicon which are the main components of the fiber to generate carbon monoxide (CO) and silicon monoxide (SiO) gases and cause thermal decomposition, thereby reducing the strength of the SiC ceramic fiber.

To avoid the oxygen-induced thermal decomposition reaction at high temperatures above 1200 °C, a curing method that does not use oxygen for cross-linking has recently been developed. This method, which is called the "electron beam curing" method, was born out of "industry-academia-government collaboration" between Nippon Carbon Co., Ltd., Osaka Prefecture University, and Japan Atomic Energy Research Institute and achieves direct cross-linking between the main chains of PCS through irradiation of silicon polymeric fibers with electron beams in a helium atmosphere. In this method, the fiber is irradiated with electron beams of about 15 MGy or more and then is subjected to thermal annealing treatment in an inert atmosphere to eliminate the large amount of residual radicals generated. This reduces the oxygen density in the SiC ceramic fiber to 0.4 mass% or less and imparts very high heat resistance to the SiC ceramic fiber which prevents it from being thermally decomposed even by heating at 1500 °C in an inert gas atmosphere for 10 h. SiC ceramic fibers produced using the electron beam curing method are already commercially available as practical products.

Thermal decomposition-based curing is an important process that allows SiC ceramic fibers that do not require heat resistance to be produced at a low cost, but thermal oxidation temperature control for synthesis of a large amount of fiber with this process is difficult as mentioned above.

For this reason, a study was also conducted to develop a method for applying oxidation treatment at room temperature by means of electron beam irradiation in an oxygen atmosphere. In this method, which is called the "electron beam oxidation curing" method, the PCS fiber is irradiated with electron beams of about 3 MGy in the oxygen atmosphere and then subjected to thermal annealing treatment in an inert atmosphere to decompose the generated peroxides and cause cross-linking. Unlike the thermal oxidation curing method, this method does not cause any heat release. However, a selective oxidation phenomenon is sometimes caused by the partial pressure of oxygen during the irradiation of the sample with electron beams, wherein only the surface layer of the sample is oxidized by the irradiation. To achieve uniform oxidation curing of a PCS fiber, it is necessary to control the partial oxygen pressure, electron beam dose rate, and/or sample temperature appropriately as will be described later. This section first describes the effects of irradiation of PCS fibers with electron beams and then theoretically explains the chemical reactions caused by the electron beam irradiation and explains in detail the phenomenon wherein only the surface of the sample is selectively oxidized by radiation.

Effects of Irradiation of Polymeric Materials with Electron Beams

Electron beam irradiation (at room temperature) of a PCS fiber that has been produced by melt spinning causes the formation of radicals in the main chains of the PCS polymeric material and the generation of hydrogen and methane gas as decomposition products, as shown in Fig. 11.3a. In the case of electron beam irradiation in an oxygen atmosphere, peroxides in the form of chemical compounds containing hydrogen generated by the irradiation and oxygen are generated from the generated radicals as shown in Fig. 11.3b, because of the diffusion of oxygen in the sample and the dissolution of the oxygen. However, in the case of irradiation at room temperature, the oxidation cross-linking does not occur. To cause the same oxidation cross-linking as in the case of thermal oxidation, it is necessary to perform several hours of heat treatment in an inert gas atmosphere at 200–300°C after the irradiation. The absorbed electron beam dose required to cause sufficient oxidation cross-linking is approx. 3 MGy. In the case of irradiation in an inert atmosphere, the radicals in the main chains of the polymeric material directly react to cause cross-linking, as shown in Fig. 11.3c.

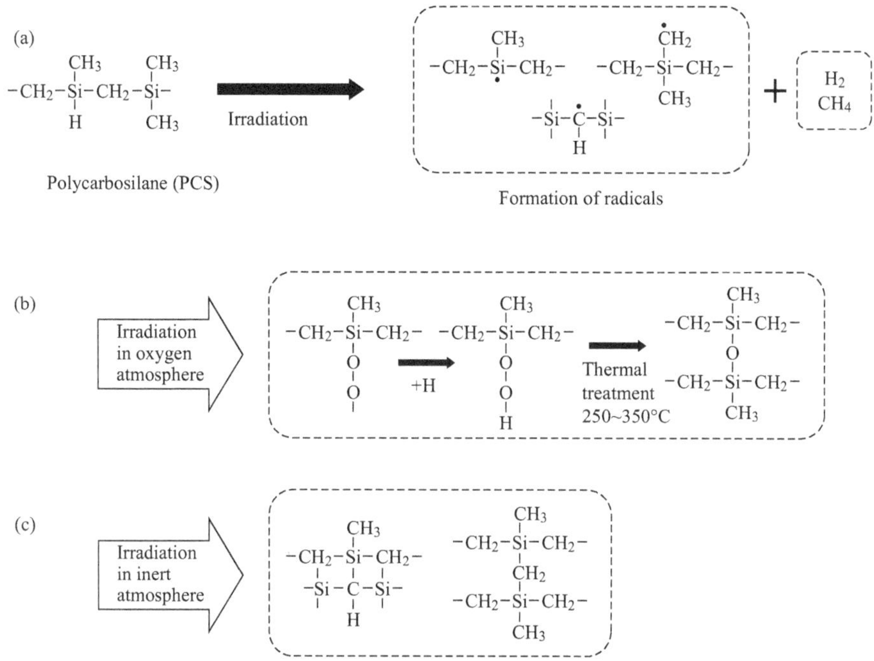

Fig. 11.3 Reaction mechanism of radiation curing

The probability of occurrence of cross-linking caused by inter-radical reactions is relatively low, and causing sufficient cross-linking requires an absorbed electron beam dose in excess of 15 MGy.

Electron beam oxidation curing can be done with about one fifth of the absorbed dose required for electron beam curing in an inert atmosphere, which means that a dramatic reduction in the time required for curing can be achieved. This is one of the merits of electron beam oxidation curing.

Unlike the case of thermal oxidation, this electron beam irradiation of PCS fiber in an oxygen-containing atmosphere is often accompanied by a phenomenon wherein only the outermost part of the PCS fiber along the circumference is oxidized by the radiation (uneven distribution of radiation oxidation). Figure 11.4 shows a scanning electron microscope photograph of a cross section of a PCS that has been irradiated with 2 MeV electron beams in air at a dose rate of 0.4 kGy/s for curing by oxidization.

The result of an energy dispersive X-ray spectroscopy (EDS) method-based analysis of the element distribution is also shown. The silicon and carbon atoms are evenly distributed throughout the cross section of the fiber, whereas most of the oxygen atoms are distributed in the outermost part along the circumference. This clearly shows that only the outermost part of the PCS fiber has been oxidized.

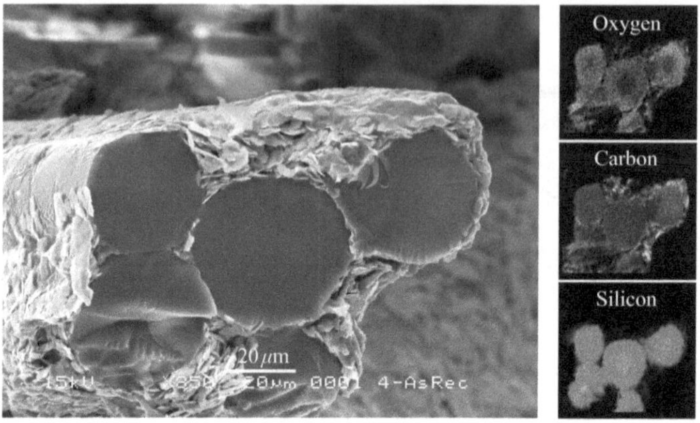

Fig. 11.4 Result of an energy dispersive X-ray absorption spectroscopy measurement-based element analysis of a cross section of a polycarbosilane fiber showing the uneven oxidation phenomenon

Analysis of the Effects of Irradiation with Radiation

Detailed studies were conducted from the latter half of the 1970s to the first half of the 1980s on the phenomenon wherein only the surface of polymeric materials is oxidized when they are irradiated with electron beams or gamma rays at a high-dose rate, as part of the efforts to evaluate the deterioration of polymeric materials caused by radiation. This is because it had been known that the state of deterioration of a polymeric material due to oxidation differs significantly between the case where it is irradiated with a certain dose of gamma rays in air, the case where it is irradiated with gamma rays at a high-dose rate for a short period of time, and the case where it is irradiated with gamma rays at a low-dose rate for a long period of time, even if the total dose is the same. This is because the number of radicals that are generated uniformly in the polymeric material by irradiation of the polymeric material with radiation varies widely depending on the dose rate of the radiation while the rate of supply of the oxygen contained in the air into the polymeric material is constant. The results of detailed experiments conducted in conjunction with the abovementioned detailed studies showed that the state of the oxidation of a polymeric material caused by irradiation of the polymeric material with radiation can be described with the following formulas:

$$\frac{d[P\cdot]}{dt} = \alpha I_R - k_1[P\cdot]^2 - k_2[P\cdot][C] \tag{11.1}$$

$$\frac{\partial[C]}{\partial t} = D_f\frac{\partial^2[C]}{\partial x^2} - k_2[P\cdot][C] \tag{11.2}$$

where $[P\bullet]$ and $[C]$ represent the radical density in the polymeric material and the oxygen density in the polymeric material, respectively, and α is the constant for the rate of radical formation in the polymeric material (mol/kg/Gy). I_R represents the

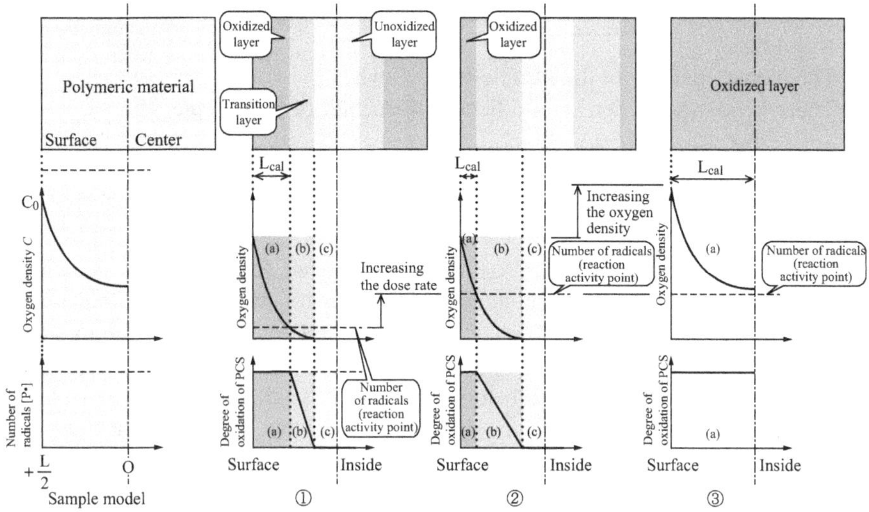

Fig. 11.5 Mechanism of oxidation caused by radiation

absorbed radiation dose rate (Gy/s), and k_1 and k_2 are the constant for the inter-radical bonding reactions and the constant for the rate of reaction between radicals and oxygen, respectively. D_f represents the rate of diffusion (m^2/s) of oxygen into the polymeric material. Formula 11.1 shows that the change in the radical density in the polymeric material per unit time is equal to the difference obtained by subtracting the number of radicals that vanish per unit time as a result of the inter-radical bonding reactions and the number of radicals that vanish per unit time as a result of their reactions with oxygen from the total number of radicals generated by irradiation per unit time. Formula 11.2 shows that the change in the oxygen density at Location x per unit time is equal to the difference obtained by subtracting the density of oxygen consumed per unit time as a result of its bonding with radicals from the density of oxygen supplied per unit time through diffusion. Figure 11.5 shows a sample model to which this differential equation applies.

The conditions for the application of this model are the following seven conditions:

1. The sample has a thickness of L and a sufficient vertical length.
2. The thickness of the irradiation sample is less than the average free path of the radiation particles, and radicals are generated uniformly in the sample. The amount of radicals does not change with time because the sample is irradiated constantly.
3. The oxidation of the polymeric material caused by the irradiation does not change the rate of diffusion.
4. The number of radicals that vanish per unit time as a result of the inter-radical bonding reactions is much smaller than the number of radicals that vanish per unit time as a result of the reactions between the radicals and oxygen ($k_1 << k_2$).

5. The oxygen density outside the sample is constant (C_0) and does not change with time.
6. The oxygen density in the sample (C) is 0 when $t = 0$.
7. There is no oxygen density gradient at the center of the sample.

$$\left(\frac{\partial C}{\partial x}\right)_{x=0} = 0$$

Conditions (1)–(4) are the approximation conditions for solving the equations, and Conditions (5)–(7) are the boundary conditions for the differential equations. Condition (1) is the condition that must be satisfied for Formula 11.2 to stand. Condition (2) is an approximation that stands in the case where PCS fibers with a diameter on the order of micrometers are irradiated with gamma rays or high-energy electron beams. Condition (4) is an assumption based on the fact that electron beam oxidation curing of PCS fibers can be achieved with about one fifth of the absorbed dose required for electron beam curing in an inert atmosphere.

Applying Conditions (2)–(4) to Formula 11.1 allows the radical density to be calculated as follows:

$$[P\cdot] \cong \frac{\alpha I_R}{k[C]} \tag{11.3}$$

By substituting Formula 11.3 in Formula 11.2, associating the oxygen density with the oxygen pressure using Henry's law and solving the rewritten equation, the following formula is obtained:

$$L_{\mathrm{cal}} = \left(\frac{2D_f M_e P_a}{\phi I_R}\right)^{1/2} \tag{11.4}$$

where L_{cal}, D_f, and M_{e} represent the thickness of the oxidized layer including the oxygen density transition layer (m), the rate of diffusion of oxygen into the polymeric material (m^2/s), and the oxygen solubility coefficient for the polymeric material (mol/kg•Pa), respectively. P_a, φ, and I_R represent the partial pressure of oxygen (Pa), the amount of oxygen consumed by the polymeric material as a result of the oxidation caused by the radiation (mol/kg•Gy), and the absorbed radiation dose rate (Gy/s), respectively. Radiation oxidation models of this formula are shown in Diagrams (1)–(3) of Fig. 11.5. The absorbed radiation dose rate during irradiation is proportional to the number of activity points of oxidation reactions that occur in the polymeric sheet (i.e., the number of radicals), but the induced activity points are distributed evenly in the PCS fiber because radiation has high penetration power. In the case where the oxygen density distribution is as shown in Diagram (1) of Fig. 11.5, sufficient oxygen is supplied to the reaction activity points in Layer (a) and the oxidation reaction in Layer (a) is limited by the rate of reaction, which results in uniform oxidation of Layer (a). However, in Layer (b), the amount of oxygen supplied to the activity points is insufficient (because of the consumption of oxygen), and the oxidation reaction is limited by the rate of supply, which results in "nonuniform oxidation" wherein the degree of oxidation decreases as the depth increases. In Layer (c), no oxygen is supplied and the irradiation effect is the same as that in a vacuum environment. Diagram (2) of Fig. 11.5 shows the result for the case where the radiation dose rate is increased without increasing the density of oxygen supplied from outside. It can be seen from the diagram that the radiation dose rate increase reduced the thickness of the oxidized layer. Diagram (3) of Fig. 11.5 shows the result for the case where the density of oxygen supplied from outside is also increased. It can be seen from the diagram that uniform oxidation was achieved even though the radiation dose rate was high. Diagrams (2) and (3) show that the thickness of the oxidized layer can be changed by changing the absorbed radiation dose rate and the oxygen density (i.e., partial pressure of oxygen). Since the rate of diffusion of oxygen is a function of the sample temperature, it can be said that the thickness of the oxidized layer can be controlled by controlling the radiation dose rate (I_R), oxygen density (P_α), or sample temperature as shown by Formula 11.4. This irradiation effect allows only the outermost part of a PCS fiber along the circumference to be cured with selective radiation oxidation of the outermost part of the PCS fiber along the circumference followed by thermal annealing to cause cross-linking.

Figure 11.6 shows an overview the electron beam accelerator the authors used for the radiation oxidation curing. This large-scale accelerator is capable of taking out current of up to 30 mA (2 MeV). The irradiation equipment used in this study was installed directly below the accelerator, and a conical beam scanner was installed to project electron beams evenly into the room. The irradiation chamber for the sample is a cylindrical stainless steel autoclave which is so designed that the irradiation atmosphere in the chamber can be altered by supplying gases from outside. Figure 11.6b is a photograph of the irradiation chamber as installed. The irradiation chamber contains a copper plate that has stainless steel cooling pipes and a heater attached to it, as shown in Fig. 11.6c. The heater and cooling pipes allow the temperature of the sample installed on the plate to be adjusted within the range between about −30 and 250 °C while irradiating the sample. This irradiation chamber can also be used in the case where PCS fibers are heated in an inert gas at 250 °C after electron beam irradiation and the generated peroxides are made chemically react to the cross-linking point. In the case of radiation oxidation cross-linking of PCS fibers, PCS fibers with a diameter between about 20 and 40 μm are cut to an appropriate length between about 5 and 10 cm, wrapped in aluminum foil and installed in the irradiation vessel, and irradiated with electron beams at a voltage of 2 MeV at a dose rate within the range between 0.40 and 2.0 kGy/s. Within this absorbed dose rate range, only the surface layer of PCS fibers is selectively oxidized by the radiation as a result of the irradiation effect represented by Formula 11.4.

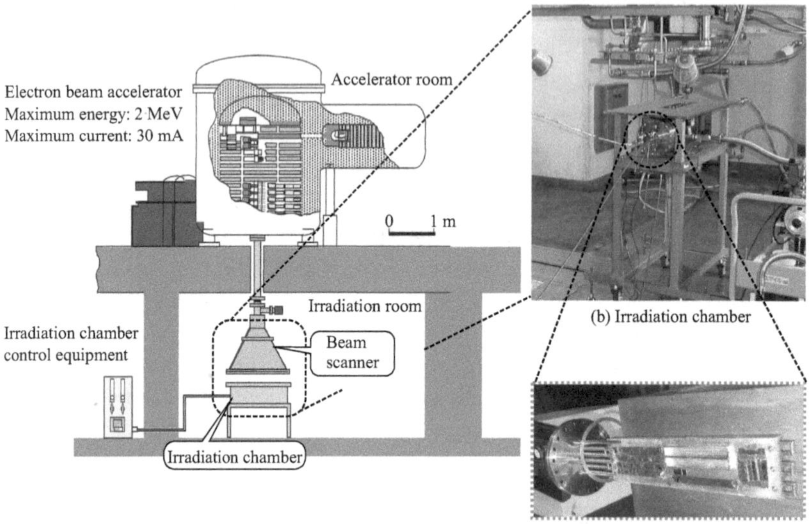

(a) Electron beam accelerator

The electron beam accelerator is capable of taking out current of up to 30 mA (2 MeV). The irradiation room is located directly below the accelerator. The irradiation chamber for the sample is installed below the conical beam scanner for projecting electron beams evenly.

(c) Inside of the irradiation chamber

This photograph shows the inside of the irradiation chamber with polycarbosilane fiber wrapped in aluminum foil installed. Liquefied carbon dioxide gas-based cooling, heater-based heating and control of the density of the oxygen in the chamber are possible.

Fig. 11.6 Electron beam accelerator and irradiation chamber

11.2.3 Processing to Make PCS Fibers Hollow and Conversion of the Obtained Hollow PCS Fibers into SiC Ceramic Microtubes Through Firing

The processing to make hollow a PCS fiber that has been oxidized only in its surface layer through radiation-based selective oxidation consists of immersing, at room temperature, the PCS fiber in an organic solvent to dissolve the core section that has not been cross-linked. For the solvent, tetrahydrofuran (THF (C_4H_8O)), cyclohexane, normalhexane, or a mixture of these is used. As shown in Fig. 11.5, a PCS fiber that has been irradiated with radiation has a transition layer which has not been oxidized sufficiently. Because cross-linking of this layer is nonuniform, the elusion cross section may vary depending on the strength of the solvent selected. Figure 11.7 presents photographs that have been taken successively to show the way the un-cross-linked core section of a PCS fiber whose surface layer has been cross-linked by electron beam oxidation is gradually eluted (extracted) in a THF solution. The un-cross-linked PCS polymeric material that forms the core section of the PCS fiber is pushed out from the ends of the PCS fiber by the internal pressure increased by the swelling due to dissolution. The time required for the extraction is about 1 min or less for a fiber whose length is about 5 cm. The solvent-based extraction process is a process that is unique to and forms the basis of the

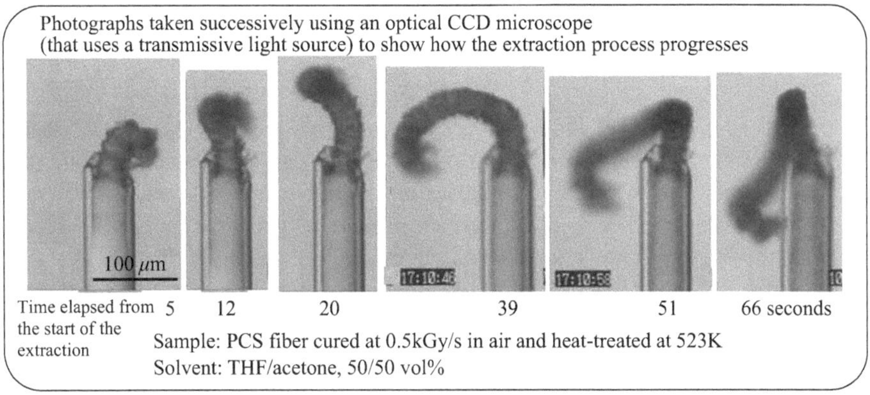

Extraction using tetrahydrofuran (THF): extraction of the un-cross-linked core section of a PCS fiber from the ends of the PCS fiber

Fig. 11.7 Treatment to make hollow a polycarbosilane fiber that has been irradiated with electron beams

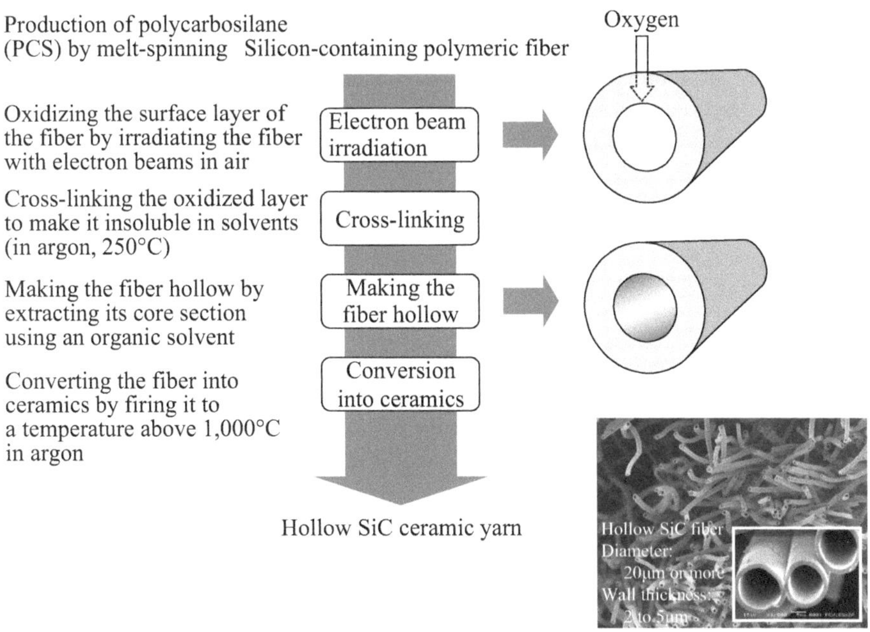

Fig. 11.8 Hollow SiC fiber production procedure and a photograph of hollow SiC fibers produced

production of SiC ceramic microtubes. By firing the obtained hollow PCS fibers in an inert gas at a temperature of 1100 °C or higher, the hollow PCS fibers can be converted into SiC ceramic microtubes. Figure 11.8 presents an outline of this procedure together with a photograph of hollow SiC ceramic fibers produced.

11.3 Future Prospect of Silicon Carbide Ceramics and Microtubes

This chapter introduced electron beam curing for polycarbosilane polymeric fibers, which are used as starting materials for producing SiC ceramic fibers. There are two different electron beam curing; the method in which the target material is irradiated with massive electron beams in an inert atmosphere to achieve direct cross-linking between the polymeric main chains and the method in which the target material is irradiated with electron beams in an oxygen-containing atmosphere to cause oxidation cross-linking. The former method is known as a method that does not cause oxidation decomposition of the fiber even when the temperature is kept at a high temperature around 1700 °C and has been used in the synthesis of ultraheat-resistant SiC ceramic fibers using PCS fibers as the starting material. The latter method has been developed into a SiC microtube production technology that utilizes the phenomenon wherein only the outermost part of the sample along the circumference is unevenly oxidized in oxidation cross-linking depending on the electron beam dose rate, the density of oxygen supplied, or the sample temperature. The SiC microtubes obtained have heat and corrosion resistance that are acceptable for use in an oxidizing atmosphere at a temperature of 1000 °C or lower. Recent studies have shown that SiC ceramics produced using such oxidation cross-linking have a hydrogen separation capability. It is probable that, if it becomes possible to impart this hydrogen separation capability to SiC microtubes, it will become possible to use them as a material for hydrogen separation SiC ceramic filters that can be used for a long period of time in harsh environments.

Exercises

1. Irradiation of a polymeric material sheet with electron beams with energy of 2 MeV at a dose rate of 0.5 kGy/s in dry air at a temperature of 293 K resulted in oxidation of the sheet down to a depth of 10 μm from the surface. State the irradiation conditions that would be required to reduce the thickness of the oxidized part to 5 μm.
2. SiC ceramics-based composite materials are recognized as materials that have high heat and corrosion resistance and allow the characteristic brittleness of ceramics to be compensated for, and development studies are being conducted to utilize them in high-temperature gas turbines, etc. They are also expected to serve as heat-resistant materials for use in radiation environments such as those in nuclear fusion reactors. What characteristics of SiC ceramics-based composite materials has this expectation stemmed from?

Chapter 12
Ion Beam Analysis of Materials

Shunya Yamamoto

In Ancient Greece, Democritus (about BC 460–370), who is said to be from Abdera located to the north of the Aegean Sea, developed the atomic theory and thought that extremely small permanent beings called atoms existed in nature and trees, animals, etc. and they would be decomposed and reduced into atoms after their death and used in another form of life. However, it was in the nineteenth century that the work to construct the concept of atoms based on the phenomenalism was started. The discovery of cathode rays led to the discovery that electrons have negative electric charges, atoms are normally electrically neutral and thus all atoms have the corresponding positive electric charges, and electrons have very small mass and thus substances that have positive electric charges represent the majority of the mass. In 1898, J.J. Thomson, a British physicist, proposed an atom model in which electrons are embedded in a positively charged substance much like raisins in raisin bread [1]. Geiger and Marsden conducted an experiment to confirm whether this model can explain the actual behaviors of atoms, based on advice from Ernest Rutherford. This is the famous experiment that used alpha particles naturally emitted by radioactive elements as a means of investigating the inside of atoms [2]. In this experiment, alpha particles were made into alpha particle beams using a piece of lead in which small holes had been made and the beams were projected onto a thin gold foil to see how the beams are scattered using a zinc sulfide screen installed behind the gold foil. As it is well known, the behavior of the alpha particles observed in the experiment was much different from the behavior predicted with the raisin bread model; some of the alpha particle beams were scattered at very large angles, suggesting that there were atomic nuclei in which positive electric charges were localized in a very small area. Through similar experiments that used different types of foils as alpha particle beam irradiation targets, the principles of the concept

S. Yamamoto (✉)
National Institutes for Quantum and Radiological Science and Technology, Takasaki, Gunma, Japan
e-mail: yamamoto.shunya@qst.go.jp

H. Kudo (ed.), *Radiation Applications*, An Advanced Course in Nuclear Engineering 7, https://doi.org/10.1007/978-981-10-7350-2_12

of atomic numbers, which are well-known facts today, were verified, including the fact that the changes in the trajectory of an alpha particle caused when it passes through the vicinity of an atomic nucleus depend on the amount of the electric charge of the atomic nucleus and that each atom has an electric charge unique to it. In addition, it was discovered that the angles of scattering of alpha particles caused by a metal foil can be accurately predicted by using Rutherford's scattering formula derived by introducing the concept of cross section, which is an important concept in physics in defining the probability of scattering, and applying basic physical principles to atomic nuclei. It is possible today to derive Rutherford's scattering formula with a quantum mechanics-based approach, but it has been confirmed that the formula derived is the same (this fact attests to the outstanding insight of Rutherford). The use of Rutherford's scattering formula leads to the discovery that the number of alpha particles with a scattering angle of θ detected per unit area is in proportion to the square of the atomic number, the foil thickness, and the number of atoms of the foil per unit area and in reverse proportion to the square of the kinetic energy of the alpha particles and $\sin^4(\theta/2)$, which is a function of the scattering angle θ. It can be considered that this discovery showed the fact that, if the kinetic energy and scattering angle of the scattered alpha particles can be measured precisely, the atomic mass etc. of the alpha particles can be determined even if the foil that interacts with and scatters the alpha particles is an unknown substance. In this sense, it can be said that the experiments showed the advantage of "observation with radiation" for the first time. However, Rutherford's backscattering phenomenon was not utilized in full-fledged analyses of substances until after the invention of ion accelerators and detectors that support technologies for analyzing the energy of scattered particles and the development of computers capable of simulating Rutherford's backscattering instantly. It was in the latter half of the twentieth century that the Rutherford backscattering spectroscopy (RBS) was widely recognized as a substance analysis method.

In the present day, which is a century after the discovery of Rutherford's backscattering phenomenon, it is no longer rare that ion beam analysis methods are used for analyzing atoms that comprise substances, thanks to the popularization of compact ion accelerators with an acceleration voltage of several MV. Nowadays, the RBS mentioned above and other methods such as the particle-induced X-ray emission (PIXE) analysis that uses the characteristic X-rays emitted when electrons fall into the inner cores of atomic nuclei are being used as standard laboratory-level analysis methods.

This section first explains the RBS, which has become a standard ion beam analysis, and then the elastic recoil detection (ERD), which is used to analyze light element impurities contained in thin-film materials, and the nuclear reaction analysis (NRA).

12.1 Principles of the Rutherford Backscattering Spectroscopy

The Rutherford backscattering spectroscopy (RBS) is a method that can be applied to thin-film samples with a thickness of several micrometers and allows the atomic composition and distribution in the depth direction to be examined in a nondestructive manner in a short period of time. Figure 12.1a shows the concept of the RBS.

Figure 12.1 shows how an incident particle with incident energy of E_1 is scattered in the backward direction as a result of its elastic collision with an atom present in the sample at a depth of t from the surface. In the case of the incidence of a fast particle beam (an ion beam with incident energy of several MeV per nucleon) onto a substance, almost all of the particles scattered at a large angle are particles scattered as a result of a single elastic collision, and the existence of particles scattered as a result of double or more the number of elastic collisions can be ignored. Therefore, for the scattering of a particle with an incident angle of θ_1, a mass of m_1 and incident energy of E_1 in the "$\theta\{=\pi - (\theta_1 + \theta_2)\}$" direction caused by a single collision of it with an atom with a mass of m_2 located at a depth of t, the energy of the scattered particle ejected from the surface of the sample (E_2) can be expressed with the following formula:

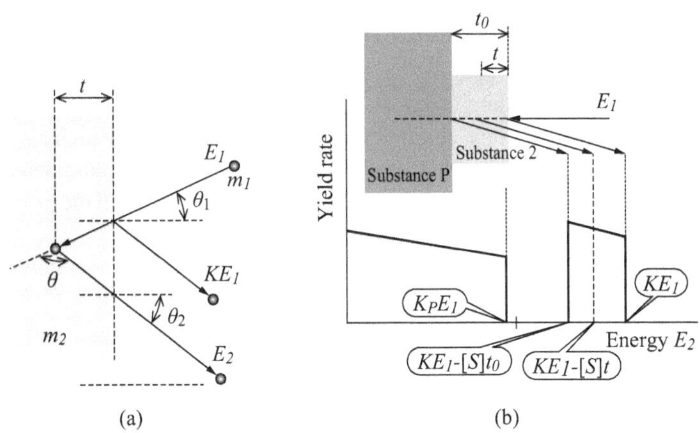

Fig. 12.1 The concept of the RBS and its spectrum (**a**) Schematic diagram showing the incidence of a particle with a mass of m_1 onto a sample with an atomic mass of m_2 and the scattering of the particle at the surface and at a depth of t (**b**) Relationship between the energy of the scattered particle generated by the incidence of a particle with incident energy of E_1 onto a layered thin-film sample composed of a substance with an atomic mass of m_p and a substance with an atomic mass of m_2 and the spectrum of the detected yield

$$E_2 = K\left[E_1 - \frac{1}{\cos\theta_1}\int_0^t S\big(E(x)_{\text{in}}\big)dx\right] - \frac{1}{\cos\theta_2}\int_t^0 S\big(E(x)_{\text{out}}\big)dx \qquad (12.1)$$

$$K = \left[\frac{\left(m_2^2 - m_1^2 \sin^2\theta\right)^{1/2} + m_1\cos\theta}{m_1 + m_2}\right]^2 \qquad (12.2)$$

The integration term of Formula (12.1) represents the electric stopping power, which is the value of the energy lost per unit length by the particle with incident energy of E_1 as a result of the inelastic scattering caused by its collision with an electron present in the sample. The first term of Formula (12.1) relates to the incident particle with a mass of m_1 that advances in the sample and the second term relates to the scattered particle with a mass of m_1 that advances in the sample. "K" in Formula (12.2), which is called the K factor, represents the energy allocation rate for an elastic collision between the m_1 and m_2 particles. If the sample is a thin film, the electric stopping power in Formula (12.1) can be approximated using the following formula because t is sufficiently small:

$$\int_0^t S\big(E(x)_{\text{in}}\big)dx \cong S(E_1)t, \quad \int_t^0 S\big(E(x)_{\text{out}}\big)dx \cong S(KE_1)(-t) \qquad (12.3)$$

Therefore, E_2 can be given by the following formulas:

$$E_2 = KE_1 - [S]t \qquad (12.4)$$

$$[S] = \frac{K}{\cos\theta_1}S(E_1) - \frac{1}{\cos\theta_2}S(KE_1) \qquad (12.5)$$

"$[S]$" in the above formulas is called the energy loss factor. Since the value of electric stopping power of any element can be calculated using Ziegler's semiempirical formula [3], the mass (m_2) of the atom with which the incident particle collided can be determined if the mass (m_1) and energy (E_1) of the particle that collided elastically are known (mass analysis). In addition, the depth (t) at which the scattering is caused can be determined from Formula (12.4) if the energy of the particle scattered at the surface of the sample (KE_1) and the energy (E_2) of the particle scattered in the sample at a depth of t are known (depth analysis). For example, if a sample made by covering the surface of a plate made of Substance P with a mass of m_p with a layer with a thickness of t_0 of Substance 2 with a mass of m_2 by means of vapor deposition is irradiated with a particle beam with energy of E_1, the incident particle will be scattered by the atoms present at the depth t ($\leq t_0$). By detecting the energy and number of the particles scattered at any given angle (θ_2) and plotting them, an RBS spectrum like the one shown in Fig. 12.1b can be obtained. By substituting the energy value that corresponds to the spectrum width of Substance 2 obtained for E_2 in Formula (12.4), the film thickness (t_0) of Substance 2 can be calculated from Formula (12.4). In addition, the yield of the scattered particles detected (scattered energy spectrum intensity [H (E_2)]) can be expressed with the following formula:

$$H(E_2) = \frac{\eta\Omega\sigma QN}{[S]} \tag{12.6}$$

where Q represents the incident particles, σ represents the differentiated scattering cross section, Ω represents the solid angle of the detector, η represents the efficiency of the detector, and N represents the atomic density.

The scattering cross section (σ (E_2)) can be expressed with the following formula:where Z_1 and Z_2 represent the atomic numbers of the incident particle and Substance 2, respectively, and e represents the charge of the electron.

$$\sigma(E_2) = \left(\frac{Z_1 Z_2 e^2}{4E_2}\right)^2 \times \frac{4\left[\left(m_2^2 - m_1^2 \sin^2\theta\right)^{1/2} + m_2 \cos\theta\right]^2}{m_2 \sin^4\theta\left(m_2^2 - m_1^2 \sin^2\theta\right)^{1/2}} \tag{12.7}$$

The spectrum area (A_2) obtained by performing integration from KE_1 to KE_1–$[S]$ t_0 along the energy axis in Fig. 12.1b can be approximated with the following formula if it is assumed that the dependence of the scattering cross section on energy is not significant:

$$A_2 = \int_{KE_1-[S]t_0}^{KE_1} \frac{\eta\Omega\sigma(E_2)QN}{[S]} dE_2 \cong \eta\Omega\sigma(E_2)QNt_0 \tag{12.8}$$

This means that the film thickness (t_0) can be calculated from the spectrum area as well. In the RBS, obtaining a spectrum in which the effect of the substrate is detached makes it easier to perform analysis, as shown in Fig. 12.1b. To achieve this, it is necessary to produce a sample that satisfies the following formula:

$$KE_1 - [S]t \geq K_P E_1 \tag{12.9}$$

where K_p represents the K factor for Substance P. If the film thickness is thin, the second term of the left-hand side can be ignored, and, as can be seen from Formula (12.2), Formula (12.9) is satisfied if m_2 is greater than m_p. Conversely, in the case where m_p is greater than m_2 or the film thickness cannot be ignored, spectrum overlapping often occurs, which makes the analysis difficult. Techniques to avoid this include the following, as can be seen from Formula (12.9):

1. Increasing the energy (E_1) of the incident particle
2. Changing the K factor by using a heavy particle as the incident particle

As an example, Fig. 12.2 shows how the K factor depends on the atomic mass for the case where the incident particle is ^{4}He and the case where the incident particle is ^{16}O. The K factor values were calculated assuming that the scattering angle is 160 degrees. If the values that correspond to m_p and m_2, that is, K_p and K, are plotted on this graph and it is assumed that m_2 is greater than m_p, the difference in the K factor becomes larger, and the RBS spectrum resolution increases as the gradient of

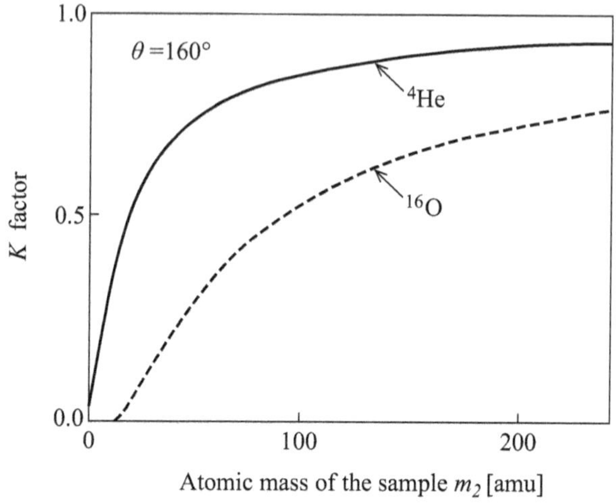

Fig. 12.2 Dependence of the K factor on the atomic mass of the sample Changes in the K factor caused by the incidence of helium or oxygen onto a substance with an atomic mass of m_2

the straight line connecting points (m_p, K_p) and (m_2, K) increases. It can be seen that the mass resolution can be improved for the mass range from around light to medium-heavy elements in the case where helium (^{4}He) is used and for medium-heavy and heavier elements in the case where oxygen (^{16}O) is used. When analyzing a multilayered film using the RBS, it is usually necessary to select the incident particle, energy, etc. to use in the analysis taking into consideration the combination of the elements that comprise the sample, the number of layers, the film thickness, etc. However, it is difficult to totally avoid spectrum overlapping. In such a case, methods are often employed in which the composition etc. is estimated by means of analysis formula-based computer simulations.

As an example of RBS for thin-film composition analysis, Fig. 12.3 shows the RBS spectrum of a thin film of tungsten oxide. Tungsten oxide is a substance that has a nonstoichiometric composition. Because the ratio of oxygen (O) to tungsten (W) (the composition ratio or the O/W ratio) varies between 1 and 3 depending on the production conditions, it is often the case that the amount of oxygen contained cannot be known precisely. The black dots in the figure show the RBS spectrum of a thin film of tungsten oxide whose composition ratio is unknown (WO_X) formed on the surface of a carbon substrate using the magnetron sputter method. This is the result of the measurement of the elastically scattered 2 MeV $^4He^+$ ions performed using a charged particle detector installed at an angle that is different from the direction of incidence by 165°. It can be seen from the figure that the RBS spectrums attributable to tungsten, oxygen, and carbon appeared separately in the order of mass because the masses of the constituent elements of WO_X are larger than the mass of the carbon that constitutes the substrate.

By applying Formula (12.8) to the obtained spectrums of W and O, the composition ratio (x) and film thickness were calculated at about 3 and 320 nm, respectively. The solid line in the figure shows the result of the simulation of the RBS

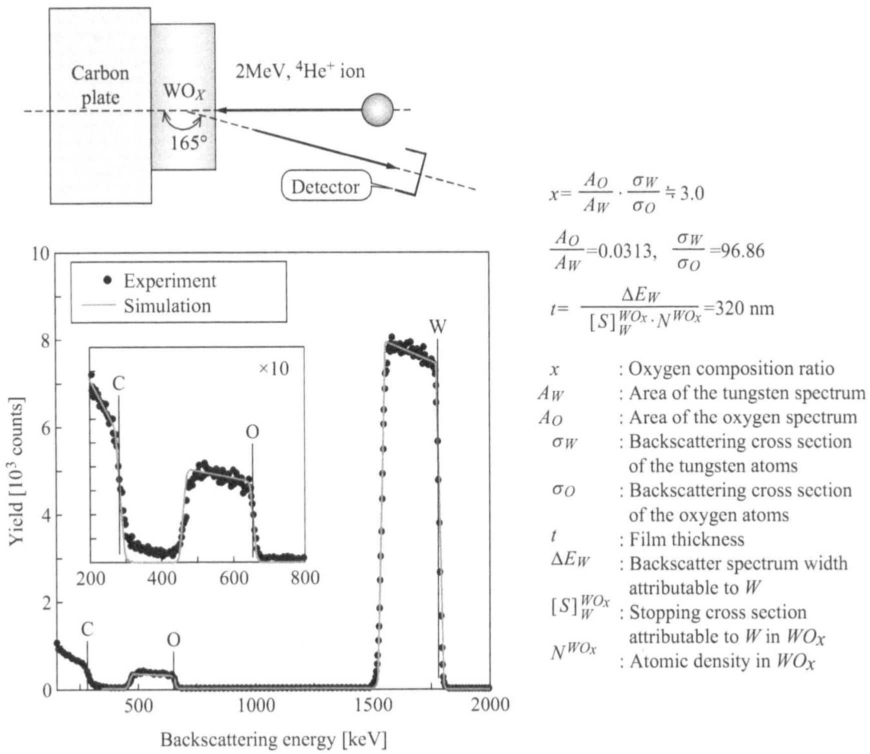

Fig. 12.3 Example of RBS for composition analysis of tungsten oxide
The inserted graph is an enlarged view of the low-energy part (200–800 keV)

spectrum performed using the obtained values to confirm the result of the calculation. The simulation result reproduced the experiment result well, and it can be seen that quantitative element analysis is possible.

12.2 Elastic Recoil Detection

The RBS detects elastically scattered incident particles, but if atoms that are lighter than the incident particles are present in the sample, lighter atoms may be ejected from the sample depending on the collision angle. The elastic recoil detection (ERD) is a method that utilizes this phenomenon to analyze elements present in substances. In the ERD, the same experiment system as that for the RBS can usually be used, and the depth distributions of light elements such as hydrogen that are present in the vicinity of the surface of the material can be measured with the isotopes discriminated. The detection target is hydrogen in the case where helium ions are used as the incident particles and hydrogen and helium in the case where

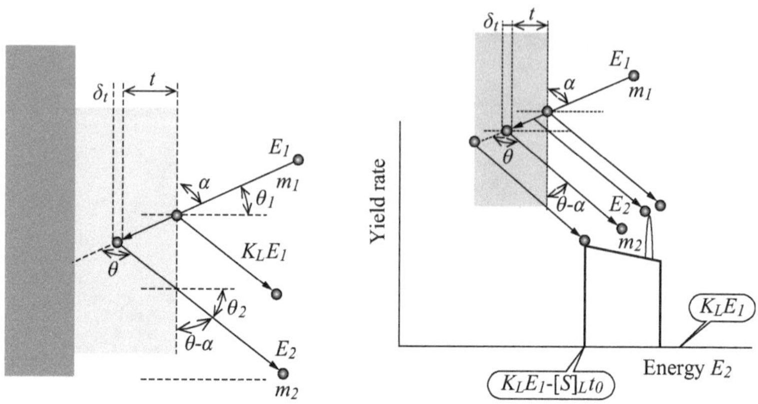

(a) (b)

Fig. 12.4 The principle of the ERD measurement and ERD spectrum (**a**) Schematic diagram showing the way a particle with a mass of m_1 collides with a particle located at a depth of t with a mass of m_2 which is much smaller than m_1 and the latter is ejected from the sample as a result of the recoiling (**b**) Relationship between the energy of the recoiling particle generated by incidence of a particle with incident energy of E_1 onto a thin-film sample and the spectrum of the detected yield. Sometimes, incident particles that have elastically collided with constituent atoms of the sample are detected. To separate these particles, a Mylar film, aluminum foil, etc. that absorbs only the elastically collided incident particles is installed in front of the detector

oxygen ions are used as the incident particles. There are also cases where nickel ions etc. are used and carbon atoms etc. are selected as the target of the elastic recoil detection. Figure 12.4a presents a schematic diagram of the ERD. This diagram shows the way an incident particle with incident energy of E_1 elastically collides with an atomic nucleus located at a depth of t from the surface of the sample and the atomic nucleus is ejected from the sample. The probability of occurrence of an elastic collision between the incident particle or ejected particle and another particle in the substance is very low, and the ejection of a particle that caused double or more the number of collisions can be ignored. Therefore, the energy (E_2) involved in a single elastic collision of a particle with an incident angle of θ_1, a mass of m_1, and incident energy of E_1 with a particle with a mass of m_2 ($<<m_1$) located at a depth of t and the resultant ejection of the latter by the former from the surface of the sample in the direction of a recoiling angle of θ_2 can be expressed with the following formula:

$$E_2 = K_L\left(E_1 - \frac{1}{\cos\theta_1}\int_0^t S\left(E(x)_{\text{in}}\right)dx\right) - \frac{1}{\cos\theta_2}\int_t^0 S_L\left(E(x)_{\text{out}}\right)dx \quad (12.10)$$

$$K_L = \frac{4m_1 m_2 \cos^2\theta}{(m_1 + m_2)^2} \tag{12.11}$$

The first term of Formula (12.10) relates to the energy passed by the incident particle to an atom (m_2) located at a depth of t through the elastic collision, and the second term relates to the energy lost by the atom (m_2) that has become a recoiling particle through interactions with electrons in the sample. K_L in Formula (12.11) is the energy allocation rate for a collision of an incident particle (m_1) with a recoiling particle (m_2) with a mass that is much smaller than the mass of the incident particle ($m_1 > > m_2$). θ is the direction of ejection of the recoiling particle expressed in terms of the angle with respect to the direction of entry of the incident particle. If the sample is a thin film, the electric stopping power in Formula (12.10) can be approximated using Formula (12.3), because t is sufficiently small. Expressing the angle of incidence and angle of ejection in terms of sample surface gradient (α) allows Formula (12.10) to be approximated as shown below. Expressing this using thickness (t) as the common parameter allows handling that is similar to that in the RBS:

$$E_2 \cong K_L E_1 - \left(\frac{K_L}{\sin \alpha} S(E_1) - \frac{1}{\sin(\theta - \alpha)} S_L(K_L E_1) \right) t \tag{12.12}$$

$$\cong K_L E_1 - [S]_L t \tag{12.13}$$

$$[S]_L = \left(\frac{K_L}{\sin \alpha} S(E_1) - \frac{1}{\sin(\theta - \alpha)} S_L(K_L E_1) \right) \tag{12.14}$$

where S (E_1) represents the stopping power of the incident particle and S_L ($K_L E_1$) represents the stopping power of the recoiling particle. For example, by plotting the particle energy and yield rate for the recoiling of a particle located at a depth of t in a sample with a thickness of t_0 in the "θin" direction caused by a particle with a mass of m_1, an ERD spectrum like the one shown in Fig. 12.4b can be obtained. The energy of the recoiling particle detected would be $K_L E_1$ if it is assumed that the particle is located on the surface and $K_L E_1 - [S]_L t_0$ if it is assumed that the particle is located at a depth of t_0, which is equal to the thickness of the sample. Determination of the starting point of the spectrum ($K_L E_1$) allows the mass of the recoiling particle to be determined from the allocation rate. In addition, determination of the starting energy allows the conversion of the observed energy (E_2) into depth (t) if $[S]_L$ is known (analysis of distribution in the depth direction). However, obtaining information for the depth direction requires obtaining precise values of $[S]_L$ and K_L, which is a difficult task. In addition, it is necessary to install an absorber (Mylar film, aluminum foil, etc.) that absorbs the incident particles scattered as a result of their elastic collision with constituent atoms of the sample in front of the detector so that the detection of the scattered incident particles is prevented and only the recoiling particles are detected.

The resolution of the elastic recoil detection in the depth direction (δt) can be calculated from Formula (12.13) as follows:

$$\delta t = \left| \frac{\delta E_2}{[S]_L} \right| \tag{12.15}$$

The resolution of the energy detected (δE_2) can be determined from the following relationship:

$$\delta {E_2}^2 = \left[\sqrt{4\pi Z_1 Z_2 e^4 N t_{ab}}\right]^2 + \left[8 \frac{m_1 m_2}{(m_1 + m_2)^2} E_1 \sin\theta \cos\theta\right]^2 \tag{12.16}$$

where Z_1 represents the atomic number of the incident particle and Z_2 represents the atomic number of the absorber installed in front of the detector. e, N, and t_{ab} represent the amount of the electric charge of the electron, the atomic density in the absorber, and the thickness of the absorber, respectively. The first term represents the energy struggling of the recoiling particle that has hit and got into the absorber with a thickness of t_{ab}, and the second term represents the corresponding geometrical spread of energy. What is necessary to improve the positional resolution is to increase the energy loss factor ($[S]_L$) in Formula (12.15). This is achieved by selecting the incident particle and energy so that the stopping power of the incident particle in the first term of Formula (12.14) becomes the largest. For example, incident energy of "$E/m = 0.5$ MeV/amu" is recommended for oxygen atoms (O) and carbon atoms (C), and incident energy of "$E/m = 2$ MeV/amu" is recommended for xenon (Xe) etc. When detecting heavy elements as recoiling particles, the effect of the geometrical spread of energy in the second term of Formula (12.16) cannot be ignored, and it is necessary to take countermeasures such as suppressing the spread by reducing the slit width of the absorber installed in front of the charged particle detector. However, when reducing the slit width, it is necessary to exercise due care in handling the results obtained, because reducing the slit width reduces the number of recoiling particles detected per unit time and increases the beam irradiation time required to obtain a smoothed spectrum, which means that in many cases the damage caused by the particle beam irradiation to the sample cannot be ignored.

As an example of ERD analysis of the distribution of elements in a thin film, Fig. 12.5 shows the result of measurement of the distribution of hydrogen in an amorphous silicon film with a thickness of 1.1 μm on a silicon substrate. In the experiment, a Mylar film with a thickness of 12.5 μm was used as an absorber that allows only hydrogen (the recoiling particles) to pass through it, as shown in Fig. 12.5a. Figure 12.5b shows the measurement result. It can be seen from the shape of the spectrum that a large amount of hydrogen is distributed in the amorphous silicon film, because the channel number on the horizontal axis is a

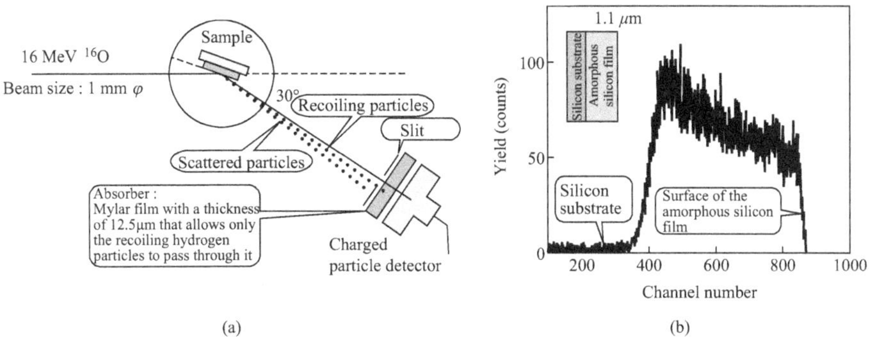

Fig. 12.5 Schematic diagram of the ERD for hydrogen distribution measurement and the measurement result (**a**) A charged particle detector was installed at an angle of 30° from the surface of the sample. A Mylar film with a thickness of 12.5 μm was used as the absorber (**b**) Recoiling particle detection spectrum obtained for incidence of 16 MeV oxygen atoms onto an amorphous silicon film with a thickness of 1.1 μm formed on a silicon substrate. Hydrogen atoms have been detected as recoiling particles

quantity that corresponds to the energy of the detected particles and thus reflects the depths of penetration of the incident particles into the sample. ERD measurement of hydrogen distribution allows nondestructive examination of the hydrogen distribution in the depth direction down to a depth of several micrometers. It has now become one of the few important methods that allow confirmation of the hydrogen distribution in the depth direction.

As can be seen from Fig. 12.5a, the detector for ERD measurement is installed at a location that is different from the location where the detector for RBS measurement is installed. Therefore, ERD measurement and RBS measurement on the same sample can be performed in parallel by adding a detector for ERD measurement. As an example, Fig. 12.6 shows the ERD and RBS results of multilayered sample comprised of alternate thin copper layers (each with a thickness of 20 nm) and thin niobium layers (with different thicknesses ranging from 10 to 60 nm) by using 16 MeV oxygen ion beam. From the result of the ERD measurement, it can be seen that many recoiling particles were ejected at locations distant from the surface. On the other hand, the result of the RBS measurement shows that, while the half-width of the spectrum of the yield of the oxygen ions scattered by the thin copper layers remained roughly the same, the half-width of the spectrum of the yield of the oxygen ions scattered by the thin niobium layers gradually increased, reflecting the progressive changes in thin niobium layer thickness. From these results, it can be said that simultaneous ERD and RBS measurement allows a clear understanding to be obtained of the state of distribution of hydrogen in the film formed and the physical structure of the film.

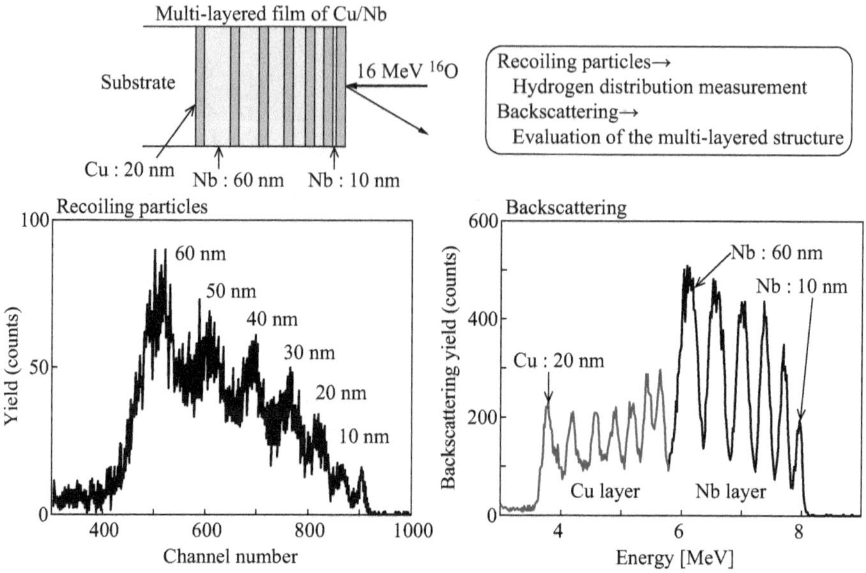

Fig. 12.6 Results of ERD and RBS measurement of thin film composed of copper and niobium layers
As shown in the figure, hydrogen distribution measurement can be made in parallel with analysis of the multilayered structure of the thin film

12.3 Resonant Nuclear Reaction

Collision of a particle with certain incident energy with a hydrogen atom can result in an increase in the nuclear reaction cross section and selective induction of a nuclear reaction ("resonant nuclear reaction") depending on the energy. This phenomenon can be utilized in ion beam analysis of the distribution of hydrogen in the sample. Figure 12.7a, b presents schematic diagrams of this method, and Fig. 12.7c presents a table showing the resonance energy, reaction cross section, resonance width and reaction yield rate of the resonant nuclear reaction, and the energy of the gamma ray released. Atoms that are known to be capable of causing a resonant nuclear reaction with hydrogen include lithium (^{7}Li), nitrogen (^{15}N), and fluorine (^{19}F). If such ion beam is projected onto a sample containing hydrogen, the incident particles will collide with hydrogen in the sample and gradually decay as a result of the loss of their energy due to the collisions. However, if the energy of the particles coincides with the energy that causes the resonant nuclear reaction, their subsequent collisions with hydrogen will result in a nuclear reaction with hydrogen, and the particles will release gamma ray. If the amount of incident particles is constant, the gamma ray yield will be proportional to the hydrogen concentration. Hydrogen concentration at any given depth can be detected by changing the location to cause the resonant nuclear reaction in the sample by changing incidence energy. As can be seen from the table in Fig. 12.7, the ^{15}N (6.385 MeV) resonant

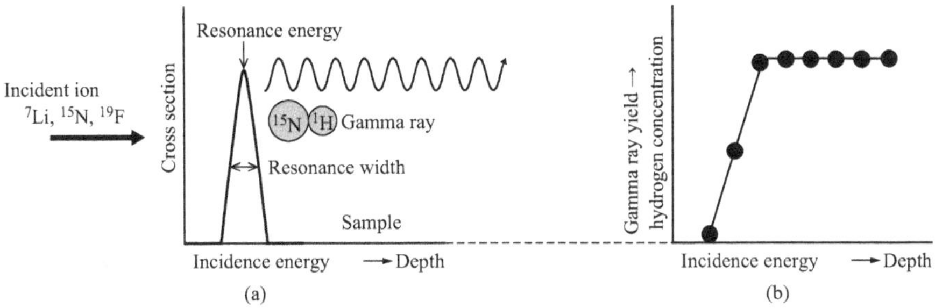

Nuclear reaction with hydrogen	(^{7}Li, γ)	(^{15}N, $\alpha\gamma$)	(^{15}N, $\alpha\gamma$)	(^{19}F, $\alpha\gamma$)
Resonance energy: E_r [MeV]	3.07	6.385	13.35	16.44
Cross section: σ [mbarn]	4.8	1650	1050	440
Resonance width: Γ [keV]	81	1.8	25.4	86
Next resonance energy: [MeV]	7.11	13.35	18.0	17.6
Gamma ray energy: [MeV]	17.7, 14.7	4.43	4.43	6.13, 6.98 7.12

(c)

Fig. 12.7 Schematic diagrams showing the concept of the resonant nuclear reaction analysis (RNRA) and a table showing types of nuclear reactions that can result in a resonant nuclear reaction with hydrogen (**a**) Schematic diagram showing hydrogen's nuclear reaction cross section in the sample for the case where Li, N or F is used as the incident particle. The incident particle collides with hydrogen in the sample and gradually loses its energy, but if its energy coincides with the energy that causes the resonant nuclear reaction, the reaction cross section will increase, as shown in the diagram. The incident particle will change into another nuclide and release gamma ray (**b**) Schematic diagram showing the hydrogen concentration distribution obtained by converting the incidence energy of ion beam onto a sample in which hydrogen is uniformly distributed into the distance in the depth direction and the converting the gamma ray yield into the hydrogen concentration. Because the energy at incidence is higher than the energy that causes a resonant reaction, gamma ray is not detected immediately after the incidence. If the energy decreases to the energy that causes a nuclear reaction, the resonant nuclear reaction will be induced. However, because the resonance energy has a certain range, the gamma ray yield will increase gradually and saturate when the energy coincides with the resonance energy (**c**) Types of nuclear reactions that can result in a resonant nuclear reaction with hydrogen. The framed nuclear reaction is often used to analyze the distribution of hydrogen in the depth direction because of the narrow resonance width

nuclear reaction allows hydrogen concentration measurement to be made with high resolution in the depth direction because it has a large cross section and a narrow resonance width. However, in the case of measurement to confirm whether hydrogen is present, ^{19}F etc., which brings a large reaction yield ($\sigma\Gamma$), may be more advantageous.

Figure 12.8 shows an example of an actual ^{15}N resonant nuclear reaction analysis of hydrogen contained in a sample. Here, the ^{1}H (^{15}N, $\alpha\gamma$)^{12}C resonant nuclear reaction that occurs at 6.385 MeV shown in Fig. 12.7c is used. This reaction releases 4.43 MeV gamma ray. However, because the resonance width is very narrow (1.8 keV) and the reaction cross section becomes about 4 orders of magnitude smaller outside the resonant reaction energy range, the nuclear reactions caused by the incident particles with energy that does not coincide with the resonance energy can be ignored. In addition, because it is outside the energy regions of the natural radioactive rays such as ^{40}K (1.43 MeV), there is no influence of the background of the gamma ray, which means that the detected yield of the 4.43 MeV gamma ray is proportional to the hydrogen concentration in the region that corresponds to the resonance energy depth. However, since ^{15}N causes the resonant nuclear reaction to generate 4.43 MeV gamma ray at 13.35 MeV as well, care must be taken to ensure that the incidence energy will not exceed 13.35 MeV.

In the actual experiment, ^{15}N ions were accelerated using the accelerator shown in Fig. 12.8a, and the diameter of the ^{15}N beam was adjusted to approx. 3 mm using an aperture slit. Then, the beam was projected onto the sample installed in the vacuum chamber shown in the figure, and the gamma ray was detected using a

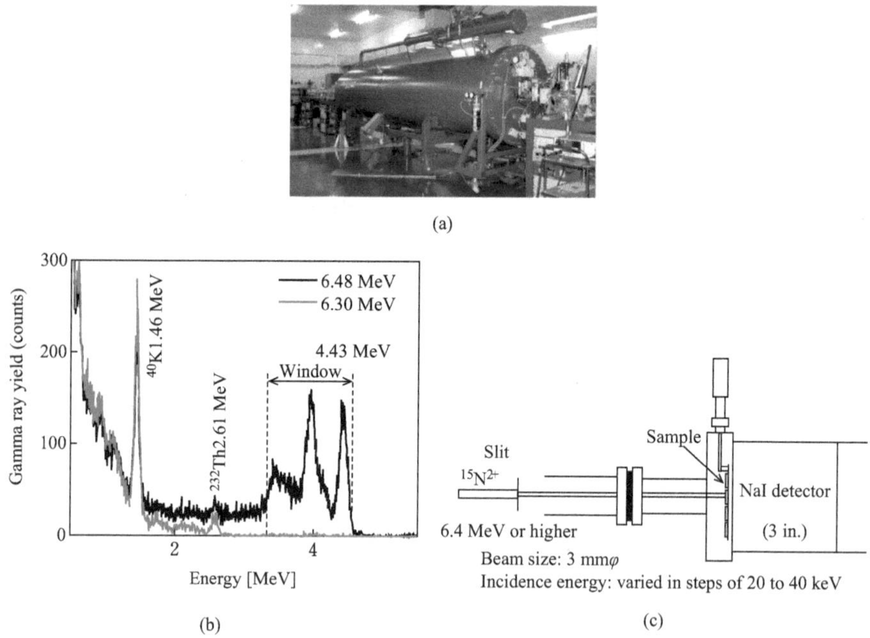

Fig. 12.8 Gamma ray energy spectra of ^{15}N irradiation of a sample containing hydrogen (**a**) Tandem accelerator installed in the Takasaki Advanced Radiation Research Institute of Japan Atomic Energy Agency (**b**) Result of measurement of the ^{1}H (^{15}N, αγ) ^{12}C resonant nuclear reaction (a ^{15}N resonant nuclear reaction) using a gamma ray detector. Because the energy of the generated gamma ray was high, a satellite peak associated with electron pair generation was detected approx. 0.5 MeV below (**c**) Schematic diagram of the equipment used to detect the resonant nuclear reaction. A 3-inch NaI detector was attached directly to the sample chamber so that as much of the generated gamma ray as possible could be captured

gamma ray detector installed behind the sample. In actual measurement of the hydrogen distribution in the depth direction, the 4.43 MeV gamma ray yield is measured while varying the incidence energy with the amount of ^{15}N ion irradiation kept constant. The relationship between the depth (t) from the surface and the incidence energy (E_i) is given by the following formula:

$$t = \frac{E_i - E_r}{S} \tag{12.17}$$

where S represents the stopping power and E_r represents the resonance energy (6.385 MeV). Measurement of the hydrogen distribution in a sample requires varying the incidence energy in steps of 10–50 keV. Because it is necessary to adjust the acceleration voltage of the accelerator and the allocation magnet as necessary according to the changes in the incidence energy of the ^{15}N ions, it usually takes about 3 h to measure one sample. The graphs in Fig. 12.9 show the spectrums of the gamma ray energy generated as a result of ^{15}N irradiation of a sample containing hydrogen.

Comparison of the energy spectrum obtained with the 6.30 MeV beam, whose energy is below the resonance energy, shown in Fig. 12.8 (the spectrum (i.e., characteristics) shown in gray) with the energy spectrum obtained with the 6.48 MeV beam, whose energy is above the resonance energy (the spectrum shown in dark black), shows that the 4.43 MeV gamma ray released as a result of

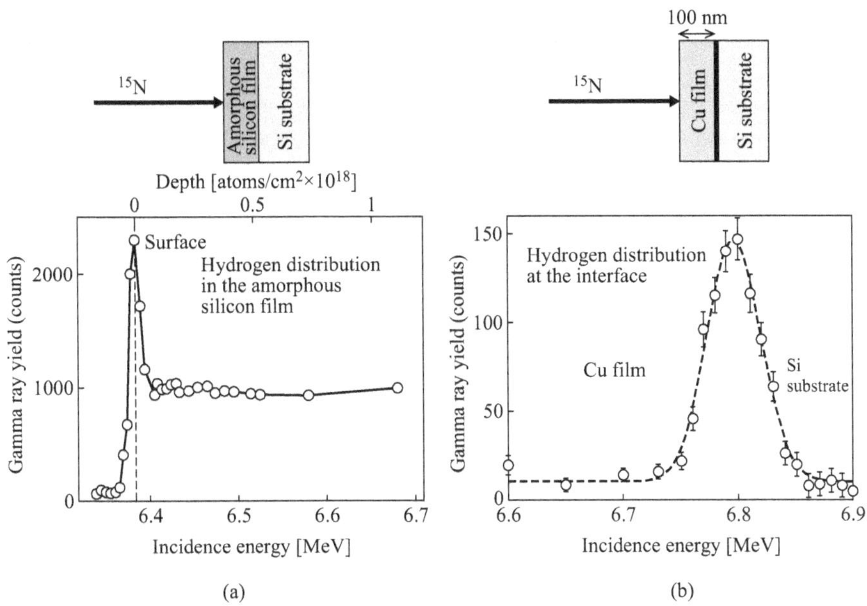

Fig. 12.9 ^{15}N resonant nuclear reaction analysis of hydrogen contained in a sample (**a**) Hydrogen distribution in the thin film (**b**) Analysis of the hydrogen distribution at the thin film interface

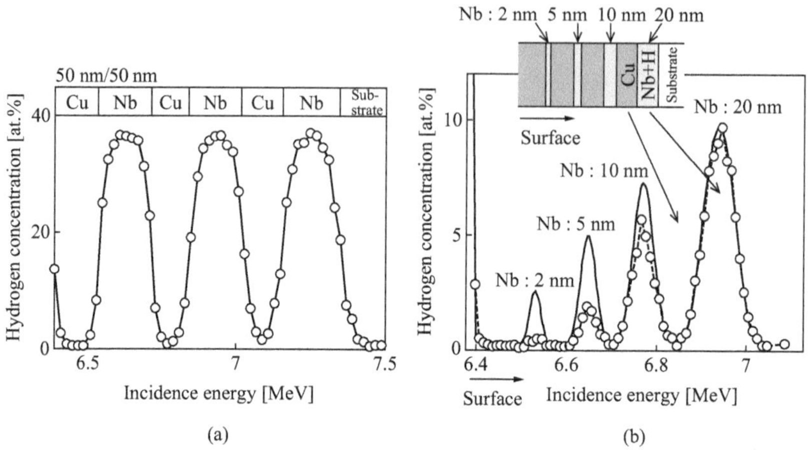

Fig. 12.10 Ion beam analysis of the distribution of hydrogen in a multilayered thin film comprised of copper and niobium layers (**a**) Result of an analysis of the concentration of hydrogen in a multi-layered sample comprised of 50 nm-thick thin layers of copper and niobium. It can be seen that a large amount of hydrogen is present in the niobium layers (**b**) Result of an analysis of the concentration of hydrogen present in a multi-layered sample comprised of alternate thin niobium layers and 50 nm-thick thin layers of copper with the thin niobium layer thickness increased progressively from 2 to 20 nm. It can be seen that the amounts of hydrogen in the thin niobium layers are proportional to the layer thickness. The solid line shows the hydrogen concentration calculated based on the assumption that all niobium layers have the same hydrogen concentration as the hydrogen concentration in the 20 nm-thick niobium layer

the nuclear reaction was detected in the spectrum shown in strong black. The value obtained by summing up these peaks is proportional to the amount of hydrogen.

Figure 12.10a, b shows the results of resonant nuclear reaction measurement of the hydrogen distribution in a multilayered sample comprised of niobium (in which solid solubility of hydrogen is high) and copper (in which solubility of hydrogen is low). Figure 12.10a shows the result of measurement of the hydrogen distribution in a multilayered film comprised of alternate 50-nm-thick layers of niobium and copper. It can be seen that hydrogen is uniformly distributed only in the niobium layers. Figure 12.10b shows the result of measurement of the hydrogen distribution in a multilayered film comprised of alternate niobium layers and 50-nm-thick layers of copper with the niobium layer thickness increased progressively from 2 to 20 nm. It can be seen that measurement is possible even in the several nanometer regions. The solid line in Fig. 12.10b shows the hydrogen concentration calculated based on the assumption that all niobium layers have the same hydrogen concentration as the hydrogen concentration in the 20 nm-thick niobium layer. It can be seen that the measured values are lower than the calculated values and that the hydrogen concentration in a niobium layer tends to decrease when the thickness is reduced. It is clear from these facts that the resonant nuclear reaction is an important hydrogen analysis method that can be applied to thin films with a thickness on the order of nanometers.

Traditionally, the X-ray diffraction method and electron diffraction method have been used in structural analyses of thin films. However, examples of X-ray diffraction method- and electron diffraction method-based analyses are very few, because it is difficult to analyze thin films in the depth direction with these methods. The analysis methods that utilize ion beams with energy of several MeV introduced in this chapter allow the distributions of hydrogen and other light elements in the depth direction and the atomic compositions of thin films to be examined at a resolution of about several tens of nanometers in a nondestructive way. The heavy ion beam analyses including the RBS, ERD, and ^{15}N–NRA methods introduced in this chapter are being used as effective means of analyses of interface structures and the behavior of light elements, which are important in studies on multilayered materials.

Exercises

1. An analysis beam that had been made by accelerating ^{4}He to 2.0 MeV was projected onto a single-element sample and backscattered He particles were detected at a scattering angle of 160°. The highest backscattering energy was 1.85 MeV. What element is the sample made of?
   (a) Carbon (C)
   (b) Aluminum (Al)
   (c) Cobalt (Co)
   (d) Gold (Au)
2. Cite two appropriate mass separability improvement methods for Rutherford's backscattering.

References

1. Beiser A translated by Sato T, Nagae M, Fukase S and Yamashita T (1984) Gendai buturigaku no kiso, Kogakusha (in Japanese); Beiser A (2002) Concepts of modern physics, 6th edn. McGraw-Hill Science/Engineering/Math, London
2. Weinberg S translated by Honma S (2006) Sinban denshi to genshi kaku no hakken – 20seiki buturi gaku wo kizuita hitobito, Chikuma shobo, Tokyo (in Japanese); Weinberg S (2003) The discovery of subatomic particles. Cambridge University Press, Cambridge
3. Ziegler JF, Biersack JP, Littmark U (1985) The stopping and range of ions in solids, vol 1. Pergamon, New York

Further Readings

4. Chu WK, Mayer JW, Nicolet MA (1978) Backscattering spectrometry. Academic, New York
5. Bird JR, Williams JS (eds) (1989) Ion beams for materials analysis. Academic, Sydney
6. Tesmer JR, Nastasi M (eds) (1995) Handbook of modern ion beam materials analysis. Materials Research Society, Warendale
7. Fujimoto F, Komaki K (kyohen) (1995) Ion bi-mu kogaku – ion· kotai sogo sayo hen, Uchida Rokakuho, Tokyo (in Japanese); Fujimoto F, Komaki K (1995) Ion beam engineering – interactions between ions and solids. Uchida Rokakuho Publishing Co., Ltd., Tokyo. (in Japanese)
8. Fujimoto F, Komaki K (kyohen) (2000) Ion bi-mu ni yoru busshitsu bunseki· busshitsu kaishitsu, Uchida Rokakuho, Tokyo (in Japanese); Fujimoto F, Komaki K (2000) Ion beam-based materials analysis and reforming. Uchida Rokakuho Publishing Co., Ltd., Tokyo. (in Japanese)
9. Ishigure K, Tominaga H, Yamaguchi H, Tateno Y, Nakazawa M (kyohen) (1990) Hosyasen oyo gijyutsu hando bukku, Asakura syoten, Tokyo (in Japanese); Ishigure K, Tominaga H, Yamaguchi H, Tateno Y and Nakazawa M (1990) Handbook of radiation utilization technologies, Asakura Publishing Co., Ltd., Tokyo. (in Japanese)

Chapter 13
Ion Beam-Induced Mutation in Plants

Atsushi Tanaka

13.1 Mutation of Plants Induced by Radiation

Living things cannot avoid mutations. Conversely, it can be said that mutations are the essence of life. Creating a diversity of variants through mutation is indispensable in the process of the birth of life and evolutions of organisms into different organisms. In this sense, it can be said that the history of life has always been accompanied by the history of mutation. However, the discovery that mutation can be caused artificially was made only after the start of the twentieth century. What lead to this discovery was the discovery by H.J. Muller [1] in 1927 that X-ray irradiation induces mutation in yellow fruit flies and the successful induction, around the same point of time, of mutations in barley and maize by L.J. Stadler [2] using X-rays and gamma rays. Since then, radiation-based plant breeding has been pursued for 80 years, and, according to the International Atomic Energy Agency (IAEA), more than 3000 plant varieties have been developed around the world to date through plant breeding that uses radiation such as gamma rays and X-rays [3]. However, because of the long history (maturity) of radiation-based plant breeding, a feeling has spread that it is no longer possible to obtain novel mutants and improved varieties, and in recent years discovery of new mutagens to obtain mutants with is being waited on.

Ion beams had been considered to affect cells and genes significantly because they deposit higher energy locally than gamma rays and X-rays and had been utilized mainly for analyses of effects of cosmic radiation since around the 1960s. However, it had not been clear what mutation ion beams can induce, and there had been no knowledge at all as to whether ion beam-induced mutation can be utilized for practical purposes such as plant breeding. In Japan, the governmental plan for

A. Tanaka (✉)
National Institutes for Quantum and Radiological Science and Technology, Gunma, Japan
e-mail: tanaka.atsushi@qst.go.jp

H. Kudo (ed.), *Radiation Applications*, An Advanced Course in Nuclear Engineering 7, https://doi.org/10.1007/978-981-10-7350-2_13

research for advanced utilization of radiation was established in 1987 as a result of efforts by organizations such as Japan Atomic Energy Research Institute (JAERI) (the predecessor body of Japan Atomic Energy Agency (JAEA)), and, as part of the plan, a plan for utilizing ion beams in inducing mutation in plants was established. As a result, an ion irradiation research facility called "Takasaki Ion Accelerators for Advanced Radiation Application (TIARA)" was established in 1991 as the world's first facility dedicated for materials and biotechnology research, and analyses of the characteristics of ion beams as a mutagen started in earnest [4]. Currently, ion beam irradiation experiments are being conducted at RIKEN, the Wakasa Wan Energy Research Center, and the National Institute of Radiological Sciences (NIRS) as well, and many projects are in progress at private sector companies, universities, and public research organizations in Japan, and these activities have started attracting attention from overseas as well.

13.2 Characteristics of the Energy Deposition of Ion Beams and the Irradiation Method

One measure of the energy deposition to a substance by radiation is linear energy transfer (LET). Gamma rays and X-rays have an LET value of approx. 0.2 keV/μm, whereas an ion beam has a much higher LET value and its energy deposition intensifies locally. In addition, an ion beam rapidly loses its speed as it deposits huge energy to the irradiation target. For example, 220 MeV carbon ions advance only about 1.2 mm in water. This is because 220 MeV carbon ions have an LET value of approx. 100 to 200 keV/μm, which is about 500 times as high as the LET values of gamma rays and X-rays (Fig. 13.1). In addition, ion beams deposit huge energy immediately before they stop in a target. This is called the "Bragg peak." The Bragg peak is the most accurate indicator of the characteristics of ion beams, but it is very narrow and difficult to control. Ions with the initial high energy (shown in the left of Fig. 13.1) have a lower LET value and a large track radius due to the generation of high-energy secondary electrons. Ions with lower energy have a higher LET value and a small track radius that causes local deposition of the impartment of energy. Therefore, it is considered that the region with relatively high LET values which is located immediately before the peak and in which LET value changes are small is the effective region for inducing mutations. Because of the physical characteristics of ion beams described above, the energy deposition to cells by ion beams is fundamentally different from that by low-LET radiation such as gamma rays.

The irradiation method used differs between irradiation facilities (such as Japan Atomic Energy Agency (JAEA) and RIKEN), but the principles are the same. In the case of TIARA's depth-controlled irradiation equipment for seeds (Fig. 13.2), scanning with ion beams of carbon ions, etc. that have been accelerated to a fraction of the speed of light by a cyclotron is made to achieve a uniform irradiation field.

Fig. 13.1 Energy deposition of ion beams

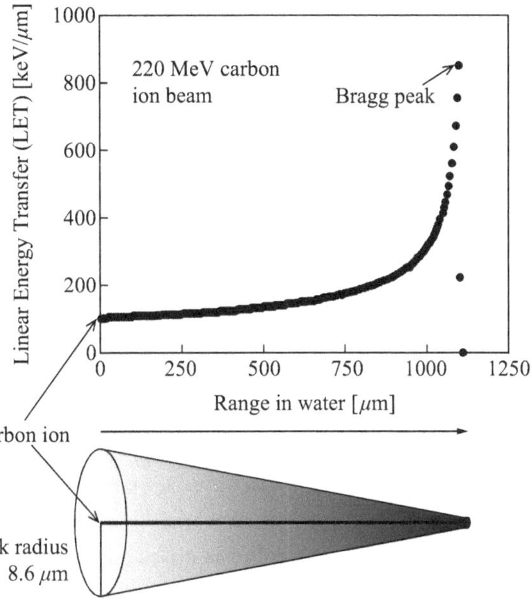

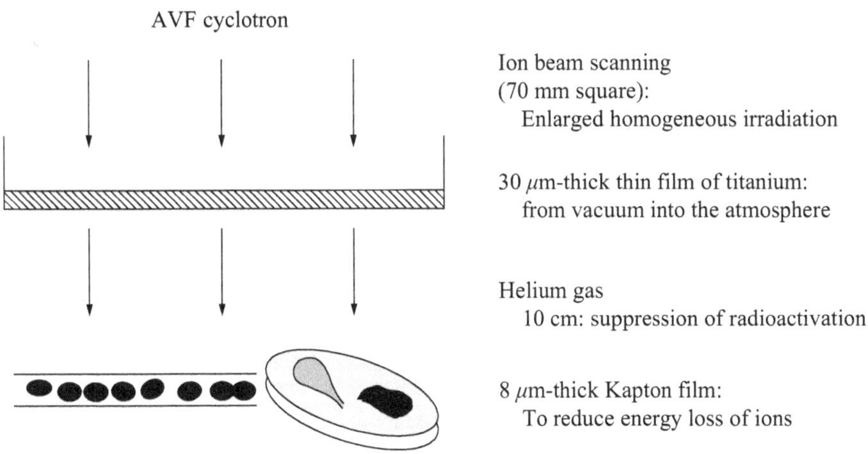

Ion	Energy [MeV]	LET [keV/ μm]	Range in water [mm]
e	2	0.20	9
He	50	19	1.6
He	100	9	6.2
C	220	122	1.1
C	320	86	2.2
Ne	350	441	0.6

Fig. 13.2 Ion beam irradiation method and physical characteristics of ions

The ion beams ejected into the atmosphere first go through the helium gas provided to suppress radioactivation and then are projected onto the sample. The irradiation sample (seeds, sterile tissue culture, etc.) is covered with a thin film such as a Kapton film and irradiated for about 30 to 120 seconds so that the ion energy loss is minimized and homogeneous irradiation is achieved.

13.3 Biological Effects of Ion Beams

To identify the characteristics of the induction of mutations in plants by ion beams, it is first necessary to know the biological effects of ion beams on plants. A lot of data has been obtained over many years on plant death and chlorophyll mutations induced by irradiation of seeds [5], and it is known that the biological effects of an ion beam are strongest when it has an LET value of 72 to 190 keV/μm [6, 7], at which the ratio of biological effects (RBE) to gamma rays can become as high as about 100 [8]. However, with the traditional ion beam irradiation, the energy deposition distribution in the sample and the effect of irradiation were nonuniform because a great deal of irradiation was made at the Bragg peak and it was difficult to determine what kinds of ions were suitable for inducing mutations. For this reason, a study for total investigation of the effect of irradiation was conducted in which the type of the ion species and LET value were varied. This study showed that the effects on the germination rate and survival rate of *Arabidopsis thaliana* (Arabidopsis) were strongest when the LET value was about 250 keV/μm [9]. In addition, an experiment was conducted to investigate this characteristic in more detail. This experiment consisted of irradiation of samples with various ion species with the energy varied for each ion species to vary the LET value. With carbon ions, the lethal effect of irradiation on Arabidopsis and *Nicotiana tabacum* (tobacco) reached the peak at an LET of approx. 200 keV/μm (Fig. 13.3) [10, 11]. The LET value of the RBE peak increased with the atomic number of the ion used. With argon, the LET value of the peak was higher than 400 keV/μm. This means that ions with an LET value between about 200 and 400 keV/μm are effective, but the optimal LET value differs between ion species. The experiment also showed that the RBE value for tobacco was more than several times higher than that for Arabidopsis, which means that ion beams are more effective on tobacco seeds than on Arabidopsis seeds. The RBE value varies widely depending not only on the target plant species but also on the tissue of the target plant species and its growth stage.

In addition, a study was conducted to determine the optimal doses for causing biological effects such as chromosome aberration and sterility. The sterile rate reached almost 100% at the shoulder end of the survival rate curve (immediately before the survival rate starts decreasing) with electron beams, while a nearly sterile state was reached at a dose of about half the dose at the shoulder end with carbon ion beams. In irradiation with tobacco seeds, no major difference was observed between gamma rays and ion beams, but the chromosome aberration rate of the root

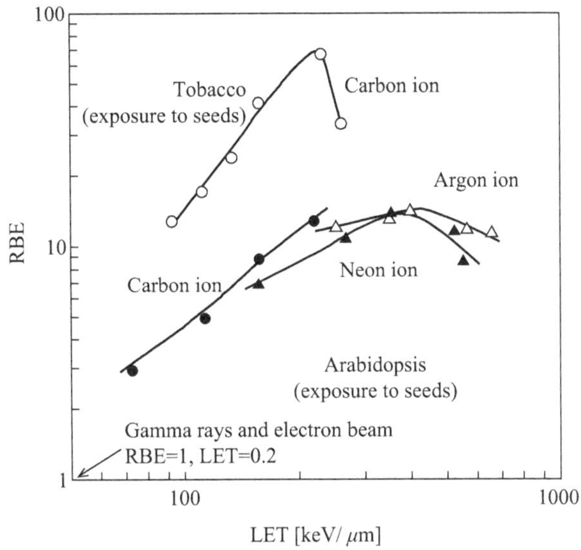

Fig. 13.3 Relative biological effectiveness (RBE) of ion beams for lethality against LET

apex reached about 80% or higher at the shoulder end of the survival curve [10]. On the basis of the results described above, it is considered that the desirable ion beam irradiation dose to induce mutations is a dose within the region in which no survival rate reduction occurs.

13.4 Characteristics of Ion Beam-Induced Mutation

13.4.1 Mutation Rate

Much of the mutation frequency data obtained in the past is based on observation of phenotypic changes such as chlorophyll-deficient or somatic cell mutation studies/ experiments, and some of this data does not accurately reflect the actual genetic mutation rates. To address this problem, an experiment was conducted to compare carbon ion irradiation with electron beam irradiation in terms of *tt* and *gl* mutation in Arabidopsis, whose characters and gene loci are clearly known. The electron beams are low-LET radiation that have the same biological effectiveness as gamma rays. As a result of irradiation of more than several tens of thousands of individuals of Arabidopsis in the experiment, it was found that the average mutation rate of carbon ions (gene locus/initial cell (diploid)/dose) was about 20 times as high as that of the electron beams used as the control (Table 13.1) [12, 13]. In the experiment, the electron beam dose was set at 750 Gy, which was five times as high as the carbon ion irradiation dose (150 Gy), in order to make the biological effects of the electron beam equivalent to the biological effects of the carbon ion irradiation. This means that the number of mutants obtained with the carbon ions

Table 13.1 Comparison of mutation rates of carbon ions and electron beams

Mutagen (dose)	Gene locus investigated	Mutation rate ($\times 10^{-6}$) (gene locus/cell/Gy)	Type of mutation (%)	
			Point-like mutation	Rearrangement
Carbon ion (150 Gy)	*tt3, tt4, tt5, tt6, tt7, tt18, tt19*	1.9	50	50
	gl1, gl2, gl3, ttg1, ttg2			
Electron beam (750 Gy)	*tt3, tt4, tt5, tt6, tt7, tt18, tt19*	0.097	75	25
	gl1, gl2, gl3, ttg1, ttg2			

(gene locus/initial cell) in actual breeding is four times greater. As the mutation rate of the 150 Gy carbon ion beam irradiation is 1.9×10^{-6} (gene locus/initial cell/dose), the average number of genes in which mutation occurs per diploid genome of the cell nucleus would be calculated at 7 (7.4 to be exact) if it is assumed that the number of genes of Arabidopsis is 26,000 ($1.9 \times 10^{-6} \times 150 \times 26{,}000 = 7.4$).

Apart from studies to determine precise mutation rates such as the experiment described above, several studies have been conducted to determine approximate mutation frequencies from characters of mutations. One study investigated the mutation frequency in the flower color of chrysanthemum with different types of radiation. In the study, cultured explants of flower petals and leaves of chrysanthemum were irradiated with carbon ions and gamma rays. The result of the irradiation showed that the frequencies of induction of mutations in flower petals and leaves were 16% and 7%, respectively. These figures were about half the corresponding figures obtained with gamma rays [14]. A study to investigate mutation frequency in cultured carnation induced by irradiation with different types of radiation showed that the mutation frequencies induced by carbon ions, gamma rays, and soft X-rays were 2.8%, 2.3%, and 1.3%, respectively [15]. A study to investigate the frequencies of the mutation that makes sugarcane hairless (trichomeless) induced by irradiation with different types of radiation showed that the frequency of the trichomeless mutation induced by helium ions was about twice as high as the trichomeless mutation induced by gamma rays [16]. In addition, studies to compare ion beams with gamma rays, etc. in terms of chlorophyll-deficient mutation in crops induced by them have been conducted frequently. A study in which rice was irradiated with nitrogen ions showed that the frequency of chlorophyll-deficient mutation was high at 11.6% [17]. For barley, a chlorophyll mutation frequency of 2.1% was observed in an evaluation of the resistance of barley individuals against yellow mosaic virus, which is a major barley disease [18]. In a study in which flowers of tobacco in the initial embryonic stage were irradiated with nitrogen ions, albino (recessive homozygote) was observed in the seeds obtained [19]. A study that compared ion beams with gamma rays in terms of the frequencies of rice chlorophyll mutation induced by them showed that chlorophyll mutation frequency induced by ion beams was up to two times greater than that by gamma rays

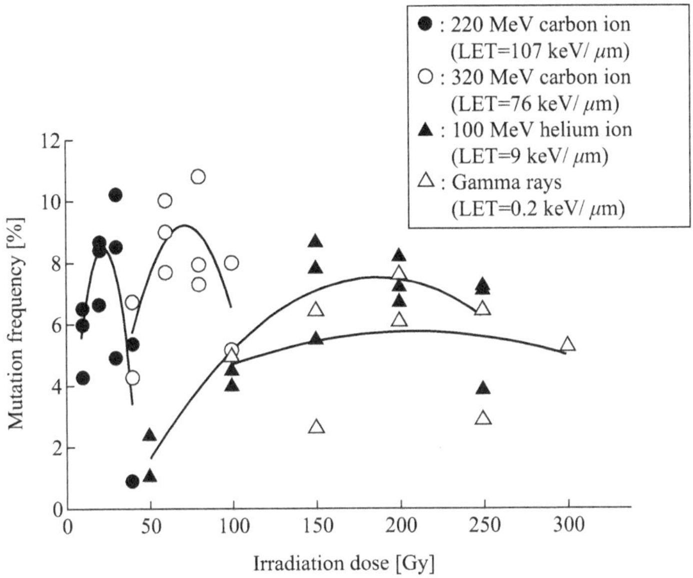

Fig. 13.4 Chlorophyll mutation frequency in rice
Each type of radiation has a peak of mutation frequency. The irradiation doses at which the peaks occur correspond to the shoulder ends of the survival curves for the irradiation of rice seeds (From Ref. Yamaguchi et al. [20], modified by the authors)

(Fig. 13.4) [20]. The study also investigated the dose dependency of mutation frequency caused by radiation and showed that the mutation frequency caused by gamma rays, helium ions, and carbon ions reached their peaks at a certain dose on the dose curve and showed for the first time that the peaks all occurred at the shoulder ends of the survival curves. This is a very important finding in mutation breeding (radiation breeding). That is, while it has traditionally been said that the mutation rate is highest around the semilethal dose (i.e., the irradiation dose at which the survival rate, etc. decrease by 50%, which is called lethal dose 50 (LD_{50})) and this has been accepted as a kind of established theory, the study clearly shows that sufficient mutations can be induced at the shoulder of the survival curve, that is, a nonlethal dose, both with gamma rays and ion beams.

13.4.2 Mutation Spectrums

Initial studies on spectrums of ion beam-induced mutations were ones that compared spectrums of ion beam-induced mutations with spectrums of gamma ray-induced mutations. In one of these studies, cultured tissues of flower petals and leaves of a pink-colored chrysanthemum variety called "Taihei" were irradiated with ion beams using a system based on a newly developed mutation induction

Table 13.2 Flower color mutations in chrysanthemum induced by carbon ions and gamma rays

	Mutation frequency (%)					
Mutagen	Light pink	Deep pink	Orange	White	Yellow	Complex/striped
No irradiation	0.3	0	0	0	0	0
Gamma rays	27.7	2.1	0	0	0	0
Carbon ions	4.6	0.3	0.3	0.3	0.2	10.2

method that combined chrysanthemum tissue culturing with irradiation, and the flower colors of the obtained regenerated individuals were examined [14]. With the gamma ray irradiation (used as the control), flower colors from light pink to deep pink were frequently obtained, while, with the ion beam irradiation, flower colors that are difficult to obtain with gamma ray irradiation including white, yellow, and orange were induced relatively easily (Table 13.2). In addition, complex color mutations which generate bicolor flowers such as ones that are yellow in the central part and pink in the surrounding outer part and striped mutations which generate flowers whose flower petals are striped with two colors were newly induced (it had not been possible to induce these mutations before the study). Thus the study showed that the spectrums of ion beam-induced mutations were different from the spectrums of gamma ray-induced mutations. Five newly obtained mutants have been registered as new varieties, and their cultivation has started in Japan. The gamma ray irradiation method used in the study is the so-called acute irradiation, in which irradiation is completed within several hours. It should be noted that there is another gamma ray irradiation method called the "chronic irradiation," in which continuous low-dose irradiation is made for a period of several days to several months or longer, and the spectrums of mutations obtained with this method would be different from the spectrums obtained in the study.

The characteristics observed in the spectrums of mutations in chrysanthemum obtained by the abovementioned irradiation were also observed in the spectrums of flower color and shape mutations in carnation induced by irradiation in another study. In the study, cultured leaves of a major European variety called "Vital" with cherry-colored, serrated flower petals were irradiated with carbon ions, and the flower colors and shapes of the obtained reproduced individuals were examined [15]. The mutagens used as the control in the study include gamma rays, X-rays, and ethylmethanesulfonate (EMS), which is the most popularly used chemical mutagen. EMS, gamma rays, etc. induced mutations that changed the flower color to light pink, pink, and red, while the carbon ions induced mutations that changed the flower color to a variety of colors including salmon pink, yellow, and cream (Fig. 13.5). The striped and complex color mutations observed in the abovementioned study on irradiation of chrysanthemum were also induced frequently. With regard to flower shape mutations, the gamma rays induced a few mutations that changed serrated petals into round ones, while the carbon ions induced many mutations that changed serrated petals into round ones and Dianthus-type flowers. In addition, the study achieved continuous flower shape changes through induced mutations. For example, serrated petals were changed into

Fig. 13.5 Carnations with various flower colors and shapes obtained by ion beam irradiation
The carnation shown in the upper left corner is the original variety "Vital" (photograph courtesy of Dr. Okamura of Kirin Holdings Company, Limited)

semi-serrated petals, semiround petals, and round petals. Out of this study, several new varieties were developed which make up the "Ion Vital Series." In addition, "Beam Cherry Series" were also recently successfully developed. "Beam Cherry Series" not only have the same disease resistance, flower life span, and productivity as those of "Ion Vital Series" but also excel "Ion Vital Series" in stem quality. It is expected that "Beam Cherry Series" will be popularized and cultivated for a long time in the future.

With regard to *tt* and *gl* mutations induced in Arabidopsis described above, a study showed that electron beam irradiation tended to frequently produce mutants in a few particular gene loci including *gl2*, while carbon ion beam irradiation tended to induce mutations uniformly in all gene loci [12]. The study discovered two new gene loci as a result of the carbon ion beam irradiation in addition to inducing traditional *tt* mutations. It is considered that the discovery of the new genes was made possible by the higher mutation rates achieved.

As a result of the studies described above, it has become clear that the characters of ion beam-induced mutations are different from those of mutations induced by gamma rays, etc. and ion beam-induced mutations have wider mutation spectrums. However, because no significant difference has been found so far between gamma rays and ion beams in terms of the spectrums of chlorophyll mutations they cause in rice described in Section 13.4.1 [20], it is necessary to make detailed analyses in the future to determine what causes differences between spectrums of chlorophyll mutation in rice induced by gamma rays and spectrums of chlorophyll mutation in rice induced by ion beams.

13.4.3 Molecular-Level Characteristics of Induced Mutation

The characteristics observed in relation to mutation rate with ion beams and to spectrum of such mutation are all attributable to molecular-level mutation that occur in the DNA. In the days when plant genomes had not been analyzed to the sufficient extent, molecular-level analysis of mutation was extremely difficult, and very few analyses of the DNA of mutants produced with gamma rays had been made. However, it became possible in the 1990s to analyze mutations on the molecular level as a result of dramatic advances in the Arabidopsis genome analysis and the development of ion beam technologies that allowed many mutants to be obtained in a single gene locus. Attempts were made to irradiate seeds of Arabidopsis with ion beams and make detailed DNA-level analyses of mutations in the transparent testa mutants *(tt)* and trichomeless mutants, glabra (*gl*), induced in progenies [12, 13, 21]. In these studies, the electron beam irradiation mostly induced point-like mutations (9 point-like mutations in genes and 3 rearrangements, i.e., large structural changes on the chromosome level), while the carbon ions induced point-like mutation and large structural change ratio of about 1:1 (14 point-like mutations and 15 rearrangements), indicating that carbon ions tend to induce large structural changes more frequently than electron beam (see Table 13.1). Analysis of the base sequences of the abovementioned mutations showed that most of the large structural changes induced by the carbon ions were inversion-, translocation-, or deletion-type changes and most of the large structural changes induced by the electron beams were inversion- or translocation-type changes. With regard to point-like mutations, the analysis showed that the carbon ions induced short-deletion-type mutations more frequently than the electron beams. Detailed analysis of the base sequences of the deleted ends showed that the point-like mutations induced by the carbon ions were short-deletion-type mutations with the DNA breakage end rejoined, while the point-like mutations induced by the electron beams were non-deletion-type mutations that had overlapping in the base sequence of the DNA break end [12]. On the basis of these results, the authors presume that ion beam irradiation tends to cause multilocation damage at the DNA break end, and, because of this, nonhomologous end joining plays an important role in the repair process. Similar findings have been obtained from a subsequent study in which an analysis was made using recombinant DNA as a marker [22].

13.5 New Mutants and Varieties Created Using Ion Beam Irradiation

Many new mutants and varieties have been created to date using ion beam irradiation. This chapter describes the characteristics of these mutants and their mutations (Fig. 13.6).

Fig. 13.6 Examples of new mutant varieties developed using ion beams
Upper left: an ultraviolet B light-resistant mutant of Arabidopsis
Upper center: a mutant with frilled flower petals
Upper right: a mutant in which pigment has been accumulated
Lower left: an "Ion No Seiko" variety with complex colored flower petals
Lower center: a "Beam Cherry" carnation variety with improved stem quality and flower color
Lower right: an "Aladdin" chrysanthemum variety with reduced axillary buds

13.5.1 Model Plants

Ultraviolet B Light-Resistant Mutants and Genes

In recent years, depletion of the ozone layer has been accelerating as a result of the continuing intensification of global warming. Because the ozone layer is a strong shield that blocks the ultraviolet light of sunlight, the amount of ultraviolet B (UV-B) light that reaches the ground has been increasing as a result of the depletion, especially in polar regions including the South Pole. It is known that UV-B light suppresses the growth of plants and causes plant leaf scorching, thereby reducing crop yields significantly. Although UV-B light-sensitive mutants have been created and studies have been conducted to determine why they are sensitive to UV-B irradiation, no report has been published yet on successful induction of an UV-B-resistant mutant. However, creating UV-B-resistant mutants using ion beams and identifying the genes responsible for the resistance allow the mechanism of UV-B resistance of plants to be clarified and UV-B-resistant crops to be developed. By irradiating 1280 seeds of Arabidopsis selected as the model plant with carbon ions and selecting well-grown individuals out of progeny individuals originating from the seeds, 4 strains of UV-B-resistant mutants (*uvi1* to *uvi4*) were ultimately obtained which grow about 1.5 to 2 times more than the wild type in an environment with a UV-B light dose that is equivalent to UV-B light doses encountered in outdoor environments in the Japanese spring and autumn [23, 24]. In addition, analyses of *uvi4* mutants led to successful isolation of the responsible gene (*UVI4*) [25]. Because the 123 base pairs (bp) had been deleted in the *UVI4* gene of the *uvi4* mutants, it is considered that the function that suppresses the endoreduplication (i.e., DNA amount increases) in the cell nucleus of the *UVI4* gene was lost as a result of the deletion, and this facilitated the endoreduplication to increase DNA contents in leaf and stem cells, which allowed the *uvi4* mutants to continue growing even under the strong UV-B light making up for the DNA wounded by the UV-B light and grow two times (or more) more than ordinary Arabidopsis. It is expected that this result will serve as an important key to understanding how plants have adapted to UV-B light and help develop technologies to increase yields of crops such as barley, wheat, and soybean and to prevent leaf scorching in garden plants.

In the *rev3* UV-B-sensitive mutants created, which were found to be totally new mutants, ion beam-induced chromosome reorganization involving base insertion was observed [26]. The existence of *rev3* mutants and analyses of *rev1* mutants, which are related to *rev3* mutants, led to the discovery that the translesion synthesis mechanism present in mammals and yeasts for overcoming damage is also present in plants and this mechanism provides a major function for recovering from DNA damage apart from the DNA repair mechanism [27, 28]. These results indicate that the UV-B light resistance mechanism of Arabidopsis is much more complex than had been predicted and is closely related to growth stages, tissues, and cell cycles of plants, and further acquisition of mutants and further gene analyses will be required to obtain the complete picture of the UV light-resistant mechanisms.

Genes that Are Important for Morphology and Function (Including Flower Colors and Shapes) of Plants

Two new mutants (*tt18* and *tt19*) that had not been previously reported were discovered in a study on mutation induction rates of ion beams [12]. These mutants (two mutants each) were obtained from 104,088 (M2) progenies originating from 26,200 seeds irradiated with carbon ions. In the two *tt18* mutants, 2 bp and 5 bp had been deleted, respectively. In one of the *tt19* mutants, approximately 1000 kbp was inserted. In the other mutant, translocation was observed in the upstream area of the gene. It was confirmed that the *TT18* gene was the gene for the pigment biosynthesis enzyme and the *TT19* gene was a gene involved in the transport of anthocyanin (the flower pigment) and proanthocyanidin (a general term for catechin, tannin, etc.) to vacuoles and their accumulation therein [29]. It is expected that control of the *TT19* gene will allow plant production systems to be developed in the future for altering the flower color tint and developing plants that have flavonoids with an anti-oxidizing effect.

In a study in which seeds of Arabidopsis were irradiated with carbon ions, the germinated plants were cultivated, and flowers of about 12,000 progeny (succeeding generation) individuals were observed, flowers whose petal tips had been mutated into serrated ones were observed [30]. This serrated mutation is attributable to the recessive character generated as a result of the deletion of one base pair in the gene named the Frill gene (*frill1*) [31]. It is considered that the Frill gene shapes flower petals properly by controlling the cell division and cell elongation in flower petals, and it was found that a deficiency of the Frill gene would result in reductions in the number of cells in the tips of flower petals and expansion of the individual cells to up to ten times the original size to make the flower petal arrangement irregular and look like frills. It is considered that the discovery of the Frill gene will lead to future findings about how plants and their flower petals are shaped from future studies on the functions of the genes involved.

With regard to other major genes that concern the plant morphology, *aar1* Arabidopsis mutants that concern the auxin plant hormone were isolated using ion beam irradiation [32]. The *aar1* mutant was obtained from about 30,000 progenies (M2) originating from 6700 seeds irradiated with carbon ions. In these mutants, a large fragment of DNA (43,895 bp) that contained the *aar1* gene had been deleted. The protein produced from the gene that causes the mutation was named "small acidic protein 1" (SMAP1) because it has small molecular weight and is rich in acidic amino acid residue. Because databases indicate that SMAP homologous genes are present in many animals and plants and have been preserved in an evolutional manner, it is presumed that they have important biological functions.

Mutants that Concern Symbiotic Nitrogen Fixation

For *Lotus japonicus* (the model plant for legume), too, ion beam-based searches for mutant genes that concern nitrogen fixation are being made. Mutants in which up to ten times as many root nodules (organs that fix nitrogen using root nodule bacteria) as those in the wild type of *L. japonicus* are formed were obtained with helium ion beam irradiation [33]. In addition to the formation of more root nodules than the wild type of *L. japonicus*, these mutants had many other altered characters, including bifurcated pistil, delayed flowering, and abnormal vein formation. It was confirmed that all of these alterations were attributable to mutations in a single gene and this gene was named *klavier*. In addition, mutants that form more root nodules than the wild type were obtained from 35,000 progenies (M2) originating from 3300 seeds irradiated with carbon ions, and these mutants were named *too much love* [34]. These facts show that ion beam-induced mutation techniques are also effective in creating legume gene resources.

13.5.2 Development of Flower Varieties

The area in which the advantages of ion beams have been utilized most is breeding of flowering plants. This is also the first area in which ion beam irradiation was put to practical use. The first practical breeding developed is several mutants obtained from the "Taihei" chrysanthemum variety described above. These practical breedings were made public in 1996, and application for registration of these varieties was made in the autumn of 2002. In the spring of 2002, sterile Hanademari named "Coral Pink" was put on the market. Three carnation varieties of the "Ion Vital Series" were also put on the market in the same year [35].

The area in which ion beam irradiation has been put to practical use most successfully is breeding of chrysanthemum with reduced axillary buds. White chrysanthemum ranks first among all flowers in Japan in terms of production, and cultivation of "Jimba," which is a major white chrysanthemum variety that is superior in white color quality and growth characteristics, has spread rapidly in recent years. However, because Jimba produces many axillary buds, bud-removal work is required. It is said that the bud-removal work accounts for one fourth to one third of all work, and it can be said that the development of chrysanthemum breeds with reduced axillary buds is the most important goal in chrysanthemum breeding. For this reason, studies were conducted wherein cultured leaves of chrysanthemum were irradiated with ion beams and more than 10,000 lines of regenerated individuals were examined. As a result, the lines that have fewer axillary buds but retain the original flower- and leaf-related characters were obtained for the first time [36, 37]. Two good mutant lines were selected from among these lines, and the selected two lines were named "Imagine" and "Aladdin" and were registered as new plant varieties in 2003. Aladdin attracted numerous requests for permission to cultivate and has been cultivated throughout Japan. However, Aladdin had one drawback. That is, its flowering is delayed by low temperatures. Accordingly, to create an evolved version of Aladdin which retains the good features of Aladdin but is free of the problem of little delayed flowering (and therefore is more economical in terms of the heating cost), Aladdin was irradiated with ion beams. As a result, the evolved version of Aladdin, which can be considered as the ultimate Aladdin, was developed successfully and named "Aladdin 2." There have been requests for license to utilize Aladdin and Aladdin 2 from over 35 agricultural cooperatives, and the total annual production in Japan is expected to exceed 40 million flowers in 2009. It can be said from the abovementioned facts that ion beam irradiation allows one-point breeding of new character that is difficult to obtain with the conventional methods.

One recent example of practical utilization of ion beam irradiation is the development of a new Osteospermum variety called "Viento Flamingo" which has novel pastel flower petals with pink stripes through irradiation of a world-famous Osteospermum variety with ion beams. In addition, a new cherry variety called "Nishina Zao" whose green part is smaller than the green part of existing variety and whose overall color is shallow yellow was developed by irradiating cherry budwood with carbon ions and grafting it. As for fragrant cyclamen, which does not have many flower color variations, a reddish purple fragrant cyclamen containing the delphinidin pigment, which had not existed in the cyclamen genus in the past, was created for the first time in the world by irradiating a fragrant cyclamen called "Kaori No Mai (meaning "fragrant dance")" with ion beams [38]. This study examined genome information including information on genes and pigment expressions in advance, found that it was possible to produce delphinidin with a single mutation, and created the reddish purple fragrant cyclamen containing delphinidin.

13.5.3 Crop Breeding

The first studies in Japan on utilization of ion beams in breeding were studies on ion beam irradiation of rice. Bacterial leaf blight-resistant and rice blast-resistant mutated individuals of a rice variety called "Koshihikari" were developed by irradiating Koshihikari with nitrogen ions, helium ions, etc. In the series of experiments conducted, ion beams, thermal neutrons, and gamma rays were compared to confirm the differences between them in terms of mutation frequency, and it was found that ion beams were the most effective mutagen in inducing mutations that enhance bacterial leaf blight resistance [39]. Currently, many studies are being conducted on irradiation of rice with ion beams, and development of new rice mutants that produce low-amylose rice, as well as new mutants that assist in environmental cleanup, is being pursued. In a study aimed at developing a rice variety that brings stable yield with cultivation with less fertilizer (i.e., less nitrogen) that has been pursued to help solve the problem of the eutrophication of Lake Biwa, 4500 unpolished grains of a rice variety called "Aki No Uta (meaning "poem of autumn")" developed in Shiga Prefecture were irradiated with ion beams, and a mutant line that brings approximately 10% more yield than "Aki No Uta" was obtained from among the progeny lines. The cultivation characteristics and nitrogen utilization efficiency of the obtained strain are currently being investigated.

In the area of barley breeding, ion beam irradiation has frequently produced mutants that are resistant to yellow mosaic virus, which is a major barley disease [40]. Ion beam irradiation is also attracting attention as a prospective means of eating taste improvement breeding such as low-amylose rice and wheat. In tobacco, combined use of anther culture with ion beam irradiation has frequently produced resistant mutant lines to potato virus Y-related diseases [41]. Because anther culture is culture that produces haploids, recessive characters also appear in the original generation in anther culture. Anther culture combined with irradiation has produced mutated haploids with resistance to potato virus Y-related diseases at a high mutated haploid production rate of 3.5%. It has been confirmed that the disease resistance is retained in the self-propagated succeeding generation of these individuals with doubled chromosomes. In addition, attempts are being made to develop wide varieties of other plants including strawberry, melon, tomato, eggplant, garlic, and sweet potato. One of the aims of these attempts is component breeding. Breeding efforts are also being made to develop labor- and energy-saving breeding such as short-nodule stem tomatoes and melons that hypertrophy at low temperature.

13.5.4 Breeding of Trees, etc.

Studies on utilization of ion beam irradiation in breeding of trees are also being conducted. In tree breeding, it takes several years or longer to obtain a progeny generation in which mutations are fixed and the selection process involves difficulties due to the large sizes of individuals. For this reason, it is necessary to devise ways to fix mutations in the original generation and make selections at the seedling stage. For white cedar and Japanese cedar, studies are being conducted to develop ultraviolet light-resistant and less pollen mutants. In studies on ion beam breeding of Japanese cedar, characteristics such as albino and wax-rich have been confirmed in the original generation [42]. Irradiation of *Ficus pumila* with carbon ions has produced a mutant with 40 to 80% higher nitrogen dioxide (NO_2) absorption capability, which was named "KNOX." KNOX production technologies will be transferred to the private sector in July 2009 for the utilization of KNOX in environmental cleanup including the greening of building walls and roads. Ion beam breeding is also being utilized in improving indigenous breeding of flowering trees and shrubs including azalea and camellia.

In recent years, utilization of ion beam irradiation in breeding of mushrooms and microorganisms has also been increasing. One example is irradiation of spores and hyphae of *Pleurotus ostreatus*, *Lyophyllum decastes*, etc. with ion beams to develop new varieties. Mushroom mutants that have been created with ion beam irradiation include mutants with extraordinary shapes such as ones with funnel-shaped pilei and ones that do not grow evenly in all directions.

13.6 Phenomena that Cannot Be Accounted for with Mutation Only

In research and development to utilize ion beams in plant breeding, several phenomena have been discovered that cannot be accounted for with gene mutation only. This section describes these phenomena.

13.6.1 Overcoming of Cross Incompatibility

It is difficult to create mature hybrids between distantly related plant species. However, it has been discovered that ion beam irradiation of tobacco pollen allows the cross incompatibility problem in hybridization of tobacco wherein seedlings grown from seeds made by hybridization die immaturely to be solved and hybrids to be obtained efficiently [44]. Seedlings grown from seeds made by hybridization of a cultivated species with a wild species resistant to green peach aphids, etc. die immaturely, but hybridization using ion beam-irradiated pollen of a cultivated species has produced mature hybrids at a yield rate of 1×10^{-3}. Because this rate is higher than frequencies of mutations caused in particular genes with ion beam irradiation, it is considered that there can be factors other than gene mutations such as genome structure changes which allow the cross incompatibility to be overcome.

13.6.2 Sex Determination Control

In a study in which an all-female strain of spinach, which is a dioecious plant, was made through breeding and it was irradiated with ion beams, individuals that form anthers in female flowers and individuals that flower mostly with male flowers appeared in the irradiated (original) generation, and such traits were fixed in progeny generations as well [44]. It is considered that this is the result of the disabling of the male flower generation suppression function of female flowers by ion beam irradiation.

13.6.3 Activation of Transposon

As it had been known from experiments with maize, drosophila, etc. that transposon, which is a movable gene, can be activated with mutation induction treatment, an attempt has been made to make the inactive transposon in rice mobile with ion beam irradiation to assist in rice gene resource development. As a result of studies conducted over many years, a sign showing that transposon in rice plants had moved was detected in the form of chlorophyll mutations [45]. It is presumed that a change in the structure of the genome DNA induced by the ion beam irradiation caused the inactive transposon to move.

13.7 Past Achievements and the Future of Ion Beam Breeding

It has been more than 20 years since studies on ion beam-induced mutation in plants started. The main objective of the initial studies was "to determine whether mutants created with ion beams are different from those created with conventional radiation such as gamma rays." No definitive answer to this question has been supplied yet, but we believe that certain characteristics have been identified through DNA-level analyses of induced mutations thanks to advances in genome analysis technologies and techniques (Fig. 13.7). Specifically, we believe that ion beam-induced mutation in plants is different from mutation using conventional radiation in that ion beam-induced mutation:

(1) Achieves higher mutation rates.
(2) Induces a wider variety of mutations (i.e., achieves wider mutation spectrums).
(3) Tends to induce definitive mutations such as deletion-type mutations.
(4) In addition, because ion beam-based induction of mutations allows sufficient mutations to be induced even with low-dose irradiation, it can be used as a means of inducing mutation in the minimum number of genes (the so-called one-point induction of mutation).

This characteristic can be utilized to achieve stepwise breeding, wherein breeding goals are attained through repeated irradiation. One example of stepwise breeding is the abovementioned development of an ideal single-flower

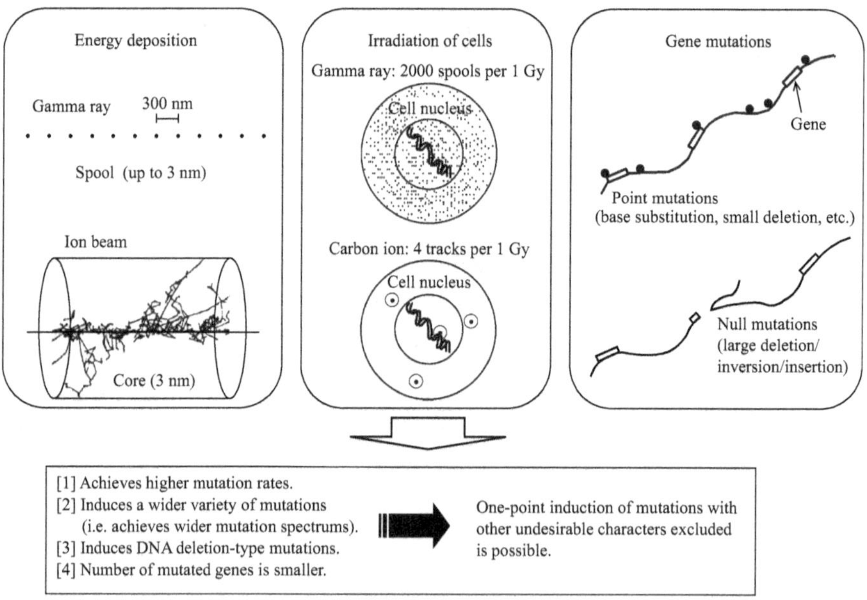

Fig. 13.7 Characteristics of ion beam-induced mutation

chrysanthemum named “Aladdin 2,” through ion beam irradiation of “Aladdin,” which is a mutant created with ion beam irradiation.

From the standpoint of basic research, it can be said that:

(5) One of the main advantages of ion beam-induced mutations is the fact that it allows induced major mutations including deletion-type ones to be utilized to isolate mutated genes. Arabidopsis genes that concern UV-B light resistance, as well as genes that concern flower petal forming and pigment accumulation, have been isolated from Arabidopsis mutants by identifying deleted parts through genome analyses. In the future, planned efficient breeding of crops may become possible as causes of mutations in crops are identified.

Ion beam breeding has been utilized mostly in the area of breeding flowering plants. However, considering the characteristics of ion beam breeding, it is considered that ion beam breeding will greatly assist in component improvement of crops, in improving disease and environmental resistance of crops and in meeting environmental remediation needs. As mentioned above, ion beam irradiation experiments for plant mutation are being conducted at JAEA, RIKEN, the Wakasa Wan Energy Research Center, and the National Institute of Radiological Sciences (NIRS), and groups at private sector companies, universities, and public research organizations in Japan are pursuing research and development in ion beam irradiation. In addition, research on radiation-based induction of mutations in breeding, i.e., radiation breeding, has become very active in other countries, especially in Asian countries. For example, chronic gamma ray irradiation facilities (such as gamma greenhouses) are being constructed one after another in Thailand, Malaysia, and South Korea. Studies on heavy ion beam-based breeding have been pursued for many years in China, and utilization of proton beam accelerators in breeding studies is being planned in South Korea. Mutation breeding is a time-consuming task, but it is expected that it will contribute to long-term environmental preservation and food security efforts by virtue of its capability to create new genetic resources.

Exercises

1. State the difference between induction of mutations using ion beams and induction of mutations using gamma rays, electron beams, etc. and the reason for the difference.
2. Explain about possible applications of ion beam irradiation to induce mutations in plants for breeding purposes.
3. Research topical studies on ion beam breeding and report the result.

References

1. Muller MJ (1927) Artificial transmutation of the gene. Science 66:84–87
2. Stadler LJ (1928) Mutations in barley induced by X-rays and radium. Science 68:186–187
3. Joint FAO/IAEA Programme, Database of mutant variety and genetic stocks. http://mvgs.iaea.org/default.aspx
4. Tanaka A, Hase Y (2010) Applications to biotechnology: ion-beam breeding of plants. In: Hatano Y, Katsumura Y, Mozumder A (eds) Charged particle and photon interactions with matter -recent advances, applications, and interfaces. CRC Press/Taylor & Francis, Boca Raton
5. Blakely EA (1992) Cell inactivation by heavy charged particles. Radiat Environ Biophys 31:181–196
6. Hirono Y, Smith HH, Lyman JT, Thompson KH, Baum JW (1970) Relative biological effectiveness of heavy ions in producing mutations, tumors, and growth inhibition in the crucifer plant. Arabidopsis Radiat Res 44:204–223
7. Mei M, Deng H, Lu Y, Zhuang C, Liu Z, Qiu Q, Qiu Y, Yang TC (1994) Mutagenic effects of heavy ion radiation in plants. Adv Space Res 14:363–372
8. Smith HH (1972) Comparative genetic effects of different physical mutagens in higher plants. In: Induced mutations and plant improvement. International Atomic Energy Agency, Vienna, pp 75–93
9. Tanaka A, Shikazono N, Yokota Y, Watanabe H, Tano S (1997) Effects of heavy ions on the germination and survival of Arabidopsis thaliana. Int J Radiat Biol 72:121–127
10. Hase Y, Yamaguchi M, Inoue M, Tanaka A (2002) Reduction of survival and induction of chromosome aberrations in tobacco irradiated by carbon ions with different LETs. Int J Radiat Biol 78:799–806
11. Shikazono N, Tanaka A, Kitayama S, Watanabe H, Tano S (2002) LET dependence of lethality in Arabidopsis thaliana irradiated by heavy ions. Radiat Environ Biophys 41:159–162
12. Shikazono N, Yokota Y, Kitamura S, Suzuki C, Watanabe H, Tano S, Tanaka A (2003) Mutation rate and novel tt mutants of Arabidopsis thaliana induced by carbon ions. Genetics 163:1449–1455
13. Shikazono N, Suzuki C, Kitamura S, Watanabe H, Tano S, Tanaka A (2005) Analysis of mutations induced by carbon ions in Arabidopsis thaliana. J Exp Bot 56:587–596
14. Nagatomi S, Tanaka A, Kato A, Watanabe H, Tano S (1997) Mutation induction on chrysanthemum plants regenerated from in vitro cultured explants irradiated with C ion beam. JAERI-Rev 96–017:50–52
15. Okamura M, Yasuno N, Ohtsuka M, Tanaka A, Shikazono N, Hase Y (2003) Wide variety of flower-color and -shape mutants regenerated from leaf cultures irradiated with ion beams. Nucl Instr Meth Phys Res B 206:574–578
16. Degi K, Morishita T, Yamaguchi H, Nagatomi S, Miyagi K, Tanaka A, Shikazono N, Hase Y (2004) Development of the efficient mutation breeding method using ion beam irradiation. JAERI-Rev 2004–025:48–50
17. Abe T, Matsuyama T, Sekido S, Yamaguchi I, Yoshida S, Kameya T (2002) Chlorophyll-deficient mutants of rice demonstrated the deletion of a DNA fragment by heavy-ion irradiation. J. Radiat Res 43(Suppl):S157–S161
18. Kishinami I, Tanaka A, Watanabe H (1994) Mutations induced by ion beam (C5+) irradiation in barley. JAERI TIARA Ann Report 3:30–31
19. Bae CH, Abe T, Matsuyama T, Fukunishi N, Nagata N, Nakano T, Kaneko Y, Miyoshi K, Matsushima H, Yoshida S (2001) Regulation of chloroplast gene expression is affected in ali, a novel tobacco albino mutant. Annals Bot 88:545–553
20. Yamaguchi H, Hase Y, Tanaka A, Shikazono N, Degi K, Shimizu A, Morishita T (2009) Mutagenic effects of ion beam irradiation on rice. Breed Sci 59:169–177
21. Shikazono N, Tanaka A, Watanabe H, Tano S (2001) Rearrangement of the DNA in carbon ion-induced mutants of Arabidopsis thaliana. Genetics 157:379–387

22. Kazama Y, Saito H, Fujiwara M, Matsuyama T, Hayashi Y, Ryuto H, Fukunishi N, Abe T (2007) An effective method for detection and analysis of DNA damage induced by heavy-ion beams. Biosci Biotechnol Biochem 71:2864–2869
23. Tanaka A (1999) Mutation induction by ion beams in Arabidopsis. Gamma Field Symp 38:19–27
24. Tanaka A, Sakamoto A, Ishigaki Y, Nikaido O, Sun G, Hase Y, Shikazono N, Tano S, Watanabe H (2002) An ultraviolet-B-resistant mutant with enhanced DNA repair in Arabidopsis. Plant Physiol 129:64–71
25. Hase Y, Trung KH, Matsunaga T, Tanaka A (2006) A mutation in the uvi4 gene promotes progression of endo-reduplication and confers increased tolerance towards ultraviolet B light. Plant J 46:317–326
26. Sakamoto A, Lan VTT, Hase Y, Shikazono N, Matsunaga T, Tanaka A (2003) Disruption of the AtREV3 gene causes hypersensitivity to ultraviolet B light and γ-rays in Arabidopsis: Implication of the presence of a translation synthesis mechanism in plants. Plant Cell 15:2042–2057
27. Sakamoto AN, Lan VTT, Puripunyavanich V, Hase Y, Yokota Y, Shikazono N, Nakagawa M, Narumi I, Tanaka A (2009) A UVB-hypersensitive mutant in Arabidopsis thaliana is defective in the DNA damage response. Plant J 60:509–517
28. Takahashi S, Sakamoto A, Sato S, Kato T, Tabata S, Tanaka A (2005) Roles of Arabidopsis AtREV1 and AtREV7 in translesion synthesis2. Plant Physiol 138:870–881
29. Kitamura S, Shikazono N, Tanaka A (2004) TRANSPARENT TESTA 19 is involved in the accumulation of both anthocyanins and proanthocyanidins in Arabidopsis. Plant J 37:104–114
30. Hase Y, Tanaka A, Baba T, Watanabe H (2000) FRL1 is required for petal and sepal development in Arabidopsis. Plant J 24:21–32
31. Hase Y, Fujioka S, Yoshida S, Sun G, Umeda M, Tanaka A (2005) Ectopic endoreduplication caused by sterol alteration results in serrated petals in Arabiopsis. J Exp Bot 56:1263–1268
32. Rahman A, Nakasone A, Chhun T, Ooura C, Biswas KK, Uchimiya H, Tsurumi S, Baskin TI, Tanaka A, Oono Y (2006) A small acidic protein 1 (SMAP1) mediates responses of the Arabidopsis root to the synthetic auxin 2,4-dichlorophenoxyacetic acid. Plant J 47:788–801
33. Oka-Kira E, Tateno K, Miura K, Haga T, Hayashi M, Harada K, Sato S, Tabata S, Shikazono N, Tanaka A, Watanabe Y, Fukuhara I, Nagata T, Kawaguchi M (2005) klavier (klv), a novel hypernodulation mutant of Lotus japonicus affected in vascular tissue organization and floral induction. Plant J 44:505–515
34. Magori S, Oka-Kira E, Shibata S, Umehara Y, Kouchi H, Hase Y, Tanaka A, Sato S, Tabata S, Kawaguchi M (2009) TOO MUCH LOVE, a root regulator associated with the long-distance control of nodulation in Lotus japonicas. Mol Plant Microbe Interact 22:259–268
35. Okamura M, Tanaka A, Momose M, Umemoto N, da Silva JAT, Toguri T (2006) Advances of mutagenesis in flowers and their industrialization. In: da Silva JAT (ed) Floriculture, ornamental and plant biotechnology, vol I. Global Science Books, Isleworth, Middlesex, pp 619–628
36. Ueno K, Nagayoshi S, Hase Y, Shikazono N, Tanaka A (2003) Effects of ion beam irradiation on the mutation induction from chrysanthemum leaf discs culture. JAERI-Rev 2003–033:52–54
37. Nagayoshi S (2003) Radiation breeding in Kagoshima prefecture -breeding of 'chrysanthemum with a few axillary flower buds' by ion beams. Radiat Indust 98:10–16
38. Kondo E, Nakayama M, Kameari N, Tanigawa N, Morita Y, Akita Y, Hase Y, Tanaka A, Ishizaka H (2009) Red-purple flower due to delphinidin 3,5-diglucosides, a novel pigment for *Cyclamen* spp., generated by ion-beam irradiation. Plant Biotechnol 26:565–569
39. Nakai H, Watanabe H, Kitayama A, Tanaka A, Kobayashi Y, Takahashi T, Asai T, Imada T (1995) Studies on induced mutations by ion beam in plants. JAERI-Rev 19:34–36
40. Kishinami I, Tanaka A, Watanabe H (1996) Studies on mutation induction of barley by ion beams. Abstracts of the fifth TIARA research review meeting, vol 165, p 6

41. Hamada K, Inoue M, Tanaka A, Watanabe H (1999) Potato virus Y-resistant mutation induced by the combination treatment of ion beam exposure and anther culture in Nicotiana tabacum, L. Plant Biotechnol 16:285–289
42. Ishii K, Yamada Y, Hase Y, Shikazono N, Tanaka A (2003) RAPD analysis of mutants obtained by ion beam irradiation to hinoki cypress shoot primordial. Nucl Instr Meth Phys Res B 206: 570–573
43. Inoue M, Watanabe H, Tanaka A, Nakamura A (1993) Interspecific hybridization between Nicotiana gossei Domin and N.tabacum L., using $^{4}He^{2+}$-irradiated pollen. TIARA Ann Rep 3:44–45

Further Readings

44. Komai F, Shikazono N, Tanaka A (2003) Sexual modification of female spinach seeds (*Spinacia oleracea L.*) by irradiation with ion particles. Plant Cell Rep 21:713–717
45. Maekawa M, Hase Y, Shikazono N, Tanaka A (2003) Induction of somatic instability in stable yellow leaf mutant of rice by ion beam irradiation. Nucl Instr Meth Phys ResB 206:579–586

Chapter 14
Experiments of Local Irradiation of Cells with Heavy Ion Microbeams

Yasuhiko Kobayashi

14.1 Effects of Heavy Ion Irradiation of Living Organisms

Ion beams have a dose distribution with a high degree of concentration because they travel nearly straight in objects and have a Bragg peak. In addition, the penetration depths of ion beams in objects can be precisely controlled by adjusting the beam energy. Furthermore, heavy charged particle beams (ion beams of elements that are heavier than carbon, hydrogen, etc.) have a high LET (linear energy transfer, the rate of energy deposition along the track of an ionizing particle, usually expressed in keV/μm) value and accordingly are called "high-LET radiation." High-LET radiation has a stronger cancer cell-killing effect than low-LET radiation such as X-rays, gamma rays, and proton beam irrespective of the cell cycle and oxygen density.

Because these physical and biological characteristics of ion beams allow them to be utilized to treat cancers effectively through ion beam irradiation from outside the body with minimal damage to the normal tissues and such treatment is effective on cancers on which the conventional radiotherapy is not effective, heavy particle beam-based cancer treatment is attracting increasing attention as a promising means of treating cancers [1] (see Chap. 17, "Medical Utilization of Radiation"). In the area of radiation-induced mutation-based breeding, too, ion beams are superior to X-rays and gamma rays in the sense that they allow mutations to be induced more efficiently, allow a wider variety of mutations to be induced, and allow mutations to be induced selectively with good characters of the parent breed retained (see Chap. 13, "Ion Beam-Induced Mutation in Plants").

It is considered that these special biological effects of ion beams are attributable to the highly localized ionization with extremely high density that occurs along the

Y. Kobayashi (✉)
National Institutes for Quantum and Radiological Science and Technology, Gunma, Japan
e-mail: kobayashi.yasuhiko@qst.go.jp

H. Kudo (ed.), *Radiation Applications*, An Advanced Course in Nuclear Engineering 7, https://doi.org/10.1007/978-981-10-7350-2_14

ion tracks in the irradiated substance, but there are many unknown aspects about the processes of induction of mutations and effective killing of cells. In the area of radiation protection, too, there is no reliable basic data that can serve as a basis of the radiation weighting factor (W_R), and systematically obtaining findings about biological effects of ion beams including heavy particle beams and cosmic rays used for medical treatment is an urgent task.

Recently, a phenomenon called the "bystander effect" was discovered wherein "bystander cells," which are cells that have not been irradiated with radiation but are located near cells that have been irradiated, show a response as if they have been irradiated [2, 3]. Details of the mechanism of this effect are yet to be known, but it is considered that the effects of the irradiation are transmitted to bystander cells through diffusion of intercellular signaling substances (such as nitric oxide (NO)) released by irradiated cells in the culture solution or through "gap junctions," which are minute channels connecting adjacent cells. It has been confirmed that the bystander effect can cause a wide range of radiation effects in unirradiated cells including apoptosis (suicide of cells), chromosomal aberrations, mutations, genetic instability (a phenomenon wherein delayed aberrations occur in cells that have survived irradiation after several times of apparently normal cell division following the irradiation), facilitation of propagation, induction of differentiation, and acquisition of radiation resistance. If the effects transmitted as a result of the bystander effect are limited to radiation damage such as cell death and DNA damage, it may be possible to call the bystander effect the "cry-in-sympathy effect" or the "forceful-double-suicide effect." However, it has recently been found that the bystander effect also contributes to vital reactions that work defensively against radiation such as the radioadaptive response [4], which is a phenomenon indicating that the harmful effects of radiation may be attenuated by a priming low radiation dose.

Because ion beams have a high LET value and impart energy to objects very dispersively both in time and space, even if a living body is irradiated with relatively high-dose (or high-dose rate) ion beams, some of the cells of the living body will remain unirradiated, which will make the effects of the bystander effect non-negligible. However, even with low-LET radiation such as X-rays and gamma rays, a situation will arise in which only part of the cells of the living body is irradiated instead of even irradiation of all cells when the dose (rate) is sufficiently low, because of the discrete nature of the ionization event. In such a case, interactions between irradiated cells and nonirradiated cells (bystander cells) will form the responses of cell population or living tissues or the response of the living body as a whole to the radiation. Therefore, to understand what effects will be caused in living bodies by a very low dose of radiation and what cellular level response phenomena are involved in such effects, it is essential to obtain cellular level and molecular level understanding of the mechanism of the bystander effect. However, with conventional random ion beam irradiation methods, it is not possible to have control regarding which parts to irradiate in individual cells or the number of cells to irradiate, and consequently the degree of irradiation achieved varies between cells. In addition, with the random ion irradiation, it is difficult to analyze irradiated and unirradiated cells separately, and consequently effects of irradiation may be mistaken for experiment errors.

If precise selective microbeam irradiation of cells is realized, it will become possible to break away from analyses of "effects of irradiation as average values," which have been the only mode of analyses possible with the conventional irradiation, and pursue analyses of true biological effects of radiation on individual cells. That is, it will become possible to separately analyze the direct effects of irradiation on the irradiated cells and the indirect effects of irradiation caused to bystander cells as a result of the bystander effect. In addition, pinpoint irradiation of specific parts of cells will allow (1) direct analysis of the molecular level mechanisms controlling the series of cell responses that cause apoptosis, etc., direct analysis of the intracellular and intercellular signal transmission, and direct analysis of the process that invokes the restoration mechanism through detection of the damage caused to molecules of living bodies by irradiation and (2) direct analysis of relationships between the distribution of the localized energy deposition in the track structure of energetic heavy charged particles and unique characteristics such as the beam-type dependency of the biological effects caused.

It is also possible to focus ion beams into microbeams with diameters equal to or less than the cell size and selectively irradiate individual tissues, organs, or cells with the microbeams from outside the living body to selectively kill or inactivate them. Such a radio-microsurgery technology can be utilized as a new living body function analysis method for the option of analyzing effects caused in living bodies by local irradiation without surgically taking out specific tissues or organs.

Currently, experiments to selectively irradiate single cells with X-ray microbeams and particle microbeams are being conducted at six facilities in Japan including the Takasaki Advanced Radiation Research Institute of JAEA and at least eight facilities in countries such as the United Kingdom, the United States, Germany, France, Italy, and China. The total number of facilities capable of conducting such experiments will exceed 30 if the facilities that are currently being planned are included [5–7]. This chapter outlines the history of past attempts to achieve selective irradiation of single cells and outcomes of recent heavy ion microbeam-based biological studies.

14.2 History of Experiments of Local Irradiation of Cells with Microbeams

Experiments of pinpoint irradiation of cells have 2 historical meanings. Such experiments were originally devised as a means of conducting radio-microsurgery experiments to examine functions of cell structures by deliberately damaging part of them and later started to be used also as a means of identifying the most radiation-sensitive parts of cells.

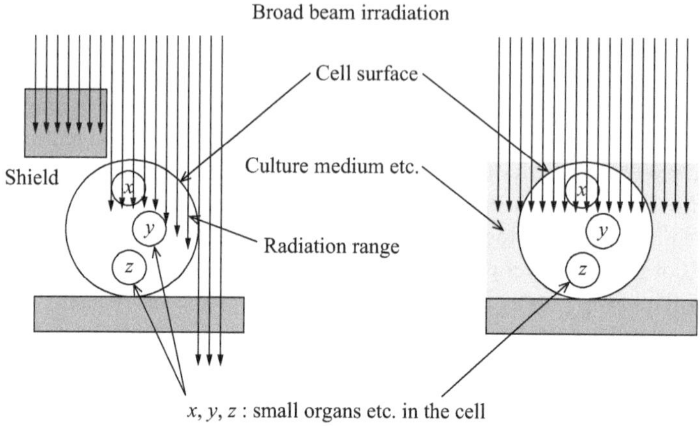

Fig. 14.1 Conceptual diagram of local irradiation with particle beams with invariable ranges (such as α-particles) using a shield

14.2.1 Early Years of Local Irradiation Experiments

According to the summary presented by R.E. Zirkle [8], the first experiments were non-microbeam-based ones in which only the surfaces of such samples as *Pteris multifida* (fern) spores, epidermal cells of onions, sea urchin eggs, and *Zygnemataceae* (charophyte) were irradiated with radiations whose ranges were known such as α-particles. These were followed by an experiment in which only portions of the samples were irradiated with α-particles, X-rays, or ultraviolet rays with the other portions of the samples shielded with an appropriate shield. This experiment showed that the effects on cell division varied depending on the irradiated portion. Figure 14.1 shows a conceptual diagram of these irradiation methods.

For example, according to P. Vintemberger's report published in 1929 on his study in which only portions of frog eggs were irradiated with X-rays and the mutation induction effect of the irradiation was examined, X-ray irradiation of only the cytoplasm induced very few mutations, while X-ray irradiation of the cell nuclei produced an effect that was equivalent to the effect of X-ray irradiation of the whole fertilized eggs.

In Japan, too, an experiment was conducted in 1939 by Mitsuo Miwa at the Japanese Foundation for Cancer Research and his colleagues in which unfertilized *Hemicentrotus pulcherrimus* (sea urchin) eggs with diameters of about 90 μm were irradiated with α-particles (with a penetration depth of 51 μm) of radium C′ (^{214}Po). In the experiment, X-ray irradiation of whole eggs delayed cleavage in all the eggs uniformly, while α-particle irradiation delayed cell division significantly in some eggs, but did not cause any cell division delay in others. In addition, by making irradiation with the α-particle penetration depth increased in steps using an appropriate shield, they discovered that the proportion of adversely affected eggs started increasing sharply as the penetration depth approached the depth that allows the α-particles to reach the cell nuclei.

Considering the fact that these discoveries were made in the days when there was little knowledge on the role of the DNA as a genetic material, there is no doubt that people at the time were very excited about them. However, it was difficult to conduct experiments of selective local irradiation of minute portions of smaller samples with these methods.

14.2.2 First Experiments of (Ultraviolet) Microbeam Irradiation of Living Bodies

Reportedly, the first person who irradiated cells with microbeams is Serge Tchakhotine at the Cancer Research Center of Heidelberg University. His paper entitled "Methodologies for Microscopic Penetrating Irradiation and Cellular Surgery (interim report)" published in 1912 contains an account of the spot ultraviolet irradiation of erythrocytes of frogs conducted by him. The paper is written in German, but he published many other papers in French and Italian on local irradiation of cells. His name, which is originally a Russian name, is spelled as Tschachotin in his papers written in German, Tchahotine in his early papers written in French, and Tchakhotine in all other papers. He developed the first microbeam irradiation equipment for local irradiation of cells. This equipment routes UV (ultraviolet) with a wavelength of 280 nm produced by separating light into spectra using a slit to the transmitted illumination system of a microscope via a prism and condenses light onto the sample using the objective lens for observation and another objective lens made of quartz glass installed in such a way that the 2 lenses face each other across the sample. He conducted many experiments in the 1920s and 1930s while continually improving the experiment system. In 1920, he reported that spot irradiation for a certain period of time of one of the cell nuclei of a fertilized sea urchin egg at the two-cell stage with a 280 nm UV microbeam with a (spot) diameter of 5 μm completely prevented further cell divisions, while irradiation of the cytoplasm resulted in no such effect even with the irradiation time doubled. In 1937, he conducted an experiment in which portions of sea urchin eggs were irradiated with UV microbeams in a respirometer and stated in the experiment report that irradiation of cell nuclei had negligible effect on the oxygen consumption, while selective irradiation of the cytoplasm resulted in dramatic reductions in the oxygen consumption when the irradiated area was large. Figure 14.2 presents the illustration providing an overview of the experiment equipment presented in his 1938 paper (we have not obtained the paper yet) as cited in the summary presented by R.E. Zirkle [8], as modified by us.

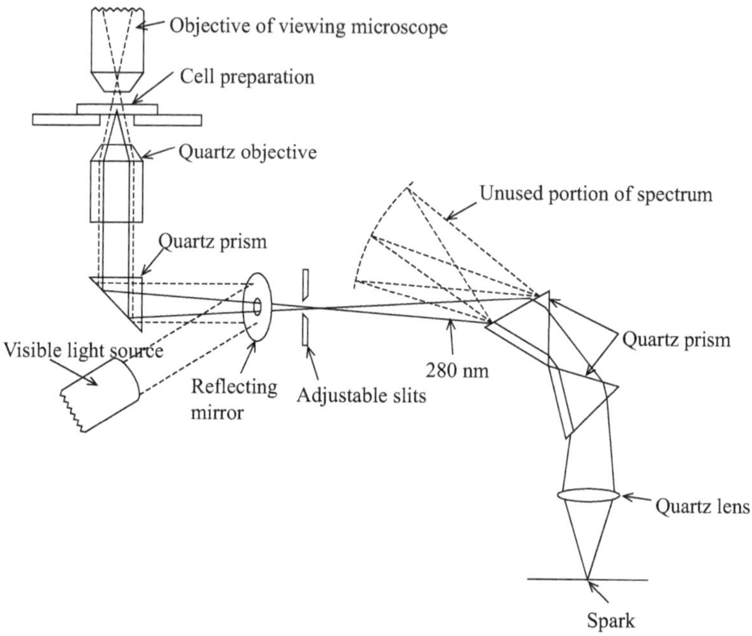

Fig. 14.2 UV microbeam irradiation equipment developed by Tchakhotine

Let us present here the interesting personal history of this great pioneering researcher to you, although this is not of direct relevance to the present topic:

Serge Tchakhotine was born in 1883 in Istanbul as a son of Russian parents, and he soon moved to Odessa with his family. After graduating from the University of Moscow, he obtained degrees in biophysics and zoology in 1907 at Heidelberg University. After completing the abovementioned paper on microsurgery, he returned to Russia in 1912 and became an assistant of Pavlov, who is famous for his discovery of conditioned reflex. Tchakhotine left Russia after the Russian Revolutions and conducted experiments of local irradiation of cells in Paris and other places while continually improving his UV microbeam irradiation equipment as described above. Meanwhile, he joined the anti-Nazi movement in Germany in 1932. From his experience as a member of the anti-Nazi movement, he thought that sociology and economics should be constructed as a human ethology based on positivistic psychology (biology) and developed his own social psychology based on the theory of conditioned reflex. In his major piece of work published in 1938 entitled "The rape of crowd by political propaganda," he called the mass mind control that takes advantage of fears and anxieties of people in tricky ways such as that used in Hitler's and Mussolini's political publicity campaigns the "psychological rape." This book was prohibited in Germany and Vichy France and was censured in the USSR as an antiestablishment book. He obtained Soviet Union nationality in 1922, but lived in Italy, France, and Germany till 1958, when he returned to the USSR. He passed away in 1973 at the age of 90.

14.2.3 Improvement of UV Microbeams

After the Second World War, R.B. Uretz at the University of Chicago of the United States and his group improved this method and conducted a study in which spindles of cells of newts and spindle fibers of spermatocytes of crane flies were locally irradiated with UV microbeams using a epifluorescence microscope capable of irradiating samples with UV through the objective and the effects of the local irradiation on cell division were examined in an attempt to determine the functions of these minute organs.

In the years that followed, laser light source-based practical microbeam irradiation equipment became available, which dramatically improved the resolution and versatility of UV microbeams. As a result, UV microbeams started to be used widely as a microsurgery experiment technology for studies to analyze functions of living bodies. For example, studies were conducted in which initial-stage embryos of *Drosophila melanogaster* (fruit fly) and *Bombyx mori* (silkworm) were irradiated with UV laser microbeams and the grown larvae and adults were examined for organ deletion, etc. in an attempt to develop a presumptive fate map.

Silkworm eggs are flat ellipsoidal eggs with a major axis diameter of approx. 1.2 mm, a minor axis diameter of approx. 1 mm, and a thickness of approx. 0.6 mm. Each egg is covered with a three-layer egg shell with a total thickness of 20 μm and has a sperm introduction hole called the micropyle at one of the ends in the major axis direction. Because it is easy to figure out the directions of both the anterior-posterior axis and the ventral-dorsal axis of silkworm eggs by their external appearance, silkworm eggs are very suitable for selective microbeam irradiation.

However, as will be described later, in an experiment conducted by Kiguchi et al. at Shinshu University in which fertilized silkworm eggs were locally irradiated with heavy ion microbeams at the "Takasaki Ion Accelerators for Advanced Radiation Application (TIARA)" ion irradiation research facility of the Takasaki Research Institute of the Japan Atomic Energy Research Institute (the present Takasaki Advanced Radiation Research Institute of Japan Atomic Energy Agency), it was confirmed that the effect of the local UV microbeam irradiation of fertilized eggs appeared in a way that was totally different from the effect of the local particle microbeam irradiation of fertilized eggs at the same growth stage. It is considered that this is because UV irradiation causes severe damage to biomembranes and protein as well and consequently hinders the movement of segmentation nuclei in fertilized eggs, while heavy charged particle irradiation only deactivates the genes of the irradiated segmentation nuclei without affecting cell structures such as protein.

14.2.4 Electron and X-Ray Microbeams

In 1957, electron microbeam irradiation equipment was developed which looked like an electron microscope and an optical microscope combined together in such a way that they faced each other. This equipment was designed to focus 30 to 150 kV electrons generated by an electron gun by means of a magnetic lens to a spot as small as 1 μm in diameter, but the equipment had shortcomings including the scattering of the electron beams within the sample which made it unusable on samples other than very thin samples and the excessively high-dose rate.

In 1967, equipment was developed which produces an X-ray microbeam with a diameter of 25 μm by guiding 100 kV X-rays through a long lead glass capillary and irradiates samples with the X-ray microbeam. However, there has been no report presenting results of an experiment of irradiation of cells conducted using this equipment.

This was followed by the development of soft X-ray microbeam irradiation equipment at the Gray Cancer Institute in the United Kingdom which selectively irradiates cultured cells of mammals with soft X-ray microbeam generated by focusing the Bremsstrahlung X-rays generated by irradiating carbon, aluminum, or titanium with electron beams (278 eV C-K, 1.49 keV Al-K, or 4.5 keV Ti-K) using a zone plate. At the Photon Factory of the High Energy Accelerator Research Organization (KEK) in Tsukuba, 5.35 keV monochromatic X-ray microbeam irradiation equipment was developed using synchrotron radiation, and an experiment of local irradiation of cells with small irradiation fields down to a 5 μm square has been started.

14.2.5 Particle Microbeams in the 1950s

Because transformation of particle beams into microbeams allows a far higher spatial resolution to be achieved than the spatial resolutions achievable with electron beams, X-rays, etc., M. Davis et al. at Cambridge University of the United Kingdom developed in the 1950s equipment that reduces α-particles from a ^{210}Po radiation source into a microbeam with a diameter of 1 to 1.5 μm using a collimator comprised of 4 perforated copper plates combined together and locally irradiates cell samples with the microbeam. They confirmed through experiments using the equipment that irradiation of spindle fibers of fibroblasts of chicken at the division stage hindered cell division. However, although combining radioactive substances with a shield is easy, it is strongly restricted by the alpha ray range. In addition, it was difficult to achieve a high-fluence rate because the distance between the radiation source and the sample to be irradiated had to be long for beams to be collimated.

In 1953, Zirkle et al. at the University of Chicago of the United States succeeded in taking out a proton beam projected vertically downward from a 2 MV Van de

Graaff accelerator into the atmosphere through a mica foil with a thickness of 5 μm and transforming the beam into a microbeam using a metal collimator. The collimator they used had a micro-aperture made by making a straight groove in one of the largest surfaces of each of 2 polished metal plates by scratching the surfaces with diamond and joining the 2 metal plates together in such a way that the scratched surfaces face each other. They locally irradiated newt heart cells in various phases of mitosis using this collimator and investigated the effects of the irradiation on the cell division process. As a result, they found that micronuclei (which are a form of chromosomal aberration) had been formed in the parts of the irradiated cells which had been hit by the microbeam.

The Brookhaven National Laboratory (BNL) of the United States succeeded in transforming proton beams, deuteron beams, and He ion beams accelerated to 11 MeV/u with a cyclotron into microbeams. The first cyclotron-based microbeam irradiation equipment was a piece of equipment designed to take out a beam that has passed through an aluminum beam scattering filter into the atmosphere through a variable diameter micro-aperture (minimum diameter, 25 μm) made of a platinum foil with a thickness of 0.4 mm. For the beam shutter, a brass plate that rotates 360 degrees every 16.6 seconds was used. It was possible to set the aperture size to a sector of this brass plate with any given angle between 0 and 345 degrees, and to set the shutter opening time at any given period of time equal to or longer than 0.2 seconds by adjusting the proportion (angle) of the aperture sector and the rotation speed. Introduction of cyclotrons dramatically raised the level of achievable particle acceleration energy and made it possible to monitor the beam strength by having the beam go through a helium gas ionization chamber before it hits the sample. Cyclotrons enabled 22.5 MeV deuteron beams to achieve a range of 2.3 mm in living samples, and it was expected that cyclotrons would make it possible to conduct local irradiation experiments on a wide variety of biological samples.

In the late 1950s, W. Zeman et al. irradiated brains of mice with 22.5 MeV deuteron microbeams with diameters of 1 mm, 250 μm, 75 μm, and 25 μm to simulate the effects of cosmic ray irradiation of brain tissues, based on the estimation that the distance from the track of primary particles of a high-energy cosmic ray to the perimeter of the high-density ionization zone surrounding the track was about 25 μm. Three to 48 days after irradiation, they anatomized the mice, dyed section samples of brain tissues, and examined them for histological changes. The result was an unexpected one. That is, the smaller the beam diameter, the higher the dose that was required to cause a histological change. The threshold for the 1 mm microbeam was 300 Gy, while that for the 25 μm microbeam was 11 kGy, a value that was 2 orders of magnitude greater than the value for the 1 mm microbeam. Incidentally, this mysterious result was recently "rediscovered" in the recovery of normal cells of rat brains irradiated with slit-like X-rays from its lethal effect. That is, a mysterious phenomenon has been reported wherein bar-code-like irradiation of rat brains with a lethal dose of SPring-8 synchrotron radiation X-rays at intervals of approx. 200 μm through slits with a width of approx. 20 μm caused almost no damage to the cells of the irradiated parts contrary to the expectation. The biological mechanism of this phenomenon has not been accounted for yet, but, because the parts irradiated with the slit-like X-rays were located alternately with unirradiated parts in the brain tissue in the irradiation experiment, it is considered that this phenomenon may have been caused by some bystander effect that makes the irradiated and unirradiated cells interact in some way.

Particle microbeams generated by these accelerators developed in the 1950s have greatly contributed to studies on biological effects of radiation including the discovery of the fact that local irradiation of cytoplasm scarcely affects the survival of cells. However, the irradiation experiments that have been conducted are experiments with relatively high doses (fluences) without exception. This is because the hit of a single charged particle could not be detected and controlled, so the beam shutter of each system had to be left open long enough that the variance in the number of particles through the cell was acceptable. For this reason, it was technically difficult to conduct low-dose irradiation experiments. In particular, it was impossible to conduct single-hit-type low-dose irradiation experiments, which are necessary to obtain a clear understanding of the biological effects of low-dose/low-dose-rate radiation.

14.3 Realization of Particle Microbeam Irradiation of Single Cells

In the latter half of the 1980s, the need to obtain a clear understanding of the biological effects of low-dose radiation started to be strongly recognized. Because realization of single-ion irradiation of cells required microbeam irradiation equipment that allows precise control in the number of target cells to hit, the movement for the development of microbeam irradiation equipment for experiments of single-ion irradiation of cells intensified backed by rapid advances in ion detection devices and computer technologies.

14.3.1 First Heavy Particle Microbeams

In the late 1980s, a heavy ion research institute (Gesellschaft für Schwerionenforschung (GSI)) in Darmstadt of the Hessen State of Germany (located near the Frankfurt Airport) succeeded in transforming various heavy ions ranging from carbon to uranium ions accelerated to 1.4 MeV/u with a linear accelerator called Universal Linear Accelerator (UNILAC) into microbeams through a nucleus track hole made in a film (an etch pit with a diameter of 0.7 to 2.0 μm in a mica foil with a thickness of 30 μm) and irradiating samples in air with single ions. This made it possible to irradiate target cells with single ions by installing an inverted optical microscope with its side down at the end of the horizontal beam line, adjusting the target position to the beam spot position while observing the sample through the objective and then turning on/off the beam while counting the penetration by each heavy particle using a semiconductor detector installed behind the target.

Experiments using this equipment were conducted to investigate the effects of irradiation of spores of *Bacillus subtilis* with high-LET heavy ions including ^{58}Ni (4240 keV/μm), ^{120}Sn (7940 keV/μm), and ^{238}U (13,700 keV/μm). The experiments showed that, contrary to the expectation, the inactivation rates were between about 40% (^{58}Ni) and 80% (^{238}U) even when the center of the target spore was hit and the local deposition of very high energy by these high-LET heavy ions did not directly connect with biological effects.

However, because this irradiation procedure was time consuming and the number of cells that could be irradiated was 400 per 10 h at best. As a result, the irradiation equipment was dismantled without being used in a full-fledged experiment of single-ion irradiation of mammalian cells.

14.3.2 Single-Ion Irradiation of Cultured Cells of Mammals

In the 1990s, the Pacific Northwest National Laboratory (PNNL) in the United States developed single-particle irradiation system which used proton, deuteron, ^{3}He, and ^{4}He ions from a 2 MV tandem electrostatic accelerator. This system was designed to transform the beam projected vertically upward into a microbeam using a knife edge-shaped slit, take out the microbeam into the atmosphere through a 5 μm-thick plastic scintillator film, and irradiate from below the sample (cells) pasted on the upper surface of a 3 μm-thick Mylar film attached to the bottom of an irradiation sample container placed in contact with the microbeam outlet. With this equipment, it was possible to replace the objective lens with a small photomultiplier tube and count the ions that have hit cells by measuring the weak light emission pulses generated as a result of the penetration of the plastic scintillator film by single particles.

An experiment was conducted in which this microbeam irradiation system was used to irradiate cultured cells taken from ovaries of Chinese hamsters (CHO-K1 cells) with 3.2 MeV ^{4}He ions at a hit rate of 1 to 5 ions per cell and the frequency of resultant formation of micronuclei, which is a form of chromosomal aberration, was analyzed. The result showed that the frequency increased in proportion to the number of hits. This was the first single-ion irradiation experiment in the world on cultured cells of mammals. This microbeam irradiation system moved to Texas A&M University in the latter half of the 1990s together with Professor L.A. Braby, who had led the experiment. It is said that there are plans to conduct further experiments using the equipment, but details of the plans are unknown.

Following the development of the PNNL single-particle irradiation system, the Gray Cancer Institute (GCI) in the United Kingdom developed microbeam irradiation equipment that used proton and He ions from a 4 MV Van de Graaff

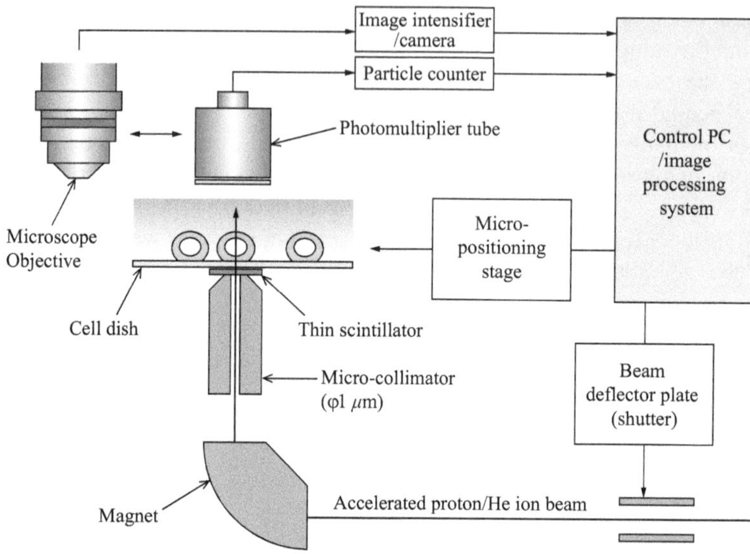

Fig. 14.3 Gray Cancer Institute's proton/He ion microbeam irradiation system

accelerator (Fig. 14.3). The basic composition of the equipment and the procedure for the single-hitting were similar to those of the PNNL. In experiments conducted using this equipment in which only the nuclei of certain cells of human fibroblasts scattered in a petri dish were selectively irradiated (hit) with 5, 10, and 15 ^{3}He ions, apoptosis was induced in unirradiated cells as well.

Columbia University of the United States also developed microbeam irradiation equipment that employed a similar system and conducted an experiment in 1997 in which nuclei of human-hamster hybrid (A_L) cells were irradiated (hit) with 1 to 8 He ions with an LET value of 90 keV/μm that had been generated to simulate radon α-particles. The experiment showed that there was high probability that even a single-hitting of a nucleus with a single α-particle can induce a mutation. The experiment also showed that the hitting of cytoplasm by an α-particle can also induce a mutation and that such a mutation may pose a greater carcinogenesis risk than mutations induced by the hitting of a nucleus, because the hitting of cytoplasm by an α-particle scarcely cause cell death.

In addition, an experiment was conducted in which 5%, 10%, 20%, and 100% of the cells of a cell group were irradiated (hit) with a single α-particle and the number of cells in which mutations were induced among 10^5 surviving cells was counted for each of the four cases. The experiment showed that the numbers for the 10%, 20%, and 100% irradiation cases were about the same (approx. 80). That is, the experiment suggested that the carcinogenesis risk brought by very-low-fluence irradiation may actually be higher than that extrapolated from high-fluence irradiation results.

14.4 Other Countries' System for Irradiating Cells with Heavy Ion Microbeams

As described above, development of particle microbeam-based systems for experiments of irradiation of cells has been actively pursued since the 1950s. Most of the system for irradiation of cells with microbeams that has been developed in the world so far is X-ray-, proton beam-, or He ion-based equipment, and it appears that the only facilities that are currently conducting experiments of irradiation of cells with microbeams of heavy charged particles beyond He ions are JAEA's TIARA, the heavy ion research institute (GSI) in Germany mentioned above, and Technische Universität München of Germany.

14.4.1 GSI's Focused Heavy Ion Microbeam Irradiation System

In the 1990s, the GSI in Germany newly installed focusing-type heavy ion microbeam irradiation system in the UNILAC beam line and started conducting experiments of selective irradiation of cells in air with various heavy ion beams with a wide range of LET values up to 14.1 MeV/u ranging from carbon to uranium. This new microbeam irradiation system was made by modifying equipment that had been used for experiments to analyze the single-event effect on semiconductor elements, etc. so that living cells can be irradiated on the atmosphere side. This system is capable of instantly irradiating many target cells pasted onto a 4 μm-thick thin film of polypropylene by taking out a heavy ion beam that has been focused into a submicron beam using a precision magnetic lens into the atmosphere through a 0.2 μm-thick thin film of silicon nitride (Si_3N_4) and scanning the beam at a high speed with computer-based beam control.

With this irradiation system, it is possible to automatically detect the locations of the target cells through fluorescence microscope-based observation of the nuclei dyed using Hoechst 33,342 and to make pseudo phase contrast image observation using transmitted illumination that uses an LED light source installed in a vacuum chamber. The typical irradiation protocol is repetitions of the following: automatically detecting several tens of cells within the field of view that can be observed through an objective with a magnification of 20, selecting the desired number of cells out of the detected cells, scanning the beam at the selected cells to irradiate them with a certain number of particles, and then moving the precision sample stage on which the irradiation sample container is installed to the next field of view to allow automatic detection of a new target cell group.

The selective single-hitting of target cells is achieved by detecting the secondary electrons generated as a result of the penetration of the thin film of silicon nitride by ions using a secondary electron detector installed in a vacuum chamber and quickly diverting the focused beam from the course concurrently with the detection using an electrostatic beam switch. As errors in the hitting of the target locations in actual cell samples are 2 μm or less including the effect of the scattering in the atmosphere window of the thin film of silicon nitride and the thin polypropylene film of the irradiation sample dish, precise separate hitting of cell nuclei and other parts of cells is possible. However, because the beam line is horizontal, it is necessary to place the irradiation sample in the "upright" position so that it becomes perpendicular to the beam. This means that there are limits to the number of cell samples and the volume of culture medium that can be used, because irradiation sample dish larger than a certain size cannot be used.

The GSI is using this system to perform visualized analyses of DNA damage in cells at locations where they have been hit by heavy ions, analyses of the dynamics of the DNA repair proteins that are accumulated at the locations where cells have been hit by energetic heavy ions, and bystander effect analysis experiments. They irradiated normal human fibroblast (AG01522) with 4.8 MeV/u ^{40}Ar and ^{12}C ions and succeeded in visualizing DNA damage in cells through fluorescent dyeing using 53BP1 (p53 binding protein 1). They also irradiated contact-inhibition-cultured AG01522 cells with broad beams and microbeams of three types of heavy ions which significantly differ from each other in the LET value, namely, 397 MeV/u C ions (LET = 11 keV/μm), 9.8 MeV/u C ions (LET = 1700 keV/μm), and 4.2 MeV/u U ions (LET = 15,000 keV/μm), and analyzed the induction of the CDKN1A protein (p21) by the bystander effect. In addition, they conducted a collaborative experiment with JAEA's TIARA in which the GSI equipment and TIARA's heavy ion microbeam irradiation equipment were used in combination on cells that had been cultured under conditions designed to allow both pieces of equipment to be used on them.

14.4.2 Focused Heavy Ion Microbeam Irradiation Equipment of Technische Universität München

Another piece of equipment in Germany for irradiating cells with heavy ion microbeams is the Superconducting Nanoscope for Applied Nuclear (Kern) Physics Experiments (SNAKE) of Technische Universität München connected to a 14 MV tandem accelerator of the university. This is heavy ion microbeam irradiation equipment capable of focusing heavy ion beams ranging from 28 MeV proton beams to 250 MeV Au ion beams into submicron beams and has been used in the development of high-energy proton beam-based deep lithography techniques and high-resolution scanning transmission ion microscopy (STIM) techniques. The method of taking out beams into the atmosphere and the accuracy of the single-hitting are about the same as those of the GSI equipment. Studies are being conducted in which HeLa cells are irradiated with various 100 MeV heavy ion microbeams of such ions as ^{16}O ions, and the accumulation of the DNA repair protein Rad51 in the parts of the cells which have been hit by ions is observed.

14.4.3 JAEA TIARA's System for Irradiating Cells with Heavy Ion Microbeams

The Takasaki Research Institute of the Japan Atomic Energy Research Institute (the present Takasaki Advanced Radiation Research Institute of Japan Atomic Energy Agency) developed equipment that collimates beams of heavy ions such as ^{12}C, ^{20}Ne, and ^{40}Ar accelerated to 10 to 20 MeV/u with the azimuthally varying field (AVF) cyclotron of TIARA using its aperture system, takes out the collimated heavy ion beams into the atmosphere using metal micro-apertures with diameters between φ5 and 250 μm and thicknesses between 100 and 500 μm, and irradiates selected parts of a sample being observed under a microscope with the extracted heavy ion beams. This equipment was originally used for studies on the development and differentiation processes of insects and on the mechanism of the gravitropism of root tip tissues of plants. Based on this technology, the world's first system for irradiating cells with heavy ion microbeams was constructed which not only was capable of hitting living cells with the beam accurately trained at them by observing them under a microscope while counting the number of ions but also allowed the locations of the cells which have been actually hit by ions and the number of cells that has been hit to be confirmed immediately after the irradiation and the effect of the irradiation to be analyzed by continuously observing selected cells during the culturing after irradiation [9].

Overview of the Heavy Ion Microbeam Irradiation System

Figure 14.4 presents a diagram that provides an overview of TIARA's ion accelerators and various microbeam irradiation systems. The equipment capable of taking out microbeams into the atmosphere and irradiating living biological samples with them is the equipment installed in the AVF cyclotron beam line. Figure 14.5 shows the external appearance of the collimation-type system for irradiating cells with heavy ion microbeams installed at Port HZ1, which is one of the vertical beam ports.

For irradiation of cells in a culture medium, equipment that projects beams vertically upward, such as the X-ray and proton microbeam irradiation equipment developed by the Gray Cancer Institute and Columbia University, is more convenient. On the other hand, for precise focused microbeam irradiation, it is desirable to make the horizontal beam line as long as possible. However, TIARA adopted a downward projection design because it conducts not only irradiation experiments on cultured cell systems but also microsurgery experiments on a variety of living bodies including insects and plants. Because it is prohibited to enter or stay in the irradiation room during irradiation, the irradiation staff leaves the irradiation room and closes the door after setting the sample under the online microscope (an inverted optical microscope) installed directly below the beam line and

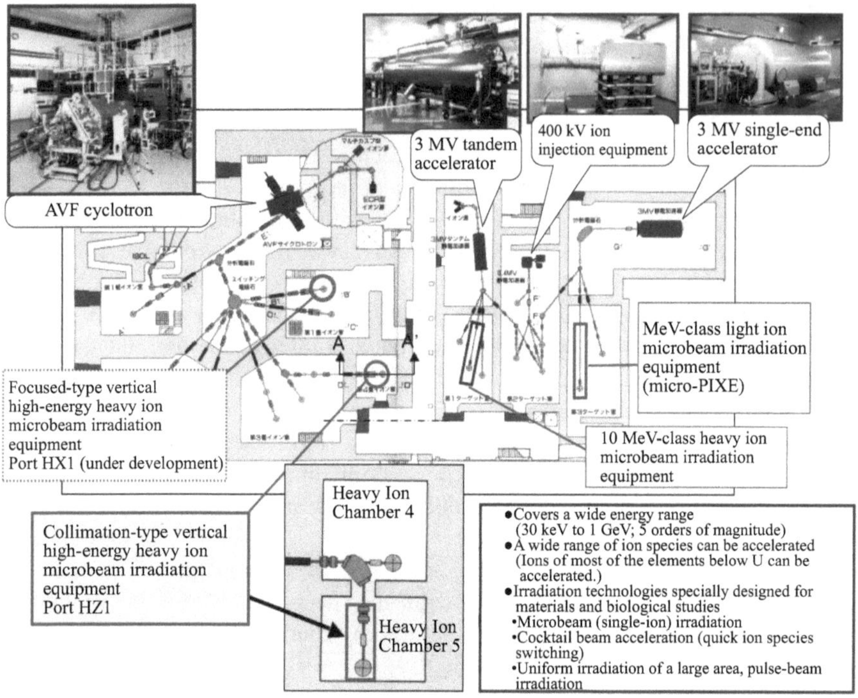

Fig. 14.4 JAEA TIARA's ion accelerators and microbeam irradiation equipment

performs all operations for microbeam training and irradiation remotely from the adjacent preparation room.

Table 14.1 shows the ion species that can be used at present together with their minimum beam diameters, ranges in water, and LET values at sample surface.

The beam diameter can be changed within the range shown in Table 14.1 by replacing the micro-aperture with the desired one selected from among micro-apertures with various diameters. Currently, micro-apertures with diameters of φ5 μm, φ10 μm, φ20 μm, φ40 μm, φ60 μm, φ90 μm, φ120 μm, φ150 μm, φ180 μm, φ200 μm, and φ250 μm are available at TIARA.

The φ5 to 20 μm micro-apertures are 100 μm- or 200 μm-thick micro-apertures, and the φ90 μm and larger micro-apertures are 250 μm- or 500 μm-thick micro-apertures. Collimating an ion beam with a long range requires a thick micro-aperture. However, achievable diameter-thickness ratios are limited because of limitations of the processing technology for making thin and straight holes in hard materials, and this limits the achievable minimum beam diameters of micro-beams of long-range ions such as He and C ions.

φ40 μm and larger micro-apertures have an 8 μm-thick polyimide film attached to them to maintain the vacuum of the beam line, but φ5 to 20 μm micro-apertures have no such film because the vacuum of the beam line is sufficiently maintained by

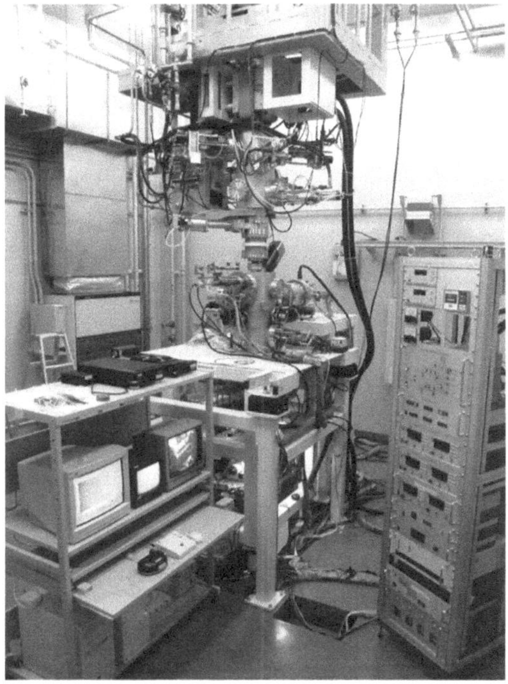

Fig. 14.5 TIARA's collimation-type heavy ion microbeam cell irradiation system

Table 14.1 TIARA's heavy ion microbeams that can be used to irradiate living cells in air

Ion species: energy	Beam diameter	Range in water	LET value at sample surface
$^{4}He^{2+}$:50 MeV	φ40–250 μm	1.8 mm	14–50 keV/μm
$^{12}C^{5+}$:220 MeV	φ20–250 μm	1.2 mm	100–300 keV/μm
$^{20}Ne^{8+}$:350 MeV	φ20–250 μm	0.7 mm	300–800 keV/μm
$^{20}Ne^{7+}$:260 MeV	φ5–250 μm	0.4 mm	370–800 keV/μm
$^{40}Ar^{14+}$:520 MeV	φ5–250 μm	0.3 mm	1130–2500 keV/μm
$^{40}Ar^{13+}$:460 MeV	φ5–250 μm	0.2 mm	1260–2500 keV/μm

the differential evacuation of the vacuum chamber at the end of the irradiation equipment.

Apart from the ion species shown in Table 14.1, the $^{14}N^{5+}$ (total energy, 195 MeV), $^{16}O^{7+}$ (335 MeV), and $^{28}Si^{10+}$ (390 MeV) heavy ions can also be used. However, TIARA has set an upper limit to the number of routinely used ion species, to use the beam time effectively by avoiding unnecessarily frequent ion species and acceleration energy switching.

Ion source switching normally takes about 3 h, but, in the case of the combination of $^{20}Ne^{7+}$ (260 MeV) and $^{40}Ar^{14+}$ (520 MeV), either of the ion beams can be taken out with a switching time of several minutes by using the cocktail beam acceleration technology, that is, by slightly changing the acceleration frequency (RF frequency) of the cyclotron while accelerating both ions simultaneously.

Heavier particles such as $^{56}Fe^{15+}$ (400 MeV), $^{84}Kr^{20+}$ (520 MeV), and $^{129}Xe^{26+}$ (560 MeV) can also be used, but detecting these ions after their penetration of cell samples is difficult because their ranges in samples are short. It is expected that use of such ions as $^{84}Kr^{28+}$ (1030 MeV) and $^{129}Xe^{37+}$ (1170 MeV) will become possible when it becomes possible through future improvements in ion sources to easily generate and accelerate multivalent ion species.

Heavy Ion Microbeam Irradiation Procedure

As shown in Table 14.1, the ranges of the currently used heavy ion beams in living body materials (i.e., their ranges in water) are short, and the beams are downwardly projected beams. For this reason, it is necessary to temporarily remove the culture medium when performing track segment irradiation on cells incubated on the bottom of a dish. Therefore, to minimize the time required to train the beam at the target cells and irradiate them, a system was constructed which creates a database of the coordinates of the positions of the cells prior to the irradiation and semiautomatically trains the beam at the target cells and irradiates them using the database. Figure 14.6 shows the procedure used in this system.

First, the coordinates of the positions of all target cells are acquired using an offline microscope installed in the preparation room adjacent to the irradiation room ((1) in Fig. 14.6 (upper left)). It is possible to automatically recognize fluorescence-dyed cells through image processing and to manually specify the desired hitting locations while observing cells with a phase contrast microscope, etc. Multiple offline microscopes including a stereoscopic microscope for specifying the local irradiation target parts of relatively large samples such as eggs and larvae of insects and sprouts of plants are connected with an online microscope installed directly below the beam line in the irradiation room via LAN, so that high-accuracy coordinates of the position of the automatic sample stage can be shared between and reproduced on these microscopes.

After setting the sample under the online microscope installed in the irradiation room, the irradiation staff closes the door of the irradiation room and returns to the preparation room and then starts the irradiation. The sample stage of the online microscope is controlled based on the database of the coordinates of the positions of the targets created in advance, the target cells are moved so that they will be hit by the microbeam in sequence, and then the target cells are automatically (or manually) irradiated with the predetermined number of ions ((2) in Fig. 14.6 (right)). The number of ions that have penetrated through the sample and their energy are detected by the photomultiplier tube and plastic scintillator installed at the location of the objective lens of the online microscope. If the sample is too thick for ions to penetrate through, irradiation is made by first measuring the particle fluence per second and then releasing the beam shutter for a certain period of time. In very-high-dose irradiation experiments in which the number of ions shot at the sample per second exceeds 10,000, too, this beam shutter release time-based control

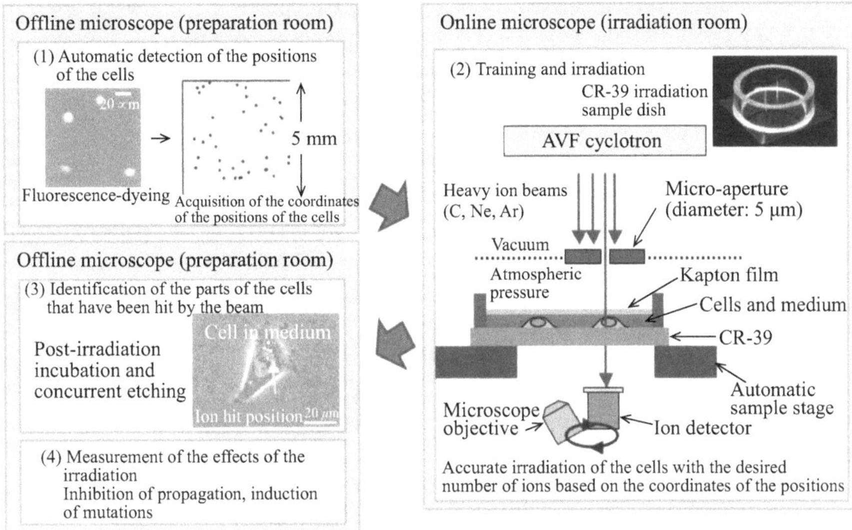

Fig. 14.6 Procedure for heavy ion microbeam irradiation of cells

of irradiation is performed, in order to avoid errors caused by failure of the plastic scintillator and photomultiplier to count some of the ions.

It is possible to acquire accurate data in the form of image on the states of all target cells at commencement of the postirradiation incubation and the locations at which the sample has been hit by ions by etching the side of the CR-39 plastic irradiation sample container (petri dish for cell culturing) for track detection which is opposite to the side on which the cells have been pasted in an alkaline solution at 37 °C for a short period of time immediately after the irradiation and then observing the microscopic images of the individual target cells and ion tracks in a single field of view ((3) in Fig. 14.6 (lower left)). It is also possible to repeatedly observe these cells under the microscope at any given point of time while continuing the postirradiation incubation ((4) in Fig. 14.6 (lower left)).

Development of New Focusing Heavy Ion Microbeam Irradiation Equipment

With the present microbeam generation methods described above, scattering of the beam in the aperture edge is unavoidable, and consequently it is very difficult to reduce ^{12}C and ^{4}He ion beams into a size smaller than a cell nucleus. Therefore, it is not likely that significant improvements in the accuracy of the hitting of cells with such ions as ^{20}Ne and ^{40}Ar ions will be achieved with these microbeam irradiation methods.

For this reason, new heavy ion microbeam irradiation equipment is being developed at TIARA which focuses beams using four-tier quadrupole magnetic lenses in another vertical projection beam line (HX1 port) of the AVF cyclotron (see Fig. 14.4). Unlike the existing collimation-type microbeam irradiation equipment that uses micro-apertures, this equipment will allow microbeams of a wide range of ion species ranging from He ions to heavy ions including long-range high-energy protons (90 MeV) to be used through the use of magnetic lenses without being restricted by the aperture material and the level of precision of processing.

First, $^{20}Ne^{7+}$ (260 MeV) was selected as the ion species to use, and, with the objective of generating a microbeam from a beam whose quality has been improved with single-turn drawing after reducing the energy width ($\Delta E/E$) to about 2×10^{-4} by flat top acceleration, confirmation of the secondary electron generation distribution at the grid edge through microbeam scanning targeting a Cu mesh for electron microscope calibration was made. As a result, it was confirmed that a beam size below φ1 μm had been achieved in vacuum.

To evaluate the shape of this converged microbeam after its extraction into the atmosphere and verify the accuracy of the single-hitting, single-hitting of the 5×5 grid point coordinates on CR-39 was attempted using $^{20}Ne^{7+}$ (260 MeV) and $^{40}Ar^{14+}$ (520 MeV) ions taken out into the atmosphere through a 100 nm-thick thin film of silicon nitride attached to a temporary chamber installed directly below the four-tier quadrupole magnetic lenses. A total of several tens of times of irradiation of the grid points was made, and the success rate of the single-hitting was about 90%. The rates of incorrect hitting, namely, the rate of the hitting of the same point with two or more ions and the rate of failure to hit, were several percent. However, measurement of the deviations from the targets for the points for which single-hitting had succeeded showed that there were dispersions of several μm or more irrespective of the ion travel distance in air. It is considered that these hit location errors are due to not only the scattering in the thin film of silicon nitride but also other factors, and ensuring that the beam is kept appropriately focused is a task for the future. In addition, TIARA is aiming at achieving high irradiation accuracy and high sample treatment efficiency at the same time by developing a system capable of instantly irradiating many target cells by integrating a system that automatically recognizes the target cells within the field of view of the microscope, trains the beam at them, and irradiates them with a high-speed focused beam scanning control system.

14.5 Utilization of Heavy Ion Microbeams in Biology

This section describes, among others, various studies that have been conducted using JAEA TIARA's heavy ion microbeam irradiation equipment.

14.5.1 Utilization of Heavy Ion Microbeams as a Living Body Function Analysis Probe

Studies to analyze the development and differentiation processes of fertilized eggs of *Bombyx mori* (silkworm) and investigate gravitropism-related functions of *Arabidopsis thaliana* roots have been conducted in which heavy ion microbeams were used as a radio-microsurgery technology that deactivates specific tissues and organs of living bodies by locally irradiating them from the outside taking advantage of the characteristic straightness of travel, high depth controllability, and strong biological effects of ion beams.

Development and Differentiation Processes of Silkworms

In the development of embryos, segmentation nuclei and cells from the fertilized nuclei gradually differentiate into cell groups and organs with different roles through repeated divisions. It is considered that locally irradiating specific parts of embryos with microbeams during this process and observing the abnormal conditions caused in the embryos (larvae before hatching) by the irradiation allows analyses to be made to determine what organs will be formed out of the cells of the irradiated parts and how the embryo development process is controlled.

Accordingly, Kiguchi et al. at Shinshu University locally irradiated various parts of silkworm eggs in the initial phase of development (at the cellular blastoderm stage 11–12 h after egg-laying) with ^{12}C ion microbeams with diameters of 60 to 250 μm (220 MeV, LET = 100 to 200 keV/μm) and examined the morphological abnormalities caused in the cuticula (external skeleton) of the embryos. Conveniently, silkworms always lay eggs in such a direction that the median planes of the embryos become horizontal. Therefore, vertically irradiating a sheet onto which a silkworm laid eggs with a beam allows the beam to penetrate through one cell layer surrounding the yolk of an egg in such a way that symmetrical penetration is achieved in the left-right direction with respect to the median plane of the embryo (Fig. 14.7). It is also possible to irradiate only one of the halves by adjusting the particle range by changing the beam energy. Figure 14.8 presents photographs of local heavy ion microbeam irradiation of silkworm eggs that have been laid onto a transparent egg-laying sheet.

As a result, it was confirmed that there were beam spot size- and dose (number of irradiation ions)-dependent close correlations between the parts that had been spot-irradiated and the parts in which morphological embryo abnormalities were induced as shown in Fig. 14.9. On the other hand, no abnormality had been caused by heavy ion irradiation in certain areas on the anterior pole (head) side, posterior pole (tail) side, and back side of the perimeter part of the eggs. It was deduced later that these were parts that were going to differentiate into serosa cells (extraembryonic areas) and the belly side and central parts surrounded by them were the germinal areas that were going to differentiate into embryos. Irradiation of the anterior pole side of

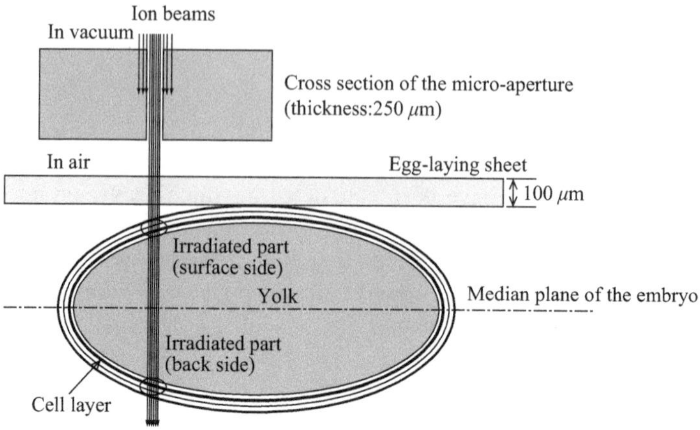

Fig. 14.7 Method of local heavy ion microbeam irradiation of silkworm eggs at the cellular blastoderm stage
One cell layer surrounding the yolk can be irradiated symmetrically in the left-right direction

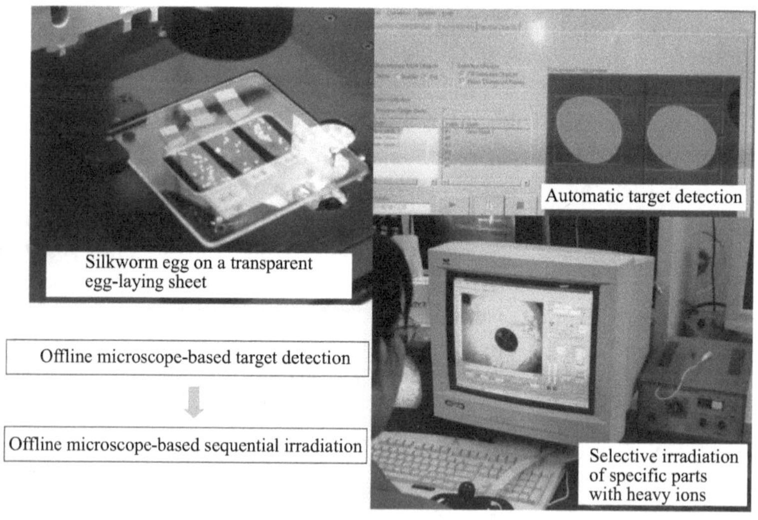

Fig. 14.8 Procedure for locally irradiating silkworm eggs with heavy ion microbeams
First, silkworm eggs laid on a transparent egg-laying sheet (upper left) are automatically detected with the offline microscope (upper right). Then, the sheet with the eggs is set in place in the irradiation chamber, and the desired parts of the eggs are selectively irradiated with heavy ions while confirming their locations with the online microscope (lower right)

these germinal areas caused morphological abnormalities such as deletion, fusion, and duplication in the stemmas, antennas, etc. Irradiation of the central part induced morphological abnormalities in the chest part, and irradiation of the posterior pole

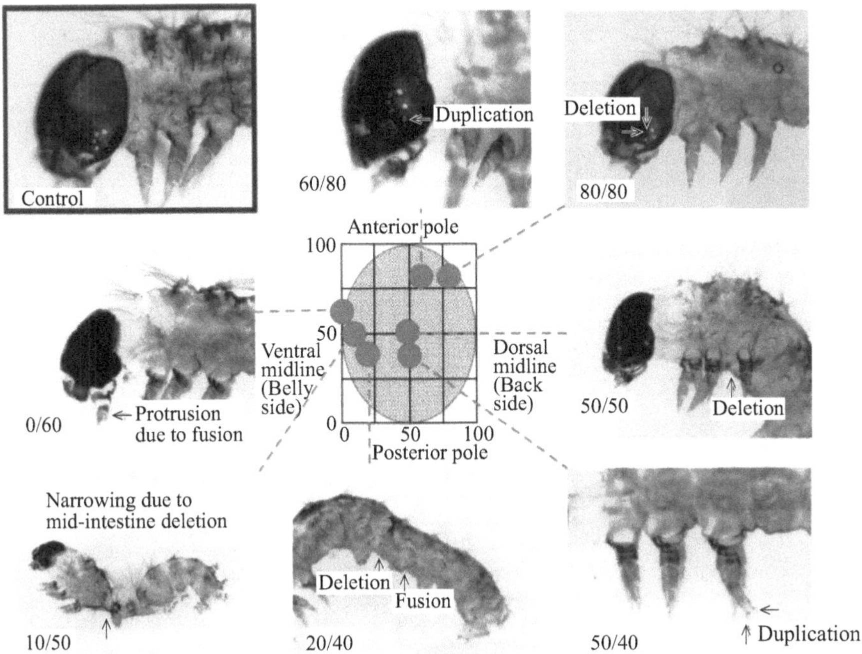

Fig. 14.9 Morphological abnormalities induced in embryos by local heavy ion irradiation of silkworm eggs at the cellular blastoderm stage

side induced morphological abnormalities mostly in the belly and tail parts. By performing local irradiation using an irradiation condition that allows these correlations to be observed most clearly, that is, irradiation of an area with a diameter of 250 μm with ^{12}C ions that correspond to 500 Gy, and analyzing the result in detail, a presumptive fate map of the silkworm eggs at the cellular blastoderm stage was created as shown in Fig. 14.10.

In addition, Fukamoto et al. at Shinshu University analyzed the effects of local heavy ion irradiation of silkworm larvae immediately after their hatching on the occurrence of the hump character (a character that makes speckles on the skin of larvae at the 4th instar and later stages protrude like humps), which is a form of mutation in silkworms, and showed that the hump character occurs as a result of abnormal divisions of dermal cells after hatching and local irradiation with ^{12}C ions higher than 250 Gy suppresses the occurrence of the hump character.

Gravitropism and Hydrotropism of *Arabidopsis Thaliana* Roots

Gravitropism is a character of plants which makes their roots extend downward (the direction of gravity), and it is said that root tips of plants have cells that detect gravity. On the basis of this knowledge, Tanaka et al. at JAEA locally irradiated

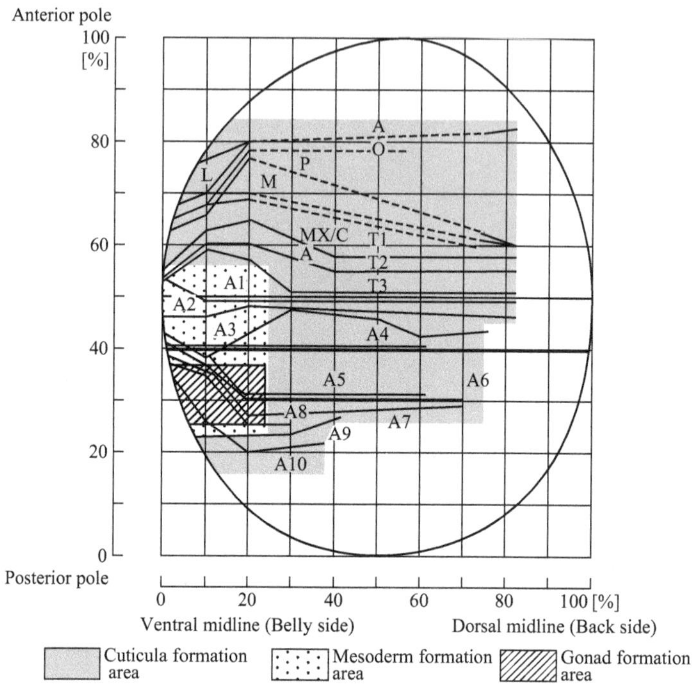

Fig. 14.10 Fate mapping of the silkworm eggs at the cellular blastoderm stage

various parts of root tips of *Arabidopsis thaliana* with ^{12}C ion microbeams with diameters between 40 and 120 μm and analyzed the effects of the irradiation on the gravitropism and showed that not only the root cap columella cells, which are considered to contain starch particles that can be capable of detecting gravity, but also cells that are located in the vicinity of the root cap columella cells and are considered to transmit the gravitropism signals to the root cells responsible for the downward extension of roots are also important for normal gravitropism.

Takahashi et al. at Tohoku University investigated the roles the root tip cells and the root extension area of *Arabidopsis thaliana* play in the hydrotropism of roots by combining local ^{12}C ion irradiation with laser ablation. Local irradiation of the root extension area with a ^{12}C ion microbeam with a diameter of 180 μm corresponding to 500 Gy resulted in temporary suppression of significance of the progression of the hydrotropism, but local irradiation of the columella cells did not suppress the He progression of the hydrotropism. Broad irradiation of the entire plant body with ^{12}C ions suppressed the expression of the *IAA5* gene, but did not suppress the expression of the *MIZ1* (*MIZU-KUSEI 1*) gene, which is vital for the hydrotropism. On the other hand, the result of 400 nm-laser ablation of root tip columella cells through the objective lens of a microscope suggested that the hydrotropism required columella cells. That is, it was suggested that the root tip and the root extension area played different roles in the hydrotropism of roots and the expression of the abovementioned gene was necessary.

Ultra-micromorphological Changes in Skeletal Muscle Cells of Mice

To clarify the mechanisms of the cell death and the cell membrane damage restoration in muscular dystrophy patients, Hiroshi Yorifuji and his group at Gunma University locally irradiated isolated skeletal muscle fibers of mice with ^{20}Ne and ^{40}Ar ion microbeams with a diameter of 20 μm and analyzed the resultant ultra-micromorphological changes based on electron microscope-based observation. Irregularly shaped protrusions and dents in the cell membrane and filament arrangement disturbances in cells were observed immediately after the irradiation, and, with the passing of time, many autophagic vacuoles were observed. This suggests that heavy ion beams alter the ultrastructure of cells and autophagic vacuoles may be playing a role in the restoration from the changes. Currently, local heavy ion irradiation of isolated skeletal muscle fibers from normal mice and various muscular dystrophy model mice and morphological and biological analyses based on the irradiation results are being made.

14.5.2 *Effects of Irradiation of Cells and the Bystander Effect*

An attempt was made to break away from analyses of "effects of irradiation as average values," which have been the only mode of analyses possible with the traditional random-type irradiation methods and directly analyze the direct effects of irradiation on individual cells and the bystander effect on unirradiated cells by irradiating selected target cells with selected numbers of microbeam particles (doses).

CHO-K1 Cells

Funayama et al. at JAEA irradiated cultured cells from ovaries of Chinese hamsters (CHO-K1 cells) they had scattered sparsely in the form of single cells with sufficient spacing between them (several ten cells in a 5 mm square) with a ^{40}Ar ion microbeam with a diameter of 5 μm and confirmed the locations at which the ion particles actually penetrated through the irradiation sample and CR-39 (i.e., the hit locations in the individual target cells) by visualizing them within 20 minutes after the irradiation. They continued the culturing up to 60 h after the irradiation without taking the cells out of the irradiation container (cell culturing petri dish) and then observed the effects of the irradiation on the cells' fissiparity ability by

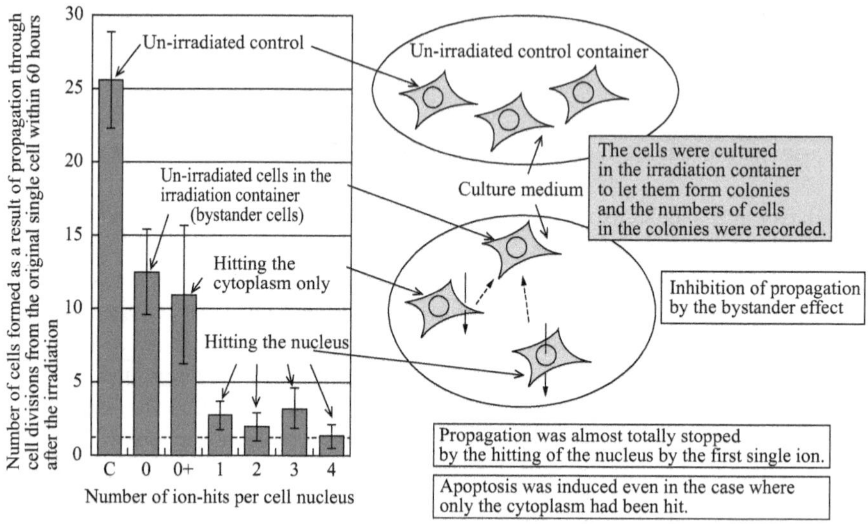

Fig. 14.11 Effects of ^{40}Ar ion irradiation of cultured Chinese hamster cells and the bystander effect

counting the numbers of cells in the colonies formed as a result of propagation through cell divisions from the original single cells (Fig. 14.11). The findings from this experiment are as follows:

1. In the cells that had been hit by ions in the nucleus, the propagation was almost totally stopped by the hitting of the nucleus by the first single ion.
2. In the cells that had not been hit by ions in the nucleus but had been hit in the cytoplasm or cell membrane, the propagation was suppressed to a certain degree.
3. The propagation of the cells that had not been irradiated at all was inhibited only in the case where an irradiated cell had been present in their culture medium in their irradiation container (petri dish), as a result of the bystander effect via the solution.

Normal Human Fibroblast

Furusawa et al. at the National Institute of Radiological Sciences (NIRS) conducted a heavy ion microbeam irradiation experiment on cultures of normal fibroblast taken from human skin incubated on the bottom of a petri dish in such a way that the cells contact each other and do not make further cell divisions for propagation so that interactions between cells in tissues of a living body can be simulated (i.e., high-density contact-inhibition cultures).

They placed cells on the bottom of a petri dish with a diameter of approx. 3 cm in such a way that the cells adhere to the petri dish bottom in contact with each other like tiles laid out without spacing and made microbeam irradiation to hit 2×2 to

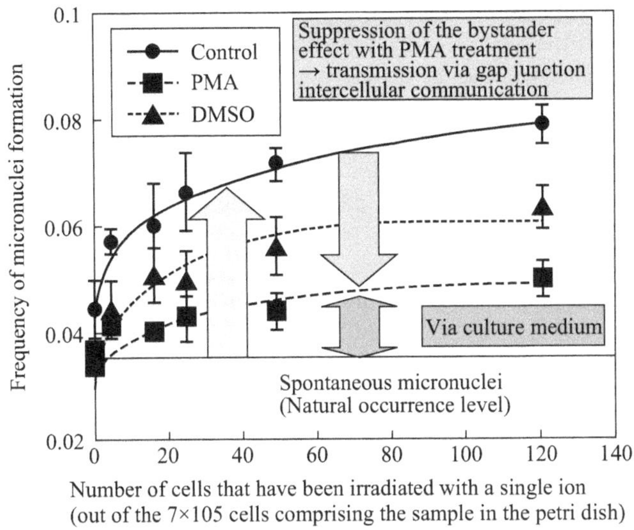

Fig. 14.12 Formation of micronuclei due to the bystander effect via gap junctions in normal human fibroblast

11 × 11 grid points with single ions of ^{20}Ne, ^{40}Ar, etc. The experiment showed that the hitting of the small numbers of grid point cells such as 5 × 5 (25) and 7 × 7 (49) grid point cells, which constituted only a very small portion of the sample comprised of approx. 700,000 cells, with single heavy ions doubled the frequency of formation of micronuclei, which are a form of chromosomal aberration (as indicated with the upward arrows in the left of Fig. 14.12). The only possible explanation for this result is that the bystander effect originating from every single irradiated cell caused DNA double-strand breaks in several hundred unirradiated cells. Increasing the number of ions to hit each cell did not change the result (no dose-response characteristic was observed). Treatment with a chemical that works to close gap junctions (phorbol myristate acetate (PMA)) significantly suppressed the bystander effect (as indicated with the downward arrows in the right of Fig. 14.12). Therefore, it is possible to consider that the part of the bystander effect which was suppressed with PMA is the bystander effect via gap junction intercellular communication and the remainder is the bystander effect due to some soluble signal molecules released into the culture medium.

Hamada et al. conducted an irradiation experiment (a joint study between Gunma University and JAEA) using high-density contact-inhibition cultures of normal fibroblast taken from human skin which were similar to the cultures used in the abovementioned NIRS experiment and confirmed the bystander effect induced by the heavy ion irradiation in relation to the clonogenic potential (colony-forming ability) of cells and the induction of apoptosis [3]. Microbeam irradiation of only several cells with ^{12}C and ^{20}Ne ions out of the sample comprised of approx. 1 million cells cultured in contact with each other without spacing (a mere 0.0003% of the cells comprising the sample) reduced the colony-forming ability of the unirradiated cells (the bystander cells) in the same petri dish (i.e., the postirradiation survival rate) by about 10%.

Interestingly, the rate of reduction of the survival rate of the unirradiated cells did not change even when the number of irradiated cells (i.e., the ratio of the irradiated cells to all cells) was increased. Because the cell cycle phases of almost all cells are the G_1 or G_0 phase under this confluent density-inhibited culturing condition, it is unlikely that there is an effect of differences in the radiation sensitivity due to the dispersion in the cell cycle phase distribution.

Another interesting point is that the changes over time of the response of the cells that had been directly irradiated with heavy ions differed significantly from those of the unirradiated cells that had been subjected to the bystander effect from the directly irradiated cells. In the directly irradiated cells, the phosphorylation of the p53 protein (which is known to be rapidly phosphorylated in response to DNA damage, etc.) increased 2 and 6 h after the irradiation to about the same level, and the induction of apoptosis continued to intensify up to at least 72 h after the irradiation. In the bystander cells, the phosphorylation of the p53 protein 2 h after the irradiation remained at the same level as that before the irradiation and only increased 6 h after the irradiation. The induction of apoptosis temporarily intensified 24 h after the irradiation, but weakened 48 h after the irradiation back to about the same level as that before the irradiation.

In conjunction with this experiment, a DNA microarray-based exhaustive analysis of the changes in the gene expression after the irradiation was made in cooperation with Iwakawa et al. at NIRS. The result showed that there were significant differences between the pattern of changes in gene expression in the directly hit cells and that in the bystander cells, including the fact that, while the expression of several hundred genes had increased in the directly hit cells, the expression of most of them had decreased in the bystander cells. For example, the expression of the interleukin gene, which is a form of cytokine, had increased in the directly hit cells, while the expression of interleukin receptor genes, etc. had increased in the bystander cells. That is, a possibility was suggested that the series of responses observed in the bystander cells in the abovementioned experiments can be the result of the activation of the intracellular signaling pathways by the effect of information transmission factors such as cytokine generated by irradiation of cells on the bystander cells. It is considered that responses of normal cells like these are a kind of self-defense mechanism that maintains homeostasis by minimizing the propagation of abnormal cells [3].

Reproductive Cells of *C. elegans* (Nematode)

Higashitani et al. at Tohoku University conducted an experiment of microbeam irradiation of gonads of *C. elegans* (a nematode species they selected as the model organism). They locally irradiated part of the hermaphroditic *C. elegans* nematode with a ^{12}C ion microbeam with a diameter of 20 μm (a beam whose size is roughly the same as the size of an oocyte) at a dose corresponding to approx. 70 Gy in the rear pachytene nucleus area (rear gonad) of each nematode (which has a front gonad and a rear gonad) and locally irradiated the remainder of the sample with the same microbeam in each nematode's tail section located far from the gonads. Then, 5 h after the irradiation, they counted the number of cells (of the front and rear gonads) in which apoptosis had occurred.

In the part of the sample which had been irradiated in the tail section of each nematode, no increase of significance was observed in the number of front and rear gonad cells in which apoptosis had occurred relative to the unirradiated sections. In the part of the sample which had been irradiated in the rear gonad pachytene nucleus of each nematode, an increase in the number of apoptosis cells was observed only on the irradiated side. These results suggest that the possibility of induction of apoptosis by the bystander effect within an individual of nematode during its gonad cell formation is not strong.

BY-2 Single-Cell Culture of Tobacco

It is known that higher plants, especially herbs, are resistant to radiation. However, there are many unknowns about the mechanism of the radiation resistance of plants. To help resolve these unknowns, Yokota et al. at JAEA developed the first microbeam irradiation experiment system in the world for floating cells. They selected protoplasts of a BY-2 cell culture of tobacco as the model for plant cells and embedded them in a distributed manner in soft agar spread on CR-39. The protoplasts were irradiated with single ions of ^{12}C using a microbeam with a diameter of 20 μm; the number of ions that have hit cells was counted and incubated after the irradiation to evaluate the colony-forming ability.

They conducted a microbeam irradiation experiment using this system on protoplasts of a single-cell culture of tobacco. The result showed that even the irradiation of each cell with 10 ^{12}C ions barely affected the colony-forming ability and the resistance of tobacco cells against radiation as measured in terms of the effect of radiation on the colony-forming ability was about 10 times as high as that of mammalian cells, thereby showing for the first time in the world that plants have radiation resistance on the level of cells. In addition, they showed (for the first time in the world) in a subsequent study on γ-rays irradiation that the yield (per DNA per dose) of double-strand breaks in tobacco cells exposed to from γ-rays was only one third of the yields of those in mammalian cells and yeasts. They also showed in the same study that the number of double-strand breaks that a single cell of tobacco can withstand was five times as large as the number of double-strand breaks that a mammalian cell can withstand and heavy ions are more efficient than γ-rays in inducing double-strand breaks at adjacent locations in the DNA.

14.6 Future of Studies on Utilization of Microbeams

In the future, microbeam-based radio-microsurgery technologies and techniques can serve as new methods for analyzing functions of living bodies by deactivating specific tissues, organs, and/or cells from outside the living body or inducing the appearance of genes from outside the living body. In addition, experiments of selective irradiation of single cells using microbeams and high-accuracy temporal and spatial analyses of the effects of such irradiation can lead to clearer understanding of the network of cell and living body responses. It is expected that such studies and experiments will contribute to the utilization of ion beams in the fields of bioscience, medical science, and bioengineering through development of new cancer treatment methods that achieve the intended effects with lower-dose irradiation, verification of the effects of low-dose or low-dose-rate irradiation of the human body with high-LET radiation such as cosmic rays, advancements in ion beam-based breeding technologies and techniques that utilize major mutations induced by ion irradiation, and development of radiation-based innovative gene modification technologies. To realize such contributions, it is necessary to break away from conventional irradiation experiments that use systems of cultured cells and design irradiation experiments that irradiate cells in states as close as possible to in vivo states in tissues and organs of living individuals. The task after the development of the system for high-speed, high-accuracy heavy ion microbeam irradiation of cells under development at TIARA which uses a focused beam is the development of microbeam technologies for precise local irradiation of the inside of tissues of living bodies and technologies for measuring or estimating the three-dimensional distribution of local energy deposition.

For detailed technical data obtained from experiments that are not presented in this chapter, refer to the respective references listed at the end of this chapter.

Exercises

1. Describe the characteristics of and applications for high-LET radiation comparing them with the characteristics of and applications for X-rays and γ-rays.
2. Cite 3 of the accelerator facilities in the world at which microbeam-based biological experiments are being conducted, and briefly describe their equipment and experiment results.
3. State your view of the biological importance of the radiation-induced bystander effect in radiation protection.

References

1. Hamada N, Hara T, Kataoka K, Sora S and Kobayashi Y (2009) Bcl-2 ha jyuryusisen chiryo no hyoteki to naruka", Hoshasen seibutsu kenkyu 44(2):216–225 (in Japanese); Hamada N, Hara T, Kataoka K, Sora S and Kobayashi Y (2009) Will Bcl-2 become a target of heavy particle beam-based treatment? Radiat Biol Res Commun 44(2):216–225 (in Japanese)
2. Hamada N, Matsumoto H, Hara T, Kobayashi Y (2007) Intercellular and intracellular signaling pathways mediating ionizing radiation-induced bystander effects. J Radiat Res 48(2):87–95
3. Hamada N, Iwakawa M, Imai T and Kobayashi Y (2008) Jyuryushisen yuhatsu baisutanda koka – syosha sareta saibo kara syosha sareteinai shui no saibo eno jyoho dentatsu. Hoshasen kagaku 51(8)31–41 (in Japanese); Hamada N, Iwakawa M, Imai T and Kobayashi Y (2008) Heavy ion-induced bystander effects-intercellular signaling from irradiated cells to their non-irradiated neighbors. Radiol Sci 51(8):31–41 (in Japanese)
4. Matsumoto H, Hamada N, Takahashi A, Kobayashi Y, Ohnishi T (2007) Vanguards of paradigm shift in radiation biology: radiation-induced adaptive and bystander responses. J Radiat Res 48(2):97–106
5. Bigelow AW, Brenner DJ, Garty G, Randers-Pehson G (2008) Single-particle/single-cell ion microbeams as probes of biological mechanisms. IEEE Trans Plasma Sci 36(4):1424–1431
6. Funayama T, Hamada N, Sakashita T, Kobayashi Y (2008) Heavy-ion microbeams – development and applications in biological studies. IEEE Trans Plasma Sci 36(4):432–1440
7. Kobayashi Y, Funayama T, Hamada N, Sakashita T, Konishi T, Imaseki H, Yasuda K, Hatashita M, Taguchi K, Hatori S, Suzuki K, Yamaguchi M, Yamashita S, Tomita M, Maeda M, Kobayashi K, Usami N and Wu L (2009) Microbeam irradiation facilities for radiobiology in Japan and China. J Radiat Res 50(Suppl):A29–A47
8. Zirkle RE (1957) Particle-cell irradiation. In: Curtis HJ, Gray LH, Thorell B (eds) Advances in biological and medical physics. Academic, New York, pp 103–146
9. Funayama T, Wada S, Yokota Y, Fukamoto K, Sakashita T, Taguchi M, Kakizaki T, Hamada N, Suzuki M, Furusawa Y, Watanabe H, Kiguchi K, Kobayashi Y (2008) Heavy-ion microbeam system at JAEA-Takasaki for microbeam biology. J Radiat Res 49(1):71–82

Chapter 15
Food Irradiation: Radiation-Based Sterilization, Insecticidal, and Inhibition of Sprouting Technologies for Foods and Agricultural Produce

Yasuhiko Kobayashi

15.1 What Is Food Irradiation?

Food irradiation is the irradiation of foods and agricultural produce with ionizing radiation to kill bacteria or insects present in them or stop the germination of agricultural produce. It is one of the physical food processing techniques currently in use, which include drying, heating, pressurizing, and freezing. The main purposes of food irradiation are:

1. Preservation of foods and agricultural produce and prevention of damage to them during storage
2. Food hygiene security and prevention of foodborne illness (sanitary)
3. Control of organisms or insects harmful to plants or plant products (phytosanitary)

The effectiveness of food irradiation in achieving these purposes and the soundness of radiation-irradiated foods and agricultural produce in terms of both safety and nutritional suitability have been confirmed since the 1980s by research organizations of countries around the world and the World Health Organization (WHO), the Food and Agriculture Organization (FAO) of the United Nations, and the International Atomic Energy Agency (IAEA) [1], and technologies to kill bacteria in spices, herbs, frozen meats, fish, and shellfish and technologies for quarantine (insecticidal) processing of tropical fruits and citrus and germination stopping technologies for garlic have been put into practical use around the world (see Table 15.1). In recent years, there have been remarkable advances in these technologies in Asia [2–5].

Y. Kobayashi (✉)
National Institutes for Quantum and Radiological Science and Technology, Gunma, Japan
e-mail: kobayashi.yasuhiko@qst.go.jp

H. Kudo (ed.), *Radiation Applications*, An Advanced Course in Nuclear Engineering 7, https://doi.org/10.1007/978-981-10-7350-2_15

Table 15.1 Amounts of irradiated foods in the world (2005) [5]

Country	Item	Amount irradiated (tons)
1. China	Garlic, spices, dried vegetables, etc.	146,000
2. United States	Spices, frozen red meats, fruits, and vegetables	92,000
3. Ukraine	Wheat and barley	70,000
4. Brazil	Spices, fruits, etc.	23,000
5. South Africa	Spices, honey, etc.	18,185
6. Vietnam	Frozen shrimp, fish, and shellfish	14,200
7. Japan	Potatoes	8096
8. Belgium	Frozen frog legs, chickens, etc.	7279
9. South Korea	Spices and dried vegetables	5394
10. Indonesia	Cocoa powder, frozen fish, frozen shrimp, etc.	4011
11. Holland	Spices, dried vegetables, frozen frog legs, etc.	3299
12. France	Frozen poultry meats, frozen frog legs, etc.	3111
13. Thailand	Spices, fermented sausages, dried fruits, etc.	3000
14. India	Spices and onions	1600
15. Canada	Spices	1400
16. Israel	Spices	1300
Others		2929
Total		404,804

The table above was developed from data contained in unpublished material (Reference [5])

However, in Japan, despite the fact that technology to irradiate potatoes with cobalt 60 (Co-60) gamma rays to stop their sprouting was put into practical use for the first time in the world as early as 1974, the scope of application of the technology has not been expanded to cover other foods, and food irradiation has been legally prohibited except for irradiation of potatoes to stop their sprouting. Irradiation of spices to kill bacteria in them, which has been in practical use in other countries, has also been prohibited in Japan.

It is considered that one of the reasons for this closed state of Japan in the area of food irradiation is the lack of exchanges of risk information between the risk management organizations, the risk evaluation organizations, researchers, business entities, and the general public. Most people in Japan do not have an opportunity to actually see irradiated foods or agricultural produce. Nor do they know how food irradiation is made. The author cannot help suspecting that this situation has allowed emotional arguments based on negative images of and prejudices against radiation presented by some citizens' groups and mass media to become dominant.

This chapter outlines the principles of radiation-based sterilization, insect disinfestation and inhibition of sprouting, present states of their application in the area of food and agricultural produce processing, their risks and benefits, methods for detecting irradiated foods, and various countries' methods of food irradiation, to help facilitate objective arguments based on scientific evidence.

15.2 Principles of Food Irradiation

Radiation-based sterilization of foods and agricultural produce takes advantage of the strong ability of radiation to penetrate matters and the vulnerability of most food-contaminating microorganisms to radiation, as does the radiation-based sterilization of medical equipment and pharmaceuticals. Irradiation at doses lower than the sterilization-level dose is effective in killing harmful insects or making them infertile as a means of quarantine processing of tropical fruits and citrus and preventing grain weevils and mites from eating crops. Irradiation at doses lower than these allows the germination or sprouting of potatoes, garlic, onions, etc. to be stopped.

15.2.1 Target Is the DNA

The component of living bodies which is most susceptible to radiation (the target) is the DNA, which is the ultra-giant molecules that bear genetic information. DNA damage is always being caused in the natural world. However, artificial intensive induction of massive DNA damage through irradiation allows divisions and propagation of cells to be stopped. The extent to which this can be achieved depends on both the nature and extent of the DNA damage caused by the irradiation and the ability of the irradiated cells to repair the DNA damage.

Of course, DNA damage can also be caused by heating-based processing such as stewing and boiling. However, while heating causes denaturing (irreversible changes in higher-order structures) or hydrolysis in all living body constituent molecules (i.e., all cell membranes, enzymes (proteins), and genes (DNA)) comprising the cells of the target substance by uniformly raising the temperature of the entire target substance, radiation imparts energy to very limited portions of the target substance in an intensive and random manner and at the atomic level as it penetrates through the target substance and causes damage to DNA molecules such as chemical changes in them and DNA strand breaks only in the cases where the portions to which it imparted energy happen to be DNA or where chemically active species generated by the irradiation act on DNA molecules.

That is, by taking advantage of the intensive and very localized energy impartment by radiation (as opposed to uniform energy impartment to all parts of the target substance) and the fact that the ultra-giant DNA molecules are the Achilles tendon of living organisms in their resistance to radiation, a sufficient level of sterilization effect can be obtained with little increase in the overall temperature of the target substance. Such non-heating-based sterilization can be used for fresh vegetables and fruits; for meats, fish, and shellfish as edibles; and for frozen foods, as well as for pharmaceuticals and natural medications which can be denatured or deactivated by heating and plastic medical equipment and food packaging and containers which are not strong against heat.

The temperature increase caused by irradiation at 10 kGy, which is a sterilization-level dose of gamma rays, is only 2.4 °C in the case of water. Therefore, irradiation at 10 kGy allows frozen foods to be sterilized without defrosting them. In other words, if the very small thermal energy we receive when we take a sip of hot tea were to be given to us in the form of acute whole-body gamma ray irradiation, we would be subjected to a dose that is several times as high as the lethal dose for humans.

15.2.2 Why Are Living Organisms Vulnerable to Radiation?

The reason for this vulnerability of living organisms to radiation lies in their adoption of DNA as the medium for recording their genetic information. The cells of a living organism contain extremely long, compactly folded DNA molecules. For example, if the 46 chromosome DNA molecules contained in a human cell nucleus (whose diameter is approx. 10 μm) were stretched out and joined together, the total length would be approx. 2 m. These ultra-giant DNA molecules record genetic information in their chemical structures. This means that the entire DNA molecule must be accurately replicated every time a cell division is made. For this reason, all living organisms have a mechanism for immediately restoring their important genetic information-carrying DNA molecules to their original states whenever they are damaged by intra-cell physiological chemical reactions, natural radiation, or ultraviolet.

The absorbed dose from irradiation with gamma rays of the same intensity (number of photons) is proportional to the number of electrons in the irradiated substance, and which of the electrons are ionized is determined randomly. Let us make some thinking experiments here. Assume that one DNA molecule that contains a total of 1 billion electrons is irradiated in one experiment and 100 million water molecules (the total number of electrons is approx. 1 billion because each H_2O molecule has ten electrons) are irradiated in another experiment [6]. The number of electrons displaced in the former experiment would be the same as the number of electrons displaced in the latter experiment if the ionizing energy and the absorbed dose are the same. If 1000 chemical bonds in a single DNA molecule were broken simultaneously as a result of irradiation at a given dose, the effect would be significant. However, in the case of decomposition of 1000 water molecules in the group of 100 million water molecules, the proportion is very small.

This situation would not change even if the water molecules with a molecular weight of 18 were replaced with proteins with a molecular weight of 10,000 or 100,000. Even if part of the pool in a cell of an enzyme protein molecule loses its function as a result of a chemical change, the effect will probably be small and the function will be immediately restored. If a protein is denatured or becomes unnecessary, it will be quickly decomposed. If the amount of a necessary protein falls short of the amount required, it will be quickly newly synthesized. However, if many strand breaks occur in a single DNA molecule that carries genetic information, the effect will be fatal to cells unless the strand breaks are repaired immediately and accurately.

15.2.3 Lethal Effects of Radiation and DNA Repair

Although the Earth's biosphere is the place with the least radiation in the galaxies, natural radiation on Earth is damaging the DNA of living organisms every day, thereby posing a threat to their lives. Acquisition of the ability to respire aerobically has made it possible to produce energy efficiently, but oxygen is essentially a dangerous chemical substance with very strong toxicity, and the DNA of living organisms is always under its attack. This is why organisms that have evolved on Earth have acquired the ability to easily restore the DNA damage that is constantly caused by natural radiation and active oxygen species generated by oxygen breathing (which is called the "DNA repair capability"). Indeed, unlike other biopolymers in cells, which are disposed of through metabolic turnovers, DNA molecules are "repaired" by various means to correct chemical changes as they occur. If massive DNA damage occurs in a DNA molecule which is beyond the DNA repair capability, it will become impossible to catch up on the repairing and the cell will die.

The dose of radiation that causes this, or the effect of sterilization by radiation in other words, is a measure of the radiation resistance of the target microorganism and depends on the DNA damage yield (which depends on the radiation chemistry conditions including the temperature, the partial pressure of oxygen, the water activity, and the dose rate) and the DNA repair capability of the microorganism. In addition, the extent to which the DNA repair capability can be utilized depends on the environmental conditions to which the food is subjected after the irradiation. In actual food irradiation, the sterilization effect varies depending on the constituents of the food and the irradiation conditions, but the causes can basically be explained based on the DNA damage yield and the DNA repair capability.

One of the irradiated food detection methods described later is a method that quantifies the magnitudes of DNA damage caused in animal and plant cells contained in the detection target food. In this method, the traces of irradiation will vanish as time passes if the food is "alive" and under conditions that are favorable for the DNA repair capability to function after the irradiation.

Lethal doses of radiation for organisms (due to acute radiation-irradiation) vary widely from several Gy for mammals to several kGy for radiation-resistant microbes, but, roughly speaking, the larger the amounts of DNA contained in individual cells of the organism, the higher the sensitivity of the organism to radiation is. That is, for random irradiations of organisms with the same G value for fatal DNA damage and similar DNA repair capability, the larger the organism, the easier it is to "down" it.

15.3 Purposes and Examples of Practical Utilization of Food Irradiation

Food irradiation is artificial irradiation of foods with radiation to achieve the following purposes:

1. Preservation of foods and agricultural produce and prevention of damage to them during storage
2. Food hygiene security and prevention of foodborne illness (sanitary)
3. Control of organisms or insects harmful to plants or plant products (phytosanitary)

The dose required in food irradiation varies widely depending on the purpose.

15.3.1 Suppression of Germination or Sprouting: 20 to 150 Gy

The tissues of potatoes and garlic which are going to develop into buds are more sensitive to radiation than their other tissues. Therefore, if a potato or garlic is irradiated with an appropriate dose of radiation at the appropriate time after harvest, only the tissues that are going to develop into buds will be prevented from making cell divisions. Such irradiation allows potatoes and garlic to be preserved in a fresh state with their germination halted. The dose required for this is 60–150 Gy for potatoes and 20–150 Gy for onions and garlic [7].

In Japan, irradiation of foods is prohibited by the Food Sanitation Act except for the sprout inhibition of potatoes with gamma rays from Co-60, and the only organization at which such irradiation of potatoes is being performed is the agricultural cooperative of Shihoro Town of Hokkaido (JA Shihorocho). Since the spring of 1974, the agricultural cooperative has been irradiating part of the potatoes harvested in the autumn of the previous year (from August to October) with radiation using their dedicated Co-60 gamma ray irradiation facility (Fig. 15.1) in the spring (from around March to April) prior to shipping them to put them on the market before newly harvested potatoes from Kyushu start becoming available on the market (i.e., in the changeover period). The purposes of this irradiation are to prevent them from sprouting while they are in show windows of retailers in the much warmer Honshu losing their value as merchandise and to prevent the production of a large amount of solanine, which is a toxic substance, in their buds and skin. The amount of potatoes that were harvested in 2007 and irradiated in 2008 to stop their germination was approx. 4000 tons, which represented 10% of the total amount of 2007 production potatoes shipped by JA Shihorocho as raw potatoes (approx. 42,000 tons) and about 0.17% of the total amount of 2007 production potatoes shipped by all agricultural cooperatives and potato farmers in Japan (2.4 million tons). “Raw potatoes” are potatoes that are supplied to retailers and

Fig. 15.1 Co-60 gamma ray irradiation facility of JA Shihorocho of Hokkaido
"1.6 m by 1.3 m by 1 m" size harvesting containers each containing approx. 1.5 tons of potatoes are irradiated while they are slowly moved and turned around the radiation source located at a distance of 5 m from them with the containers. The irradiated potatoes are stored in the containers till their shipment

restaurants via fruits and vegetables markets. None of the irradiated potatoes are shipped as potatoes to be processed into things such as potato chips.

The annual amount of potatoes shipped after being irradiated to inhibit their sprouting was between approx. 7000 and 8000 tons up to 2006 (2005 production potatoes) but decreased to less than 4000 tons in 2007 (2006 production potatoes) as a result of the decision of agricultural cooperatives to ship such potatoes to only those customers who promised to provide labeling on the potatoes that have been irradiated to inhibit their sprouting to allow the consumers to distinguish them (Fig. 15.2). However, according to agricultural cooperatives, there were such reactions from consumers as "they were delicious," "they are convenient because they do not develop buds," and "availability of potatoes that do not develop buds is a big help at this time of the year for small croquette shops like us," although there were also consumers who asked whether they were harmless. These reactions indicate that the practice of labeling irradiated potatoes at shops has rooted and consumers have recognized advantages of high-quality potatoes that do not develop buds. As a result, some wholesalers were motivated to start selling potatoes that have been irradiated to inhibit their sprouting, and the annual amount of potatoes

Fig. 15.2 Potatoes that have been irradiated to stop their sprouting are sold in small bags with a seal attached on which the words "Potatoes treated to inhibit sprouting" and "Irradiated with gamma rays" are printed

shipped after being irradiated started increasing steadily again and reached 4112 tons in 2008 (2007 production potatoes) and approx. 4500 tons in 2009 (2008 production potatoes) [8].

In 2003, the use of maleic hydrazide, which had been sold in Japan under such product names as Erno Liquid and had been used as the agricultural chemical (plant growth regulator) to prevent garlic from sprouting, was suddenly prohibited because it was discovered that the hydrazine contained in it as an impurity was a carcinogen, and it was reported that this caused panic in garlic-producing areas such as Aomori Prefecture in a certain period after the banning. Because garlic totally loses its value as merchandise once it develops buds, the use of agricultural chemicals to prevent it from developing buds was essential for garlic producers to ship garlic steadily throughout the year. In addition, most of the garlic farmers had been doing the pre-shipment conditioning work during the agricultural off-season in the winter after the harvesting and drying, but this became difficult. Because there was no substitute agricultural chemical for maleic hydrazide, it was expected that the banning of the use of maleic hydrazide would result in the concentration of the shipment of garlic in a certain period and resultant sharp drops in garlic prices. In addition, garlic farmers feared that, if they are forced to have a changeover period (the period during which no garlic is shipped between the end of one shipment season and the start of the next shipment season), market shares of imported garlic can increase sharply. The resultant creation of a changeover period would result in sharp increases in the market share of imported garlic.

Irradiation treatment is a very effective means to inhibit sprouting of garlic and has already been put into practical use in China and South Korea. In China, approx. 80,000 tons of garlic were irradiated in 2005 to inhibit their sprouting [2–5]. The abovementioned changeover period problem would have been solved if irradiation of garlic had been approved in Japan soon after the banning of maleic hydrazide. However, because there was no prospect of such approval and there was concern that consumers may fear or protest against "irradiated garlic," garlic producers did not even make an application for approval. Instead, they constructed large refrigerating warehouses and have been using them to prevent harvested and dried garlic from germinating till their shipment by storing them at a temperature of −2 to −3 °C. However, with this approach, the garlic starts developing buds as soon as their temperature returns to the air temperature during the physical distribution process after shipment to deliver them to retailers. The authors are worried that it may take an eternity before irradiation of garlic is approved in Japan, even though garlic producers and consumers in Japan can benefit from this excellent technology.

The authors think that the lesson that must be learned from this example is that it is important to evaluate the effectiveness and safety of new technologies including food irradiation technologies in advance and establish a system in advance for exchanges of information and opinions on risks between administrative organizations, producers, physical distribution companies, and consumers based on mutual trust so that, even if there is no prospect that the technology can be immediately put into practical use, they can be used as soon as they become necessary.

15.3.2 Killing Harmful Insects or Making Them Infertile: 0.1 to 1 kGy

Killing harmful insects present in crops, fresh fruits, and vegetables or making them infertile is an effective means of preventing damage to crops, fresh fruits, and vegetables in storage due to propagation of mites and weevils including grain weevils that eat them and in preventing the entry of exotic quarantine target fruit flies into Japan (agricultural produce quarantine). However, achieving these purposes does not necessarily require the immediate killing of such harmful insects. Preventing their propagation (i.e., making them infertile) suffices for achieving these purposes. Because the life spans of adult insects that get into crops, etc. are not so long, preventing the hatching of the eggs laid and the emergence of the larvae suffices. Immediate killing of adult insects that got into crops, etc. requires a dose of 3–5 kGy, but making them infertile requires only a dose of 1 kGy or less, which is sufficient for deactivating the highly radiation-sensitive reproductive cells [7].

When crops such as rice are bitten by weevils, mold will develop from the bitten parts. Irradiation of crops to prevent weevils, etc. from eating crops also helps in preventing the development of mold and resultant contamination with mold poisons such as aflatoxin.

Traditionally, methyl bromide had been used to kill these harmful insects. However, because the Montreal Protocol designated it as an ozone layer-depleting substance whose use must be totally abolished, developed countries stopped producing it by 2005 and developing countries are also going to do so by 2015. In Japan, the use of methyl bromide has already been prohibited except in the cases where it must be used for plant quarantine to control harmful insects that have not previously occurred in Japan or in the cases where an application for approval of its

Fig. 15.3 Chestnut weevil larvae that have come out of chestnuts by biting holes in the chestnut skin after eating away the contents

use as an "indispensable use" has been made to the competent international body because of the absolute absence of a substitute means and that use has been approved.

At present, the only use of methyl bromide in Japan which has been internationally approved as an "indispensable use" is its use in the fumigation of harvested chestnuts to kill harmful insects called chestnut weevils in them. As many of the chestnuts before harvest already contain laid chestnut weevil eggs, if they are not fumigated with methyl bromide to kill the chestnut weevils, it is almost certain that white fat larvae will come out from them one after another by biting holes in the chestnut skin after eating away the contents (Fig. 15.3). Because the excrements of the larvae do not come out of the chestnuts, there is no way to know whether there are chestnut weevils from the outside until they come out of the chestnuts. Gamma ray irradiation at about 500 Gy is an effective means to kill these troublesome chestnut weevils. With regard to non-radiation-based methods, a method has been developed in which chestnut weevils are killed by subjecting chestnuts to high-pressure carbon dioxide at 30 atms for 30 min. This carbon dioxide gas has already been registered as an agricultural chemical (under the name "high-pressure carbon dioxide gas"), but the method has not been put into practical use because it requires a large pressure vessel and carbon dioxide gas is more expensive than the methyl bromide fumigant [9, 10].

For killing harmful insects in imported crops, fumigation with phosphine (aluminum phosphide agent-based fumigation) is being used in lieu of methyl bromide, but it is anticipated that variations of insects which are resistant to the chemical will occur in the near future.

For killing harmful insects in imported fresh fruits and vegetables, the following treatment methods are being used aside from the phosphine-based fumigation:

1. Low-temperature treatment: 12–20 days at a temperature between about −0.6 and 3 °C

2. Steam heat treatment: 10–50 min in saturated steam at a temperature between 43 and 50 °C
3. Warm water immersion treatment: immersion in warm water at about 46 °C for about 10 min
4. Dry-heating treatment: about 1 h in dry air at about 80 °C

However, all of these methods degrade the quality of fresh fruits and vegetables.

The radiation treatment, which allows harmful insects to be killed without degrading the quality of crops, fresh fruits, fresh vegetables, etc., has recently been recognized internationally not only as a means of killing harmful insects but also as a means of plant quarantine treatment. At the April 2003 conference of the Interim Commission on Phytosanitary Measures (ICPM) held under the International Plant Protection Convention (IPPC), the "Guidelines for the Use of Irradiation as a Phytosanitary Measures" (ISPM#18) were adopted. The Guidelines were further improved and expanded in April 2009 (ISPM#28). International guidelines on the use of radiation in treatment of plants to protect them from the designated harmful insects and items, such as a guideline recommending an absorbed dose of 70 Gy or higher for the prevention of the emergence of Mexican fruit flies, have been established, and irradiation-based quarantine treatment has been put into practical use in the United States for mangoes, papayas, etc. that are brought into the United States from India, Southeast Asian countries, Hawaii, etc. (Figs. 15.4 and 15.5).

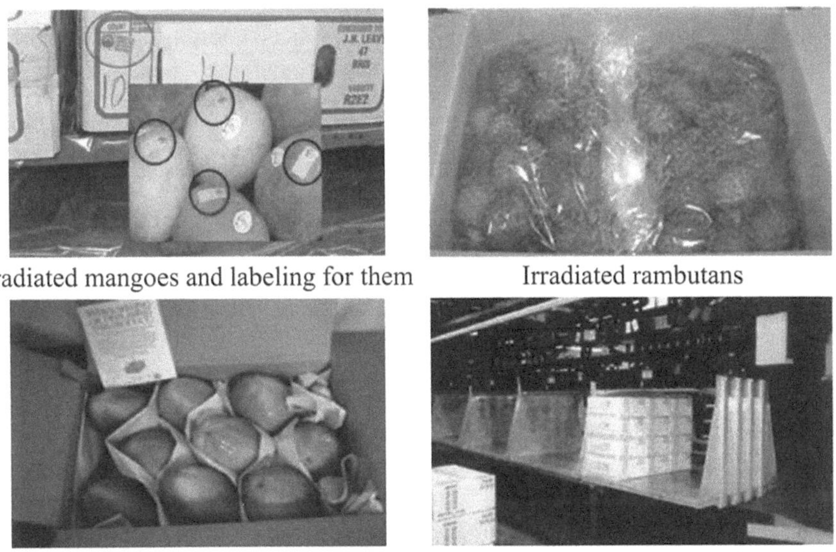

Irradiated mangoes and labeling for them

Irradiated rambutans

Irradiated papayas

X-rays irradiation facility for fruits in Hawaii

Fig. 15.4 Irradiation of tropical fruits for quarantine treatment in the United States

Fig. 15.5 Indian mangoes for export to the United States

15.3.3 Sterilization and Disinfection: 1 to 50 kGy

Radiation-based sterilization of foods and raw materials for processed foods is categorized as follows according to the purpose:

1. Sterilization to reduce bacteria in foods to prolong the periods for which they stay good to eat
2. Disinfection to kill food-poisoning microorganisms
3. Sterilization of medical equipment, pharmaceuticals, hospital food, and sterilized experiment animal feeds (complete sterilization)

The numbers of putrefying microorganisms present in meats, raw fish, raw shellfish, and processed foods made from them can be reduced significantly with irradiation at 1–3 kGy. In addition, by combining such irradiation with storage at 10 °C or lower temperature, the periods for which such foods stay good to eat can be extended to two to several times the periods they stay good to eat without irradiation. The dose required to kill off the pathogenic microorganisms present in foods such as *Salmonellae* is 2–4 kGy for irradiation at a temperature between 0 °C and room temperature and 3–7 kGy for irradiation at a temperature below −20 °C (i.e., irradiation of foods in the frozen state). In particular, such microorganisms as enterohemorrhagic *Escherichia coli* O157, *Campylobacter*, and *Vibrio parahaemolyticus* can be easily killed off with irradiation at 1 kGy [7].

Among the microorganisms that survive this absorbed dose, *Psychrobacter*, decomposing yeasts and lactic bacteria are not pathogenic and are slow to propagate, although they decompose foods. Because killing off spore-forming microorganisms such as *Bacillus cereus* and *Clostridium botulinum* requires an absorbed dose of 10 kGy or higher, these spore-forming pathogenic microorganisms can propagate at room temperature. However, it is considered that the food poisoning they cause can be prevented by using irradiation at approx. 3 kGy and

Fig. 15.6 Irradiated frozen mince of beef (beef burger patties) being sold in the United States

low-temperature storage in combination. On the other hand, food-poisoning *Norovirus* can be deactivated by subjecting them to a high temperature between 60 and 70 °C, but they are highly resistant to radiation, and it is considered that deactivating them with irradiation requires an absorbed dose of 10 kGy or higher [7].

Based on these findings, systems for irradiation-based sterilization of frozen shrimp and frozen frog legs have been put into practical use in France and Belgium, and systems for irradiation-based sterilization of frozen mince of beef (beef burger patties) to prevent food poisoning caused by enterohemorrhagic *Escherichia coli* O157 (Fig. 15.6) and systems for irradiation-based sterilization of frozen poultry meat (chicken and turkey) to prevent food poisoning caused by *Salmonellae* have been put into practical use in the United States [2–5]. Systems for irradiation-based sterilization of frozen shrimp, fish, and shellfish and dried fish and shellfish have also been put into practical use in Indonesia, Thailand, and Vietnam [2–5] (Fig. 15.7).

The items that are internationally traded between many countries for which irradiation-based sterilization treatment is being most widely used throughout the world are spices, herbs, and dried vegetables, as shown in Fig. 15.8. In many cases, natural spices are heavily contaminated with microorganisms. The number of microorganisms present in such contaminated natural spices is 10^4–10^8 per gram. The majority of these are decomposing microorganisms such as *Bacillus pumilus* and *Bacillus subtilis* that form heat-resistant spores, but *Bacillus cereus* and

Fig. 15.7 Irradiated foods purchased in Bangkok, Thailand
Seasoned dried fish, dried mangoes, dried tamarinds, Chinese-style sausages, fermented sausages, and powdered paprika (from left to right in the top row and right to left in the bottom row)

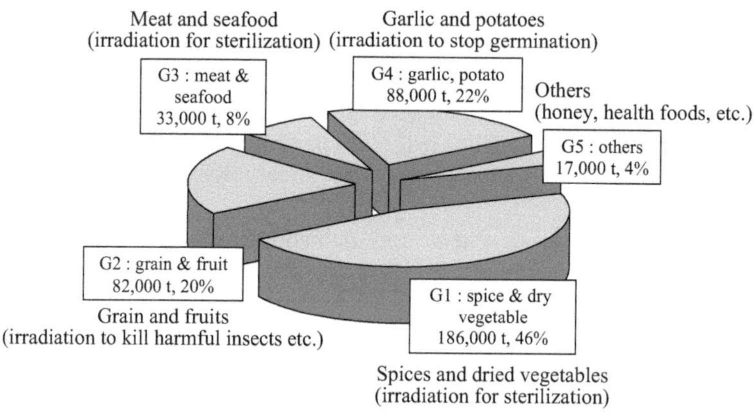

Fig. 15.8 Breakdown of the foods that were irradiated in the world in 2005 and the irradiated amounts [2–5]

Clostridium botulinum that form pathogenic spores are also present. The use of unsterilized contaminated natural spices in small amounts such as their use in cooking at kitchens of restaurants and private homes does not pose a significant risk of food poisoning. However, in the cases where they are used as raw materials for processed foods or where foods to which they have been added are heat-cooked and then preserved, pathogenic microorganisms contained in them in the form of heat-resistant spores can propagate and cause food poisoning. For this reason,

Japan's Food Sanitation Act requires that the number of microorganisms contained in spices for use for processed foods be 10^3 or less per gram.

Because the troublemaking spore-forming microorganisms are resistant to heat and practically none of them can be killed with cooking at 100 °C, high-temperature airstream sterilization treatment that uses superheated dry steam (2–5 s at 140–180 °C) is being used in Japan. However, subjecting spices to high temperature degrades the fragrance and color tone, which are very important properties of spices. On the other hand, irradiation of spices with radiation at 7–10 kGy achieves the equivalent sterilization effect without degrading the fragrance or color tone. For this reason, irradiation has become the mainstream in the world for the sterilization of spices, and it is expected that its use in the sterilization of spices will increase further. Among the 57 countries in which food irradiation has been approved, the only countries in which sterilization of spices by irradiation has not been approved are Japan and Uruguay.

15.3.4 Ionizing Radiations Used for Food Irradiation

Internationally, countries are required to ensure that their measures relating to import and export of foods and agricultural produce comply with the Agreement on the Application of Sanitary and Phytosanitary Measures (SPS Agreement). The international hygiene standards for foods (intended for consumption by humans) used under this SPS Agreement are the Codex International Standards for Foods established by the commission for international standards for foods that was set up jointly by the United Nation's FAO and WHO (Codex Alimentarius Commission). Table 15.2 presents an outline of the Codex International General Standard for Irradiated Foods adopted in 2003 (CODEX STAN 106-1983, REV.1-2003). Only the three types of radiation listed below are permitted for food irradiation. The purpose of the upper limit values for the energy is to prevent foods from being activated as a result of the irradiation:

1. Gamma rays from Co-60 (1.17/1.33 MeV) or gamma rays from cesium 137 (Cs-137) (0.66 MeV): These gamma rays have high penetrating power and allow foods covered with thick packing material to be irradiated. Because the dose rate is relatively low, irradiation of foods that require a high dose is made in the form of batch processing that requires time on the order of hours.
2. Electron beams generated from machine sources operated at or below an energy level of 10 MeV: The penetrating power is weak, but, because the dose rate is very high, foods can be irradiated while they are being moved on a treatment line conveyor. Because the generation of electron beams can be stopped by turning off the power, irradiation can be made safely. It is also possible to take advantage of the low penetrating power to selectively irradiate the surface part only.

Table 15.2 Outline of the Codex international general standard for irradiated foods

1. Scope	
2. General	
2.1. Radiation sources	Gamma rays from the radionuclides ^{60}Co or ^{137}Cs, X-rays generated from machine sources operated at or below an energy level of 5 MeV, electrons generated from machine sources operated at or below an energy level of 10 MeV
2.2. Absorbed dose	The maximum absorbed dose delivered to a food should not exceed 10 kGy, except when necessary to achieve a legitimate technological purpose
2.3. Facilities and control of process	Approval of facilities, securing safety, conformance with the good hygiene practice, operator training, keeping operation and dose measurement records, disclosing records in audits, and management in conformance with the Codex implementation rules (CAC/RCP19-1979, Rev. 2-2003)
3. Hygiene of irradiated foods	
3.1. Hygienic handling of foods	The irradiated food should be prepared, processed, and transported hygienically in accordance with the provisions of the *Recommended International Code of Practice – General Principles of Food Hygiene* (CAC/RCP 1-1969, Rev. 3-1997), including the application of the seven principles of Hazard Analysis and Critical Control Point (HACCP) system where applicable for food safety purposes. Where appropriate, the technical requirements for the raw materials and end product should comply with applicable hygienic codes, food standards, and transportation codes
3.2. Consideration for public health	Any relevant national public health requirement affecting microbiological safety and nutritional adequacy applicable in the country in which the food is sold should be observed
4. Technological requirements	
4.1. General requirements	The irradiation of food is justified only when it fulfills a technological requirement and/or is beneficial for the protection of consumer health. It should not be used as a substitute for good hygienic and good manufacturing practices or good agricultural practices
4.2. Food quality and packaging requirements	The doses applied shall be commensurate with the technological and public health purposes to be achieved and shall be in accordance with good radiation processing practice. Foods to be irradiated and their packaging materials shall be of suitable quality, acceptable hygienic condition, and appropriate for this purpose and shall be handled, before and after irradiation, according to good manufacturing practices taking into account the particular requirements of the technology of the process
5. Re-irradiation	1. Re-irradiation shall not be made except for re-irradiation of grain, beans, or dried vegetables with low moisture content which have been irradiated for sterilization purposes
	2. The following is not considered as re-irradiation:
	(a) Irradiation of foods whose raw materials have been irradiated at low dose levels for purposes other than food safety, e.g., quarantine control, prevention of sprouting of roots and tubers

(continued)

Table 15.2 (continued)

	(b) The food, containing less than 5% of irradiated ingredient, is irradiated
	(c) The full dose of ionizing radiation required to achieve the desired effect is applied to the food in more than one increment as part of processing for a specific technological purpose
	3. The cumulative maximum absorbed dose delivered to a food should not exceed 10 kGy as a result of re-irradiation except when it is necessary to achieve a legitimate technological purpose and should not compromise consumer safety or wholesomeness of the food
6. Postirradiation verification	When required and where applicable, analytical methods for the detection of irradiated foods may be used to enforce authorization and labeling requirements. The analytical methods used should be those adopted by the Codex Alimentarius Commission
7. Labeling	
7.1. Inventory control	For irradiated foods, whether prepackaged or not, the relevant shipping documents shall give appropriate information to identify the registered facility which has irradiated the food, the date(s) of treatment, irradiation dose, and lot identification
7.2. Labeling for packaged foods	General Standard for the Labelling of Prepackaged Foods (CODEX STAN 1-1985, REV1991): The name of the food and the fact that it has been irradiated must be indicated with words. The use of the logo mark (shown in the lower left of Fig. 15.7) is optional. This labeling requirement also applies to foods that contain an irradiated raw material
7.3. Labeling for foods in bulk containers	The declaration of the fact of irradiation should be made clear on the relevant shipping documents. In the case of products sold in bulk to the ultimate consumer, the international logo and the words "irradiated" or "treated with ionizing radiation" should appear together with the name of the product on the container in which products are placed

3. Converted X-rays generated from machine sources operated at or below an energy level of 5 MeV: The Bremsstrahlung X-rays generated by projecting an electron beam onto a plate of heavy metal such as tungsten or tantalum are used. Bremsstrahlung X-rays provide high penetrating power equivalent to the penetrating power of gamma rays, and its generation can be stopped by turning off the power as in the case of electron beams, but it is said that the energy conversion efficiency is low and the cost is high. It may become the mainstream in the future if improvements in endurance and cost reductions are achieved through technological innovations.

15.4 Advantages and Disadvantages of Food Irradiation

Advantages of the irradiation of foods to sterilize foods, kill harmful insects present in foods, and/or inhibit the sprouting of foods or agricultural produce over other sterilization technologies such as heating- and chemical-based sterilization technologies include the following:

1. Because it is possible to treat foods to sterilize them or kill harmful insects present in them without heating foods, it is possible to apply such treatment to fresh fruits, fresh vegetables, meats, and seafood in the chilled or frozen state with minimal degradation in the color tone and fragrance and minimal loss of nutrients.
2. Unlike the fumigation treatment using the potentially carcinogenic ethylene oxide gas or the ozone layer-depleting methyl bromide, there is no risk that toxic substances may remain in the food or any environmental impact may be caused.
3. The high penetrating power of radiation allows efficient and reliable treatment to be achieved with the food (including its insides) evenly sterilized regardless of the shape. Food irradiation is especially effective in sterilizing powdery foods, which are difficult to heat evenly. Another major advantage is that food irradiation allows foods to be sterilized after packing with them covered with airtight packaging, thereby allowing them to be kept in a clean state until they are taken out of the packaging before use.

Disadvantages of food irradiation differ depending on the type of the food and the purpose but, generally speaking, include the following:

1. The cost is high. It is said that the cost of irradiation to inhibit sprouting of potatoes is approx. 2–3 yen per kilogram, but, since the irradiation cost increases roughly proportional to the dose, the cost of sterilization-dose irradiation is 10–100 times as high as the cost of sprouting stopping-dose irradiation.
2. Food irradiation is not suitable for certain foods. The taste and/or suitability for processing may change depending on the food or irradiation condition. For example, high-dose irradiation of a meat or dairy product may cause it to emit a strange smell or change its color tone, and high-dose irradiation of flour may reduce its viscosity. In addition, reductions in the amounts of certain nutrients such as the water-soluble vitamins B_1 and C may occur, although the extents of such reductions are about the same as or smaller than the extents of the corresponding reductions caused by heating-based treatment.
3. Consumers may misunderstand food irradiation and avoid irradiated foods. Food manufacturers, physical distribution companies, retailers, and restaurants may suffer corporate image degradation and financial damage caused by misleading reports by the mass media or encounter boycotts and personal denunciation by anti-food irradiation groups.

As the cost of food irradiation is higher than the costs of conventional methods for sterilization or killing harmful insects, food irradiation is used only in the cases where the value of the product is high and the benefits of irradiating it are greater than the cost or the cases where there is no other appropriate method. Because food irradiation is not suitable for certain foods, it is not likely that all existing food processing and food hygiene technologies (including heating, freezing, canning, retort pouching, agricultural chemicals, and food additives) will be replaced by food irradiation in the future.

The strange smell meats, etc. emit when they are subjected to high-dose irradiation is called the "irradiation smell." This smell, which is similar to the smell meats emit when they are subjected to sunlight, basically consists of the smells of the volatile substances generated as a result of the decomposition of lipids and proteins. It has been reported that the main components of the irradiation smells from beef, pork, etc. are hydrocarbons, carbonyl compounds, alcohols, and mercaptans (sulfur compounds). High-dose irradiation in the air at room temperature tends to generate these irradiation smells, while irradiation under a deoxygenated, low-temperature, or frozen condition suppresses the generation of these smells. Even if a weak irradiation smell is generated, it will be further weakened to a non-sensible level by heating. The characteristics of the irradiation smells generated depend largely on the original components of the food and the type of the food. In the case of spices and tea leaves, for example, irradiation does not alter their original good fragrances, but sometimes strengthens them.

Of course, not all of the technically feasible food irradiation utilization methods will be put into practical use in the future. The extents to which individual methods will be commercially accepted will be determined by their evaluations regarding their practical advantages and cost effectiveness, as for other technologies. In a sense, it can be said that the systems for irradiating various foods which have been put into practical use in countries around the world are rare successful examples in which their values have been successfully demonstrated to the society as a result of efforts to make up for or eliminate various disadvantages.

In Japan, however, food irradiation is prohibited by law except for the irradiation of potatoes to inhibit their sprouting (which is being made to provide potatoes to consumers in a fresh and delicious state during the changeover period between the end of one shipment season and the start of the next shipment season), and import of irradiated foods is also prohibited by law. For this reason, there is no opportunity for general consumers to actually eat irradiated foods and confirm their good quality themselves. In addition, Japan is filled with ungrounded opinions that irradiated foods are dangerous (as can be easily seen by making a search on the Internet), and such opinions are fueling anxiety of consumers. The present situation where the majority of Japanese consumers do not have an opportunity to obtain correct information on food irradiation and consequently have not understood the concept of "irradiating foods with ionizing radiation" has been a major hindrance to the development of practical technologies and systems for food irradiation by enterprises. It will not be possible for Japanese consumers to benefit from food irradiation, which is a new means of food processing for Japan, unless consumers' understanding and trust about the safety and usefulness of food irradiation are obtained through resolute actions by the risk management organizations based on scientific data and tenacious efforts of people concerned in various areas to communicate correct information on risks.

15.5 Safety of Irradiated Foods and Risks Associated with Irradiated Foods

Potential hazards associated with food irradiation include the following:

1. Generation of toxic substances (including substances with chronic toxicity, carcinogenicity, or teratogenicity)
2. Generation of induced radioactivity
3. Induction of mutations in survived microorganisms and resultant occurrence of harmful microorganisms, facilitation of toxin production, and/or increases in the radiation or chemical resistance

Aside from the potential hazards described above, some people are concerned about the possible degradation in the quality of foods such as the possible reductions in the amounts of certain nutrients, the possible loss of certain nutrients, and the risk of accidents at irradiation facilities, although these are not food safety problems.

15.5.1 Food Irradiation Is a Sufficiently Verified Food Processing Technology

Numerous studies that relate to these concerns have been conducted over many years both in Japan and overseas, but none of them have presented evidence suggesting that eating irradiated foods causes any adverse effects to the human body. The following facts have been confirmed by past studies [1, 11, 12]:

1. Results of animal feeding experiments, etc. show that irradiated foods have no acute toxicity, chronic toxicity, carcinogenicity, mutagenicity, teratogenicity, or genotoxicity and food irradiation does not cause a change in food components which adversely affects human health.
2. The energy of the Co-60 gamma rays, electron beams of 10 MeV or less, and X-rays of 5 MeV or less used for food irradiation is below the nuclear reaction threshold, and no induced radioactivity is generated under proper irradiation conditions.
3. The risks associated with survived microorganisms are the same as those in other sterilization methods. Irradiation does not increase pathogenicity or toxicity.

Reductions in the amounts of nutrients such as vitamins and the loss of such nutrients caused by irradiation are not much more than those caused by heating-based treatment, etc. The results of past studies conducted around the world indicate that irradiation at 1 kGy or less causes negligible reductions in the amounts of nutrients and that irradiation at a dose above 1 kGy and below 10 kGy causes very minor reductions in the amounts of nutrients provided that the irradiation is made in a deoxygenated environment. It has been shown that the vitamins that are relatively easily decomposed as a result of irradiation are the water-soluble vitamins B_1 and C and that the extents of the decomposition are about the same as or smaller than the extents of decomposition caused by heating-based treatment [1, 12].

Needless to say, radiation-based disinfection technologies for medical equipment and food packaging for aseptic packaging are well-established technologies. As of 2006, there are a total of 24 industrial irradiation facilities (Co-60 and electron beam accelerators) throughout Japan, and the ratio of the 2005 sales in Japan of the medical equipment disinfected with radiation to the 2005 sales in Japan of all disinfected medical equipment was as high as 45%.

Therefore, it can be said that food irradiation is effective and irradiated foods are safe as long as the requirements of the good manufacturing practice (GMP) for foods are satisfied and that the irradiated food-related risks that result from failure to observe the GMP are essentially not different from the nonirradiated food-related risks that result from the incorrect use of nonirradiation-based treatment methods such as canning methods, freezing methods, and heating-based sterilization methods.

15.5.2 Safety Evaluations in Japan

In Japan, studies on the effects of food irradiation, on irradiation technologies, and on the soundness of irradiated foods (safety plus nutritional suitability) were conducted from 1967 to 1983 as Designated General Atomic Studies in a nationwide effort promoted by the Japanese government. In these studies, seven foods were irradiated: potatoes and onions (irradiation to inhibit sprouting), rice and wheat (irradiation to kill harmful insects), sausages and steamed fish paste (irradiation for sterilization), and satsuma mandarins (irradiation for anti-molding). The satsuma mandarins were irradiated with electron beams in the surface part only, and the other foods were irradiated with Co-60 gamma rays. These studies showed that food irradiation was an effective means to prevent damage caused by germination, insects, and decomposition. The studies on the effects of food irradiation were conducted by the National Food Research Institute and the National Institute of Animal Industry of the Ministry of Agriculture, Forestry and Fisheries, and the studies on irradiation technologies were conducted by the Takasaki Research Institute of the Japan Atomic Energy Research Institute. The basic research on the effects of radiation was conducted by RIKEN, and the studies on the soundness of irradiated foods were conducted by the National Institute of Health Science of the Ministry of Health and Welfare, the National Institute of Health, the National Institute of Health and Nutrition, and the Food and Drug Safety Center. (The organization names are the names used at the time.)

These studies included an experiment in which the effects of irradiated foods on mice, rats, etc. were investigated by feeding them with irradiated foods and unirradiated foods (the control). Because this experiment initially encountered various problems such as deterioration of the nutritional balance in experiment animals caused by feeding them only with the selected food in excessive amounts (which occurred irrespective of whether the food had been irradiated or not) and disorders caused by original components of foods such as toxic components of onions, additional experiments, etc. using revised methodologies were conducted. As a result, it was ultimately concluded in all of the following tests that none of the abovementioned seven irradiated foods caused any adverse effects on experiment animals [11]:

1. Chronic toxicity tests: Body weight, mortality rate, hematologic examination, and histopathologic examination.
2. Generation tests: Animals that had been fed with irradiated foods were cross-bred, and female animals in the pregnancy and lactation periods and weaned infants were also fed with irradiated foods. This was repeated to cover three generations, and teratogenicity and genotoxicity tests were conducted.
3. Mutagenicity tests: Mutation induction tests using microorganisms and chromosomal aberration tests using mice and cultured cells of mammals.

In consideration of the results of these safety verification tests, the Ministry of Health and Welfare (the present Ministry of Health, Labour and Welfare) approved the irradiation of potatoes to inhibit their sprouting in 1972 and tentatively prohibited all other irradiations of foods with radiation. As a result, JA Shihorocho started irradiating potatoes to inhibit their sprouting and shipped their first irradiated potatoes in the spring (i.e., the changeover period) of 1974.

In addition, the Food Irradiation Research Committee, which had been organized by the Japan Radioisotope Association (JRIA) and was comprised of JRIA and about 15 other organizations including universities and national and public research institutes, conducted various studies from 1986 to 1991. These studies showed that there were no problems in irradiated foods relating to induced radioactivity, nutritional changes, mutagenicity, or microbiological safety by showing that no induced radioactivity is detected in spices irradiated with a 10 MeV electron beam at 30 kGy, the digestibility and immunochemical characteristics of proteins are not changed by irradiation, the extent of reduction in the amount of vitamin C in irradiated potatoes is not different from that in unirradiated potatoes, irradiation of spices does not cause generation of mutagenic substances, and irradiation does not induce mutations in harmful microorganisms, thereby attesting to the soundness of irradiated foods [11].

The Ministry of Health and Welfare initially intended to approve irradiation of other foods (including the irradiation of onions to inhibit their sprouting) in sequence following their approval of the irradiation of potatoes in 1972, but the will to push forward with putting food irradiation into practical use waned at both the Ministry of Health and Welfare, which was in the position to grant approval, and enterprises and the government ministries and agencies concerned, which were in the position to seek approval, because of persistent anti-food irradiation movements by some consumer groups such as a boycott of irradiated potatoes. Since then, no step forward has been taken up to the present.

15.5.3 Safety Evaluations That Have Been Made at International Organizations

The World Health Organization (WHO) and the Food and Agriculture Organization (FAO) of the United Nations noticed the potential of food irradiation technologies as a means to reduce outbreaks of diseases via foods and help solve the food resources depletion problem and established in 1961 a joint expert committee with the International Atomic Energy Agency (IAEA) and initiated in 1971 an international project to conduct soundness verification tests to evaluate the safety and nutritional suitability of irradiated foods. A total of 24 countries including the United States, the United Kingdom, France, West Germany, and Japan participated in the project and conducted studies till 1980 [11].

In the toxicity tests conducted in studies before the 1970s, experiment animals such as rats were fed with foods irradiated at a dose 100 times as high as the standard dose, as in safety evaluations of such chemicals as food additives and pharmaceuticals, because the effects of radiation on food components had not been clearly known. For this reason, the abovementioned deterioration of the nutritional balance in experiment animals caused by feeding them only with the selected food in excessive amounts and disorders caused by original components of foods were not uncommon in these tests. Animal test experts say that, to evaluate test results correctly, it is necessary to confirm the following points [13]:

1. Whether dose dependency is observed in the phenomenon under consideration such as weight loss
2. Whether trends that are constant throughout the feeding period are observed
3. Whether trends that are constant throughout the generation test are observed
4. Whether or not the abnormal conditions detected are pseudo-abnormal conditions due to other factors

In the abovementioned international project, too, it was initially thought that irradiating foods at a dose 100 times as high as the standard dose was an appropriate way of conducting tests as in the case of safety evaluations of food additives. However, in consideration of study reports that showed what types of decomposition products are generated by radiation in what amounts, the joint FAO-IAEA-WHO expert committee announced in its 1976 conference that "food irradiation is a physical processing method similar to heating and freezing, and it is not appropriate to treat it in the same way as that for food additives."

On the basis of the results of the international project, the joint FAO-IAEA-WHO expert committee announced its conclusion in 1980 that "irradiation of foods does not cause any toxicological, nutritional or microbiological problems as long as the total average dose is 10 kGy or lower, and it is not necessary from now on to conduct soundness verification tests on individual foods that have been irradiated at a dose of 10 kGy or lower."

Based on this conclusion, the Codex Alimentarius Commission adopted the "International General Standard for Irradiated Foods" (Table 15.2) and the "International Standard for the Conduct of Food Irradiation" in 1983 and recommended the member countries to accept these standards.

In 1992, the WHO convened a conference of the expert committee in response to a request from the Australian government. In this conference, the expert committee reevaluated data from systematic evaluations conducted by the Food and Drug Administration (FDA) of the United States after 1980 and carefully analyzed past reports of dubious credibility (including the Indian Nutrition Research Institute's report claiming that feeding irradiated wheat to undernourished children caused increases in polyploids (a type of chromosomal abnormality) in their lymphocytes (this report has been denied by about 20 reports of double checks made throughout the world and has also been denied by the governments of the United States, the United Kingdom, and Canada, as well as by the report of the investigation conducted by the Indian government itself, but some people are still using it as a basis for their anti-irradiated foods movements) and refuted them. On the basis of this analysis, the expert committee re-concluded that "foods that have been irradiated at a dose of 10 kGy or lower are sound foods" [1]. The "10 kGy or lower" proviso does not mean that there is data showing that toxicity was observed in foods that had been irradiated at a dose above 10 kGy. It was added because the expert committee believed that a dose of 10 kGy or lower was sufficient for practical purposes.

After that, the need to make evaluations at high doses above 10 kGy for hospital food, spices, and other special purposes was recognized, and a soundness evaluation was conducted again. Based on the result of this evaluation, the WHO expert committee declared in its 1997 conference that the use of a dose above 10 kGy would not cause any problems and recommended that the dose limit should be abolished [14]. This recommendation has remained valid up to the present.

15.5.4 Are 2-Alkylcyclobutanones Dangerous?

2-Alkylcyclobutanones were reported as products of radiolysis of neutral fats in 1972, and studies were conducted to confirm whether they can be used as markers for detecting irradiated foods. As a result, it was reported in 1993 that they were characteristic products of irradiation.

In 1998, an experiment was conducted in which 2-dodecylcyclobutanone (2-dDCB) generated from palmitic acid, which is a major fatty acid that represents 20–25% of the lipids of beef (a 2-alkylcyclobutanone), was added to cultured cells of humans and rats at a high concentration (5.2 mM). The comet assay method (single-cell electrophoresis method)-based observation made in the study showed that DNA strand breaks had occurred, suggesting that 2-dDCB may have weak mutagenicity. Therefore, double checks were made by a joint German-French team from 1999 to 2001. The results of the reverse mutation tests conducted (using the Ames test method, etc.) were all negative (no mutagenicity was observed), and the results of the rat feeding experiments conducted showed that 2-dDCB had no carcinogenicity, but the tests and experiments suggested that giving 2-dDCB together with a carcinogen to rats can enhance the carcinogenicity of the carcinogen. The amount of 2-dDCB given to the rats in the study corresponds to the intake into the human body that would result if a human weighing 60 kg continued eating 4.8 tons of irradiated American beef burger patties (containing approx. 6.0 μg of 2-dDCB per 125 g) every day.

In a study that followed (which was reported in 2006), 2-dDCB was applied to cultured cells at a concentration of 53 μM, which is a high concentration but is approx. one hundredth of the 5.2 mM used in the abovementioned 1998 experiment. The comet assay method-based observation and micronuclei (a type of chromosomal aberration) generation test performed in the study showed that 2-dDCB at this concentration had a weak DNA damage effect that was roughly the same as that of palmitic acid (the precursor of 2-dDCB) (a DNA damage effect that is two to three times as strong as that at zero concentration). However, it was concluded that it was meaningless to unnecessarily emphasize the risks posed by the very small amounts of 2-dDCB generated in foods by irradiation, which can be considered as negligible risks compared to the risks posed by the palmitic acid contained in beef burger patties in far larger amounts. The lowest 2-dDCB concentration at which a difference of significance from the control section was observed in this experiment corresponds to the intake that would result if a human weighing 60 kg ate 15.8 tons of irradiated American beef burger patties.

In short, the current conclusion is that it is not necessary to view 2-alkylcyclobutanones as a problem, as shown by the fact that the amounts of 2-alkylcyclobutanones generated in actual foods by their irradiation (which are determined by the lipid content and the dose) are very small and the "toxicity" levels of such 2-alkylcyclobutanones are not much different from those of nutrients contained in large amounts in unirradiated foods as well as by the fact that no abnormality (such as carcinogenesis) was observed in the past long-term animal feeding experiments in which mice, rats, hamsters, rabbits, and beagle dogs were fed with frozen chicken that had been irradiated at 59 kGy [12].

15.6 Irradiated Food Detection Methods

Irradiated foods are relatively difficult to detect because irradiation inherently leaves no residual chemical substance in foods unlike agricultural chemical- or additive-based treatment and the changes caused in food by irradiation are not much different from the changes caused by heating-based treatment, etc. However, there are several established practical irradiated food detection methods [15].

15.6.1 Principles of Irradiated Food Detection Methods

Irradiated food detection methods are categorized into the following three types according to the principle of the method:

1. Physical detection methods: detection methods that detect the metastable states caused by a surplus or shortage of electrons that have accumulated in inorganic matters such as sand and dust adhering to or present in such foods as spices and the long-life organic radicals generated in cellulose, bones, and crystalline glucide contained in foods by the electron spin resonance (ESR) method, thermoluminescence (TL) method, photo-stimulated luminescence (PSL) method, etc.
2. Chemical detection methods: detection methods that detect the radiolysis products (such as hydrocarbons and 2-alkylcyclobutanones) generated from lipids, etc. contained in foods by such methods as the gas chromatography-mass spectrometry (GC/MS) method.
3. Biological detection methods: detection methods that deduce whether the food has been irradiated or not by means of comparative observation to determine the ratio of the living microorganisms to the dead microorganisms and identify their types (with a focus on such point as whether or not there have been unusual reductions in the numbers of microorganisms that are vulnerable to radiation) or the comet assay method that subjects animal and plant cells contained in foods to single-cell gel electrophoresis (the method was named the "comet assay method" because short DNA fragments look like comets with tails) to analyze the extents of DNA strand breaks in the cells.

15.6.2 Differences Between Japan and Other Countries in the Roles Given to Irradiated Food Detection Methods

In February 1999, the European Union (EU) issued two food irradiation-related European Commission (EC) Directives as part of its efforts to establish unified regulations for the handling of irradiated foods.

One of them is EC Directive 1999/2/EC, which specifies the general principles for food irradiation, the conditions for approval of food irradiation, the radiation source requirements, and the irradiated food labeling requirements. The labeling requirements include a requirement that words be used in the labeling and a requirement that labeling be provided on products containing irradiated raw materials as well, irrespective of the amounts of the irradiated raw materials contained. In addition, the Directive requires that approval be obtained prior to operating a food irradiation facility, irradiation facilities in EU countries be audited by government agencies of the respective countries, and irradiation facilities in non-EU countries be registered after receiving an inspection by an auditor organization of the European Commission and passing that inspection.

The other directive is EC Directive 1999/3/EC. This directive provides a list of the food items permitted to be irradiated in all EU countries (a "positive list"). The list specifies dried aromatic herbs, spices, and vegetable seasonings as the food items that are permitted to be irradiated in all EU countries. The Directive states that the individual food items that are permitted to be irradiated in individual EU countries under the current domestic laws of the respective countries (such as frozen frog legs which are permitted to be irradiated in France and Belgium and poultry meats which are permitted to be irradiated in the United Kingdom) will continue to hold that status until an expanded list of the food items permitted to be irradiated in all EU countries is provided in the future.

The EU considered that the establishment of such common regulations for all EU countries would require standard irradiated food analysis (detection) methods for obtaining supporting data for labeling. Accordingly, the Comité Européen de Normalisation (CEN) initiated a research project in 1990 to identify or develop such analysis (detection) methods. As a result, ten methods including the ESR method, TL method, PSL method, and GC/MS method were adopted as standard European methods (standard CEN analysis methods) in the period from 1996 to 2004. Nine of them were also adopted as standard analysis methods for the Codex Alimentarius Commission international standards for irradiated foods (standard Codex analysis methods) in the period from 2001 to 2003 [15] (Table 15.3).

However, the Food and Drug Administration (FDA), which is the United States' authority responsible for evaluating the safety of irradiated foods and approving them in the United States based on the evaluation results (Table 15.4), takes the position that irradiated foods are essentially safe foods and such detection methods are not necessary. Accordingly, the United States has not introduced any official detection method.

The WHO takes the position that violation of irradiated food labeling rules is not a public health problem but a problem concerning brand value-related economic actions similar to the problem of disguised labeling of the area of production and disguised "organic food" labeling, which should be treated as a compliance-related problem.

In Japan, the "Notification Method of the Ministry of Health, Labour and Welfare" was established on July 6, 2007, as the first official detection method in Japan (FOODSEC No.0706002) [16]. This is a TL method-based detection method, but employs unique evaluation criteria that differ in various aspects from the evaluation criteria of the standard CEN analysis method (EN 1788), which is also a TL method-based method [17]. However, many researchers specializing in detection methods contested this detection method asserting that there is no rationality in specifying

Table 15.3 Irradiated food detection methods adopted for the Codex standards and standard European irradiated food analysis methods (standard CEN irradiated food analysis methods)

CEN analysis method number	Description of the method	Targets (detection limits (in kGy))	Codex classification
EN 1784 (established in 1996 and revised in 2003)	Gas chromatography-based hydrocarbon measurement	Chicken (0.5), pork (0.5), beef (0.5), avocados (0.3), mangoes (0.3), papayas (0.3), Camembert cheeses (0.5)	Type II (note 1) (adopted in 2001)
EN 1785 (established in 1996 and revised in 2003)	GC/MS-based analysis of 2-alkylcyclobutanones	Chicken (0.5), pork (1), liquid whole egg (1), Camembert cheeses (1), salmon (1)	Type III (note 2) (adopted in 2001)
EN 1786 (established in 1996)	ESR measurement of bones	Chicken (0.5), bony chops (0.5), trout (0.5), frog legs (0.5)	Type II (adopted in 2001)
EN 1787 (established in 1996 and revised in 2000)	ESR measurement of cellulose	Powdered paprika (5), nutshells (2), strawberries (1.5)	Type II (adopted in 2001)
EN 1788 (established in 1996 and revised in 2000)	TL measurement of inorganic silicates	Herbs and spices (6), shrimp (1), fresh vegetables/fruits (1), shellfish (0.5), dried vegetables/fruits (8), potatoes (0.05)	Type II (adopted in 2001)
EN 13708 (established in 2001)	ESR measurement of sugar crystals	Dried papayas (3), dried mangoes (3), dried figs (3), raisins (3)	Type II (adopted in 2003)
EN 13751 (established in 2002)	Photo-stimulated luminescence (PSL) measurement	Herbs and spices (10), shellfish (0.5)	Type III (adopted in 2003)
EN 13783 (screening) (established in 2002)	The difference between the total number of microorganisms including the number of dead microorganisms (as measured on a filter after fluorescent-dyeing them) and the number of living aerobic microorganisms (APC) after plate culturing is used as the index (DEFT/PAC method)	Herbs and spices (5)	Type III (adopted in 2003) (screening method)
EN 13784 (screening) (established in 2001)	Comet assay-based detection of DNA strand breaks	Chicken (1), pork (1), plant seeds (1)	Type III (adopted in 2003) (screening method)
EN 14569 (screening)	Whether sterilization treatment has been applied or not	Chicken (2.5)	

(continued)

Table 15.3 (continued)

CEN analysis method number	Description of the method	Targets (detection limits (in kGy))	Codex classification
(established in 2004)	is deduced based on a comparison of the amount of endotoxin determined by the limulus test (LAL) and the number of living Gram-negative bacteria (LAL/GNB method)		

(Note 1) Type II (reference methods): These methods are selected from the type III methods and used for calibration and dispute-solving
(Note 2) Type III (alternative affirmation methods): These methods are used for control, inspections, and regulation analyses

Table 15.4 Irradiated foods that have been approved by the Food and Drug Administration (FDA) of the United States

Item	Purpose	Upper limit for dose	Approved year
Wheat and wheat products	Killing harmful insects	1.0 kGy	1963
Potatoes	Inhibit sprouting of potatoes	1.0 kGy	1965
Pork (raw)	Parasite control	1.0 kGy	1985
Fresh fruits and vegetables	Suppressing growth	1.0 kGy	1986
All foods	Killing harmful insects	1.0 kGy	1986
Enzyme preparations	Sterilization	10 kGy	1986
Dried spices/seasonings	Sterilization	30 kGy	1986
Poultry meats	Pathogenic microorganism control	3.0 kGy	1990
Frozen meats (NASA space food)	Disinfection	Minimum dose 44 kGy	1995
Chilled red meats	Pathogenic microorganism control	4.5 kGy	1997
Frozen red meats	Pathogenic microorganism control	7.0 kGy	1997
Raw eggs with shell	Pathogenic microorganism control	3.0 kGy	2000
Seeds for bean sprouts	Pathogenic microorganism control	8.0 kGy	2000
Raw or frozen shrimp, crabs, and shellfish	Pathogenic microorganism control	5.5 kGy	2005
Fresh lettuce and spinach	Pathogenic microorganism control	4.0 kGy	2008

analysis procedures and judgment criteria that differ from those of the standard CEN analysis method when it was not necessary to do so. In addition, quarantine office personnel did not welcome this detection method because it imposed a heavy burden on them [18]. Furthermore, there was concern that discrepancies may occur between judgment results in Japan (the importing country) and judgment results in other countries (exporting countries) because the detection method was introduced without adjustment to make it consistent with international standard methods such as CEN and Codex methods, and such discrepancies may lead to international disputes. For these reasons, revisions to make the analysis conditions that concern the irradiation judgment criteria closer to those of EN1788 were made in the Notification dated March 30, 2010 (FOODSEC No. 0330).

Many EU countries are conducting surveillance of foods on the market using the standard CEN analysis method. They have exposed noncompliant foods such as South Korean instant noodles with irradiated seasonings (spices and dried vegetables) through their surveillance efforts. Most of the exposed noncompliant foods were foods permitted to be irradiated but were in violation of the requirement that irradiated foods must be labeled as such. In addition, the EU explains that violations of the food irradiation rules and regulations in connection with irradiation of items that are not on the list of items permitted to be irradiated (such as health foods and supplements which are not permitted to be irradiated) are not a food safety problem but a problem concerning compliance.

However, because irradiated food monitoring inspections in Japan are based on the official position that irradiated foods are not allowed in Japan (except for potatoes irradiated to inhibit their sprouting) because their safety has not been confirmed, they are in effect exposing noncompliant import foods as ones in violation of the Food Sanitation Act to "secure the safety of import foods," despite the assertion that their main purpose is to grasp the current situation of distribution of irradiated foods in Japan. That is, a situation exists in Japan where the problem of labeling noncompliance in connection with failure to correctly indicate how foods have been processed and treated has been mixed up with the problem of distribution of dangerous and harmful foods which must be prevented to ensure food safety, and the risk management organizations have abandoned communicating information on risks associated with irradiated foods.

15.7 Current Situations Surrounding Food Irradiation in Japan and Overseas

As of 2005, food irradiation has been approved in 57 countries around the world. According to the latest estimates [2–5] based on questionnaire-based and on-site surveys, the annual total amount of foods irradiated around the world in 2005 was 405,000 tons, and the scale of economy was 1610 billion yen (Table 15.1).

By country, China ranked first in the amount of foods irradiated in 2005. The total amount of foods irradiated in China in 2005 was 146,000 tons including 80,000 tons of garlic irradiated to stop their germination and 52,000 tons of dried vegetables and spices irradiated for sterilization and represented 36% of the total amount of foods irradiated in the world in 2005. The total annual amount of foods irradiated in China has reportedly reached 200,000 tons. In the second place was the United States, in which a total of 92,000 tons of foods were irradiated including 80,000 tons of spices, 8000 tons of chicken and minced beef, and 4000 tons of fruits and vegetables. In the third place was Ukraine, in which a total of 70,000 tons of wheat and barley were irradiated using electron beam accelerators to kill harmful insects in them. By monetary value, the United States ranked first and represented 53% of the world total. This is because the majority of the irradiation made in the United States was irradiation of high-priced spices to sterilize them.

The amounts of irradiated foods are sharply increasing in Asian countries including China but are on the decrease in EU countries because of the recent strengthening of the labeling violation control. However, because the estimates described above are estimates of "the amounts of foods irradiated" in the respective countries and areas and not statistical data on the amounts of irradiated foods imported/distributed, whether the amounts of irradiated foods consumed in EU countries have increased or decreased and how they have increased/decreased are not clear.

The only food irradiation permitted in Japan is irradiation of potatoes to inhibit their sprouting. For example, turmeric that has been irradiated for sterilization is permitted to be used as a herbal medicine provided that certain requirements are satisfied, but using it as an ingredient of curry is a violation of the Food Sanitation Act. On the other hand, irradiation of food and drink containers for sterilization is permitted, and gamma ray and electron beam irradiation is competing with ultraviolet irradiation, hot air-based drying, high-pressure vapor, hydrogen peroxide water (immersion/spraying), ethylene oxide gas, etc. as means of food and drink container sterilization.

Because irradiating spices to be exported to developed countries for sterilization has become a standard international practice, the All Nippon Spice Association made an application for approval of irradiation of spices with radiation for their sterilization in 2000 [19]. However, the Ministry of Health and Welfare (the present Ministry of Health, Labour and Welfare) has not responded to date.

In 2006, the Japan Atomic Energy Commission (JAEC) emphasized the need for the Ministry of Education, Culture, Sports, Science and Technology; the Ministry of Health, Labour and Welfare; the Ministry of Agriculture, Forestry and Fisheries; and other relevant organizations to pursue the following efforts [21] in the JAEC Decision dated October 3, 2006, that was made based on JAEC's consideration of the report of the Technical Committee on Food Irradiation entitled "About Irradiation of Foods with Radiation" [20]:

1. Consideration and evaluation based on the Food Sanitation Act and the Food Safety Basic Act of food irradiation which is considered useful, to judge the appropriateness of irradiation from the standpoint of food safety administration
2. Promotion of the collection of scientific information on the soundness of irradiated foods and promotion of food irradiation-related research and development
3. Discussions about the way irradiated food labeling should be
4. Promotion of research and development for establishing official detection methods for inspections by administrative organizations, promotion of research and development aimed at achieving technological advances, and consideration/evaluation of actions relating to the monitoring of irradiated foods and the provision of guidance on irradiated foods
5. Promotion of public hearings and public relations activities and of the disclosure of information for increasing the degree of social acceptance of not only food irradiation but also radiation utilization as a whole and enhancement of education and training to provide Japanese citizens with basic knowledge on radiation

The Food Safety Commission of Japan, which is one of the risk evaluation organizations of Japan, indicated its position in its 74th meeting held on December 16, 2004, that there was no problems to be addressed about potatoes irradiated to stop their germination which were being distributed in Japan and it was not necessary to conduct risk evaluations for them but the Commission would initiate risk evaluations for other foods upon request from risk management organizations (such as the Ministry of Health, Labour and Welfare) [22, 23]. In addition, the Commission decided in its 182th meeting held in March 2007 that it would not conduct assessments of the effects of irradiated foods on human health because there was no request from the Ministry of Health, Labour and Welfare but it would continue its information gathering efforts [24, 25].

On the other hand, the Ministry of Health, Labour and Welfare, which is one of the risk management organizations of Japan, chose not to consider whether to accept irradiated food safety-related judgments of the WHO and other organizations (thereby choosing to stay left behind by the global standard) and started on July 6, 2007, an import food monitoring inspection to detect spices that have been irradiated with radiation, with the Pharmaceutical and Food Safety Bureau of the Ministry declaring their position in an official notice entitled "About the Implementation of the Monitoring Inspection (Foods that Have Been Irradiated with Radiation)" (FOODSAFIM No. 0706003) [26] that they will treat noncompliant import foods detected with the TL method (which had been designated as the "Notification Method of the Ministry of Health, Labour and Welfare") [16] as foods in violation of Article 11 of the Food Sanitation Act. Since then, the Ministry has added inspection target items one after another, and noncompliant paprika from Germany; powdered Maca from the United States, Peru, etc.; and oolong tea leaves, dried Shiitake mushrooms, dried leeks, and red peppers from China have been exposed. However, if the judgments of the WHO and other organizations are correct and there is no problem about the safety of irradiated foods, it follows that these inspections being conducted by Japan's quarantine offices are inspections that are unnecessary from the standpoint of the original purpose of securing food safety. If this is the case, it can be said that these inspections are disadvantaging Japanese citizens not only by wasting tax money but also by taking human and financial resources from the inspections that are really necessary to secure import food safety.

On the other hand, companies in the food industry are conducting voluntary inspections of spices, herbs, and dried vegetables they bought overseas to make sure that they do not bring irradiated foods into Japan, and the cost of these voluntary inspections and the cost of collection and disposal of inadvertently imported irradiated foods are ultimately borne by the consumers. For example, in January 2009, products named "Orthodox Indian Masala" and "Easy Chai" that had been imported from India in a fair trade project were voluntarily collected by the importer because they had been irradiated with gamma rays for sterilization [27].

For more information on the current statuses and results of discussions on and evaluations of food irradiation technologies in Japan, refer to Reference [28], which presents a comprehensive time series list of proceedings, reports, and notification documents of the councils and committees concerned. It is expected that, if irradiation of spices for sterilization is permitted in Japan in the future, utilization of irradiation to inhibit sprouting of garlic and utilization of irradiation as an alternative to the methyl bromide-based fumigation treatment to kill harmful insects in grain and imported tropical fruits (quarantine treatment) will be considered as the next step.

15.8 Prospect (or Hope) of Social Recognition of Food Irradiation

The present abundance of foods in Japan is supported by the working people who are constantly making efforts and technological innovations in all areas of the food-related industries including agriculture, food production, trade, and physical distribution. Food irradiation is one of the technologies that have been created out of the need to ensure that Japanese citizens can continue to enjoy this abundance of foods and is advantageous in that it causes less degradation of the quality of foods and is environmentally friendly because it does not use chemicals. However, it can be said that Japan has been far from successful in communicating information on risks associated with irradiated foods to its citizens. It is considered that this situation is partly attributable to the so-called radiation allergy or atomic energy allergy.

However, the risk management organizations of Japan, which are responsible for ensuring food safety for the Japanese citizens, have done nothing since the 1980s, while the United Nations organizations and the governments of the United States and European countries were continuously making serious discussions on how to utilize food irradiation in securing food safety. As a result, food irradiation has been put into practical use in many countries around the world, and many of them have already started trading irradiated foods internationally, while Japan has been left outside the mainstream in this area. Even now, the risk management organizations of Japan are just waiting for a national consensus to be reached as if they are thinking that all responsibility resides in private sector companies. As a result, the Japanese consumers have been left without even the option of buying irradiated foods, which are high quality and safe, to try them. Indeed, the Japanese consumers have been given little information about the high quality and safeness of irradiated foods.

After the discovery by Pasteur, Koch, and other scientists in the mid-nineteenth century that decomposition, fermentation, and infectious diseases are caused by microorganisms, commercial-scale sterilization of cow's milk based on Pasteur's low-temperature sterilization method (pasteurization) was started in Denmark and Sweden in 1885, and sterilization of cow's milk was started in New York around 1890 [29]. After that, the producers of cow's milk in the United States and Japan were legally obligated to sterilize their cow's milk by heating in 1909 and 1933, respectively. It is said that fierce anti-heating-based sterilization movements took place at the time. Proponents of these movements asserted that "sterilization is blasphemy of Nature," "sterilization is nothing but a cheating technique for selling dirty milk as clean milk," and "sterilization destroys Vitamin C and deactivates minerals such as calcium so that they are not utilized in the human body." These assertions are very similar to the assertions of people today who are against irradiated foods. It took several decades for cow's milk sterilized by heating to be socially accepted and firmly establish its position as safe and good food. The authors wonder how long it will take for irradiated foods to be socially accepted and firmly establish its position as safe and good food.

The present situation surrounding food irradiation in Japan is that (1) the administrative organizations say that they will not change their position that food irradiation should be prohibited except for special cases because there has been no request to approve food irradiation from the food industry and the consumers apparently have not agreed on the introduction of food irradiation, (2) the consumers think that irradiated foods are dangerous because they are prohibited by law, and (3) the food industry is forced to refrain from making requests to the government because the consumers are showing a negative attitude about the introduction of irradiated foods, even if the negative attitude is based on misunderstanding. It is considered that parting from this situation and making rational choices based on scientific rationales would require a system for constructive communication which allows the administrative organizations responsible for securing food safety, the food industry, and the well-informed consumers to exchange information and opinions on risks associated with irradiated foods while building mutual trust as equal stakeholders.

Several attempts have already been started to achieve such constructive communication. For example, the "Roundtable for Food Communication" [30] is a citizens' group that is comprised of people in various positions including ordinary housewives, office workers, researchers, educators, and people from the mass media and administrative organizations interested in food safety who gathered to obtain correct, scientific information on food safety by learning from each other. They selected food irradiation as one of their topics and made field trips and hands-on experiments to learn about food irradiation. They conducted tests on foods that are familiar to consumers they selected from the viewpoint of "curiosity-driven housewives" including steamed fish paste, cheeses, canned tuna, bacon, cow's milk, tofu, raw udon (a Japanese noodle), cucumbers, broccoli, cut vegetables, ginger, garlic, potatoes, chestnuts, grapes, unmilled rice, spices (black pepper, white pepper, cassia, cardamom, turmeric, coriander, and red pepper), strawberries, and oranges to confirm how the sterilization effect depends on the dose and irradiation conditions, whether the irradiation smell is left, whether they have the appropriate quality and soundness as foods, and whether irradiation has extended the periods for which they stay good to eat, and exchanged opinions about the results of the hands-on experiments and presented the experiment results at conferences of academic societies [31, 32]. Experience of enjoyable collaborative work with experts as equal partners through such hands-on experiments may allow citizens to have a feel as to what they should know about food irradiation, which is not very easy to understand scientifically, and to what extent they should know about them. This should lead to the recognition among citizens that food irradiation is a technology that they can choose for their own benefits. This example may be one answer to the questions, "What roles should science play for citizens?" and "How the provision of information to citizens by experts should be made?", not only for food irradiation but also for other new technologies such as nuclear power generation and gene recombination.

In this chapter, we chose Japanese general summary papers as references over ones written in other languages wherever possible. Other useful literature includes a general summary paper [33] that provides an overview of the utilization of radiation in food processing and agriculture covering a wide range of topics from the basics to the latest applications, the "Food Irradiation Q&A Handbook" [34] published in 2007 by the Japan Atomic Industrial Forum Inc. (which can be downloaded in the form of a PDF document), the "Food Irradiation Database" [35] available for browsing on the web site of the Takasaki Advanced Radiation Research Institute of Japan Atomic Energy Agency, and the "Related Information" [36] section of the web site of the Japanese Research Association for Food Irradiation.

Exercises

1. Food irradiation is being used as a means to (1) preserve foods and agricultural produce and prevent damage to them during storage, (2) secure food hygiene and prevent foodborne illness, and (3) control organisms or insects harmful to plants or plant products (phytosanitary). Cite examples of the actual use of food irradiation in these three applications, and describe the characteristics of food irradiation in these three applications comparing them with other technologies.
2. Aside from irradiation of potatoes to inhibit sprouting, radiation is being used in a wide range of applications around us including sterilization of food containers, cosmetics, contact lens conditioning solutions, manufacturing of semiconductor

devices, foamed plastics, automobile tires, and electrical wiring in television cathode ray tubes. However, this fact has hardly been recognized by the general public. State your view about the reasons for this.

3. Study the rationales of anti-food irradiation groups (such as the Public Citizen and the Japan Liaison Against Food Irradiation) on their web sites, and state your view about the legitimacy of the rationales.

References

1. WHO Sekai hoken kikan (1996) Syousha syokuhin no anzensei to eiyou tekisei, Kopu syuppan, Tokyo (in Japanese); World Health Organization (WHO) (1996) Safety and nutritional adequacy of irradiated food. Co-op Books, Tokyo (in Japanese)
2. Kume T, Furuta M, Todoriki S, Uenoyama N, Kobayashi Y (2009) Quantity and economic scale of food irradiation in the world. Radioisotopes 58(1):25–35
3. Kume T, Furuta M, Todoriki S, Uenoyama N, Kobayashi Y (2009) Status of food irradiation in the world. Radiat Phys Chem 78:222–226
4. Kume T (2008) Sekai ni okeru syokuhin syousha no syoriryou to keizai kibo. syokuhin syousha 43:46–54 (in Japanese); Kume T (2008) Quantity and economic scale of food irradiation in the world. Food Irradia 43:46–54 (in Japanese)
5. Dokuritu gyousei houjin Nihon genshiryoku kennkyu kaihatu kikou (2007) Heisei 19 nendo houshasen riyou no keizai kibo ni kansuru cyousa houkokusho – Syokuhin syousha kaigai cyousa – (Naikakufu itaku jigyou), Heisei 19 nen 12 gatu (in Japanese); Japan Atomic Energy Agency (2007) Report of the research on the economic scale of the utilization of radiation in FY 2007 (research on food irradiation overseas) (A project conducted in December 2007 on behalf of the cabinet office) (in Japanese)
6. Ohno S (2007) Genshiryoku syoutotu to houshasen sayou – Nichijo seikatu ni kakawari no fukai houshasen rikai no tameni. Houshasen kyouiku 11(1):3–12 (in Japanese); Ohno S (2007) Atomic collision and effects of radiation – for better understanding of applications of radiation that are closely related to our daily lives. Radiat Educ 11(1):3–12 (in Japanese)
7. Itoh H (2006) Syokuhin no syousha kouka to eiseika. Houshasen to sangyou 112:36–41 (in Japanese); Itoh H (2006) Effects of food irradiation and food hygiene. Radiat Ind 112:36–41 (in Japanese)
8. http://biotech.nikkeibp.co.jp/fsn/kiji.jsp?kiji=2599
9. Miyanoshita A (2007) Yakkai na kuri gaicyu Kurisigizourimushi – kuri ni musikui ga nai riyuu -. Kaoku gaicyu 28(2):167–168 (in Japanese); Miyanoshita A (2007) Chestnut weevils – mean insects that harm chestnuts; the reason why chestnuts do not have wormholes. Harmful House Insects 28(2):167–168 (in Japanese)
10. Miyanoshita A, Imamura T (2009) Kouatu tansan gasu o mochiita Kurisigizourimushi saccyu gijutu. Shokubutu boueki 63(5):302–306 (in Japanese); Miyanoshita A, Imamura T (2009) High-pressure carbon dioxide gas-based insecticidal technology to kill chestnut weevils. Plant Prot 63(5):302–306 (in Japanese)
11. Itoh H (2006) Waga kuni ni okeru syokuhin syousha gijutu no kaihatu. Houshasen to sangyou 110:36–42 (in Japanese); Itoh H (2006) Development of food irradiation technologies in Japan. Radiat Ind 110:36–42 (in Japanese)
12. Itoh H (2007) Shousha syokuhin no kenzensei hyouka to houshasen bunkai seiseibutu. Houshasen to sangyou 133:26–31 (in Japanese): Itoh H (2007) Evaluation of the soundness of irradiated foods and radiolysis products. Radiat Ind 113:26–31 (in Japanese)
13. Itoh H (2007) Tokutei sougou kenkyuu deno doubutu shiken no kekka ni tuite. Houshasen to sangyou 115:6–11 (in Japanese); Itoh H (2007) About the results of the animal experiments conducted in the designated general studies. Radiat Ind 115:6–11 (in Japanese)

14. WHO Technical Report Series 890 (1999) High-dose irradiation: wholesomeness of food irradiated with doses above 10 kGy. World Health Organization, Geneva
15. Itoh H (2007) Shokuhin no syousha gijutu, Kenchi hou to kongo no tenbou. Houshasen to sangyou 114:33–39 (in Japanese); Itoh H (2007) Food irradiation technologies, irradiated food detection methods and the future prospects of food irradiation. Radiat Ind 114:33–39 (in Japanese)
16. http://www.mhlw.go.jp/topics/yunyu/hassyutu/dl/263.pdf
17. Todoriki S, Saito K, Tsujimoto Y (2008) Netu ruminessensu hou ni okeru TL hakkou kyokusen no sekibun ondo hani no TL hi ni taisuru eikyou – Kousei roudou shou tuuchi hou to CEN hyoujun bunseki hou no hikaku -. Shokuhin syousha 43:25–36 (in Japanese); Todoriki S, Saito K, Tsujimoto Y (2008) Effects of the temperature integration range of the TL luminescence curve on the TL ratio in the thermo – luminescence method – comparison of the standard CEN analysis method with the notification method of the Ministry of Health, Labour and Welfare. Food Irradiat 43:25–36 (in Japanese)
18. Yamazaki M, Sekiguchi M, Nakagawa S, Yuzuki S (2009) Netu ruminesensu (TL) hou no genre to sokutei no jissai. Shokuhin syousha 44:24–31 (in Japanese); Yamazaki M, Sekiguchi M, Nakagawa S, Yuzuki S (2009) Principles of the Thermo-Luminescence (TL) method and practical application of TL method-based measurement. Food Irradiat 44:24–31 (in Japanese)
19. http://www.ansa-spice.com/M08_SpiceQandA/SpiceQandA.html
20. http://www.aec.go.jp/jicst/NC/iinkai/teirei/siryo2006/siryo40/siryo11.pdf
21. http://www.aec.go.jp/jicst/NC/iinkai/teirei/siryo2006/kettei/kettei061003_2.pdf
22. http://www.fsc.go.jp/iinkai/i-dai74/dai74kai-siryou2-2.pdf
23. http://www.fsc.go.jp/iinkai/i-dai74/dai74kai-gijiroku.pdf
24. http://www.fsc.go.jp/iinkai/i-dai182/dai182kai-siryou3-1.pdf
25. http://www.fsc.go.jp/iinkai/i-dai182/dai182kai-gijigaiyou.html
26. http://www.mhlw.go.jp/topics/yunyu/monitoring/dl/02-070706a.pdf
27. http://www.peopletree.co.jp/news/notice/masara1.html
28. Nihon shokuhin syousha kenkyuu kyougikai jimukyoku (2009) Waga kuni ni okeru syokuhin syousha gijutu no kentou joukyou. Syokuhin syousha 44:32–42 (in Japanese); Secretariat of the Japanese Research Association for Food Irradiation (2009) Current statuses and results of discussions on and evaluations of food irradiation technologies in Japan. Food Irradiat 44:32–42 (in Japanese)
29. Tsuchiya F (2001) Gyuunyuu dokuhon, NHK syuppan, Tokyo (in Japanese); Tsuchiya F (2001) Introduction to cow's milk. NHK Publishing, Inc., Tokyo (in Japanese)
30. http://food-entaku.org/
31. Nakagawa S (2009) Dai 46 kai aisotopu/houshasen kenkyuu happyoukai sanka houkoku. Houshasen kagaku 88:49 (in Japanese); Nakagawa S (2009) Report of my participation in the 46th meeting for presentation of isotope and radiation research. Radiat Chem 88:49 (in Japanese)
32. http://food-entaku.org/food-entaku20091113happyou.pdf. (Ichikawa M (2009) Syokuhin shousha no risuku komyunikeshon, Dai 13 kai houshasen purosesu sinpojiumu kouen yousisyuu, 2009 nen 11 gatu 13 nichi, Tokyo). http://food-entaku.org/food-entaku20091113happyou.pdf (in Japanese). (Ichikawa M (2009) Communicating information on risks associated with food irradiation. Summaries of presentations made at the 13th symposium on radiation processes held on November 13, 2009 in Tokyo) (in Japanese)
33. Hayashi T (hen) (2008) Syokuhin/Nougyou bunya no houshasen riyou. Saiwai shobo, Tokyo (in Japanese); Hayashi T (ed) (2008) Utilization of radiation in food processing and agriculture. Saiwai Shobo, Co., Ltd., Tokyo (in Japanese)
34. http://www.jaif.or.jp/ja/sangyo/qa-handbook_intro.html
35. http://foodirra.jaea.go.jp/
36. http://jrafi.ac.affrc.go.jp/index.html

Chapter 16
Radiation-Based Plant Diagnostics: Positron Imaging-Based Studies of Plants

Shinpei Matsuhashi

16.1 Positron Imaging-Based Studies of Plants

Positron imaging is a new technology for visualizing materials transfer in living plant under non-invasive condition. In the positron imaging, positron-emitting RI itself or labeled compound is used as a tracer. In the field of medicine, positron imaging is being utilized in the form of PET (positron emission tomography) for image-based diagnosis of cancers and is greatly helping in detecting cancers. Plant positron imaging is a positron imaging technology specially developed for studies of plants.

16.1.1 Discovery of Positrons and Positron Imaging

Nearly 40 years after the discovery of the X-ray by Roentgen in 1895, C.D. Anderson, while conducting a study of cosmic rays using a cloud chamber placed in a magnetic field, discovered in 1932 that some of the electrons showed a track that turned to the direction opposite of the direction of normal electrons when the cosmic ray energy was sufficiently dampened with lead, even though their mass was the same as that of normal electrons. This was the discovery of positrons (positive electrons). The existence of positrons itself had been predicted by P.A.M. Dirac in 1928 by developing equations for the particulate and wave characteristics of electrons that showed electrons and positrons could exist in two energy states (positive and negative).

S. Matsuhashi (✉)
National Institutes for Quantum and Radiological Science and Technology, Chiba, Japan
e-mail: matsuhashi.shimpei@qst.go.jp

H. Kudo (ed.), *Radiation Applications*, An Advanced Course in Nuclear Engineering 7, https://doi.org/10.1007/978-981-10-7350-2_16

Medical utilization of positron-emitting radionuclides started in 1951 with the attempt by F.R. Wren Jr. et al. to achieve ^{64}Cu-based rendering of brain tumors by the simultaneous measurement method using a pair of detectors made by attaching photomultiplier tubes to cylindrical NaI crystallines. Then, a prototype positron camera equipped with detectors that were deployed so that they face each other was developed in 1959, and a prototype equipment equipped with 32 NaI detectors that were deployed so that they surround the head was developed in 1960. In 1969, an equipment equipped with detectors deployed in a circle, which is the origin of present PET (positron emission tomography) equipment, was developed. In 1972, PET equipment equipped with polycrystalline-type detectors that were deployed so that they face each other and were designed to rotate around the measurement target was developed. Initial development of PET equipment was mainly the development of PET equipment for two-dimensional imaging, but subsequent development of X-ray computed tomography (CT) led to the development of PET equipment dedicated for tomography in 1973. Initial PET equipment employed NaI detectors. However, BGO ($Bi_4Ge_3O_{12}$) started attracting attention as the new detector material because of its non-deliquescent and the higher detection efficiency expected to be achieved with it, and, in 1977, PET equipment equipped with BGO detectors was developed.

In Japan, development of positron imaging technologies started in 1974 with the creation of a prototype positron camera by Okano et al. at RIKEN. The first PET equipment in Japan was developed in 1980 by the National Institute of Radiological Sciences in a joint project with Hitachi Medical Corporation and by the Research Institute for Brain and Blood Vessels-Akita in a joint project with Shimadzu Corporation. Following the development of PET equipment with a single layer of detector ring, whole-body PET equipment with multiple detector ring layers (such as three-layer, five-slice-type and four-layer, seven-slice-type PET equipment) was developed one after another [1–6].

Development of PET technologies and equipment evolved from the development of technologies to acquire tomographic images using a single layer of detector ring to the development of equipment with multiple detector ring layers to the development of measurement sensitivity improvement technologies and then to the development of spatial resolution improvement technologies. Searches for new crystalline and improvement of photomultiplier tubes, etc., made as part of the efforts to develop component technologies for these, coupled with advances in computer technologies, led to the development of three-dimensional PET equipment that acquires data directly as stereoscopic images rather than reconstructing three-dimensional images from stacked two-dimensional tomographic images and the development of PET-CT equipment that combines X-ray CT with PET, which makes diagnosis easier and more reliable by acquiring function images and shape images at the same time and stacking them. Technological development to improve the spatial resolution, which is a measure of the performance of PET equipment as measuring equipment for diagnostics, has constantly been pursued. The spatial resolution of whole-body PET equipment was only slightly finer than 20 mm in the mid-1970s but was improved to 7–8 mm in 10 years (i.e., by the mid-1980s) and to 4 mm in 20 years (i.e., by the mid-1990s). The spatial resolution of PET equipment for the head was improved to a little finer than 3 mm by the mid-1990s.

This section outlined the history of the development of positron imaging, which is a radiation-based measurement technology, with a focus on the history of the development of "detectors." Needless to say, to develop and improve technologies, activities to develop positron-emitting radionuclides which are measurement targets and develop marker compounds that use them and develop technologies to analyze images obtained by measurement, as well as activities to develop higher-quality measurement technologies utilizing feedback of various findings, are being pursued in parallel with them affecting each other.

16.2 Plant Positron Imaging Technologies

16.2.1 Overview of Plant Positron Imaging

In the 1990s, research and development of plant positron imaging technologies was started with the objective of developing new research methods to investigate functions of living bodies (especially on the individual level) in a noninvasive manner. This research and development was started as a joint study between Japan Atomic Energy Agency (JAEA; the former Japan Atomic Energy Research Institute), which had been conducting studies on accelerator-based production of RIs using the Takasaki Ion Accelerators for Advanced Radiation Application (TIARA) (an ion beam irradiation research facility that had been completed before the start of the research and development and was comprised of four accelerators including an AVF cyclotron) of the JAEA Takasaki Advanced Radiation Research Institute, and Hamamatsu Photonics K.K., which had been conducting studies to develop more compact and higher-performance PET equipment.

This research and development selected plants, whose dynamics had scarcely been studied, as the measurement target and aimed at developing imaging technologies to observe the ways plants live as they are, including from where to where substances controlling functions of plants on the individual level (including trace substances capable of doing so such as hormones) move, and what happens as they move. For achieving this required high-sensitivity detection, it was decided to use radioisotopes (RIs; It seems that these are more often called "radionuclides" than "radioisotopes" outside Japan) as tracers by labeling them. The research and development aimed at extending the frontiers of plant research by using positron-emitting radionuclides, which had already been utilized in the field of medicine and emit positrons (positive electrons) that generate a pair of gamma rays through annihilation with electrons. The developed equipment is shown in Fig. 16.1 [7, 8].

These new measurement technologies for living plants included the following three types of technologies (Fig. 16.2):

Fig. 16.1 Plant positron imaging equipment developed jointly by JAEA and Hamamatsu Photonics K.K. The equipment shown in the photograph is equipment 1, which is comprised of two pairs of two stacked detector units. Equipment 1 has already been decommissioned and is currently being used as demonstration equipment for explaining the structure of plant positron imaging equipment and the mechanism of the measurement at an exposition museum

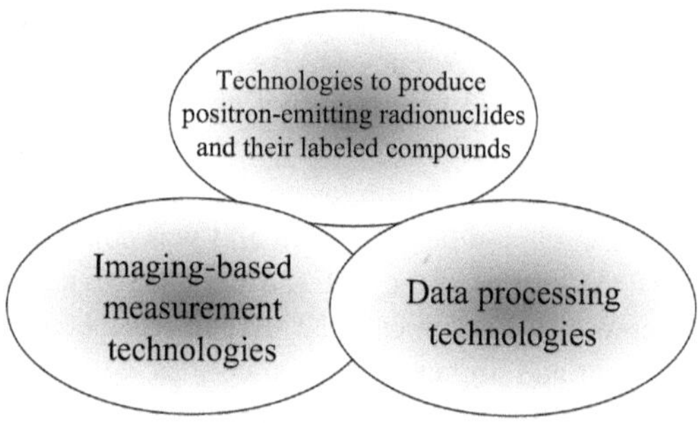

Fig. 16.2 Three types of plant positron imaging technologies

1. Technologies to produce and develop positron-emitting radionuclides and their labeled compounds
2. Imaging-based measurement technologies to measure the dynamics of RI tracers inside plants in real time
3. Image data analysis technologies to deepen the understanding of physiological functions of plants

The research and development initially focused on the development of positron-emitting radionuclides and their radiolabel compounds and the development of imaging equipment. Because the new imaging equipment which formed the core of the imaging-based measurement technologies was imaging-based measuring equipment that used positron-emitting radionuclides as tracers, Kume et al. named it "plant positron imaging equipment (positron-emitting tracer imaging system (PETIS))."

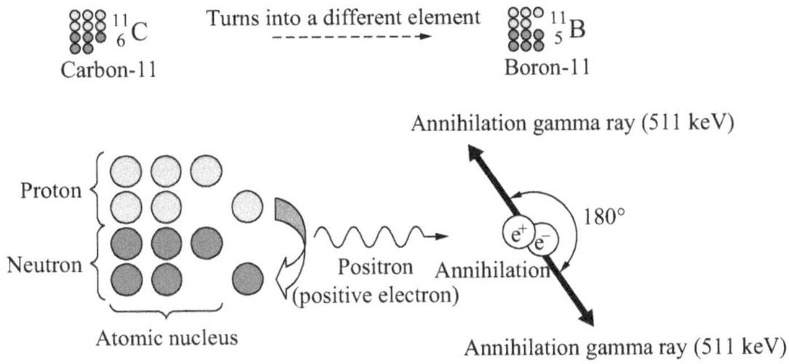

Fig. 16.3 Decay of a positron-emitting radionuclide (^{11}C) and resultant generation of annihilation gamma rays

16.2.2 Plant Positron Imaging Equipment (PETIS)

Principles of Positron Imaging Measurement

PETIS-based positron imaging measurement uses positron-emitting radionuclides as tracers. The term "positron-emitting radionuclides" is a generic term for radioisotopes that β^+ decay. In β^+ decaying, one of the protons in the atomic nucleus changes into a neutron. During this change, a positron (positive electron) is emitted as the differential electric charge between the proton and neutron. As the name implies, positrons are positively charged electrons (antiparticles of electrons). A positron emitted from an atomic nucleus has energy. It will travel around in random directions until the energy is expended and ultimately annihilate together with an electron. This annihilation generates a pair of 511 keV gamma rays (annihilation gamma rays). One of these gamma rays corresponds to the mass of the positron and the other to the mass of the electron. These two gamma rays travel in opposite directions (i.e., in directions 180 ($\pm$0.5) degrees apart from each other). In positron imaging measurement, the pair of annihilation gamma rays is measured using the simultaneous measurement method.

Figure 16.3 presents a schematic diagram showing what happens as a positron-emitting radionuclide (^{11}C (carbon-11)) decays. One of the protons in the atomic nucleus changes into a neutron, and the ^{11}C turns into stable ^{11}B (boron-11). The positron released from the atomic nucleus as a result of this annihilates with an electron, and a pair of annihilation gamma rays is projected with the two 511 keV gamma rays traveling in opposite directions (i.e., in directions 180° apart from each other).

Because the measurement target of plant positron imaging measurement is plants rather than animals, deployment on a single plane is possible. Therefore, the development aimed at developing a system that does not require the reconstruction of images from two-dimensional images, which is required in the case of PET equipment. Positron imaging is based on simultaneous measurement of light signals from scintillator crystals placed on the detectors. Because PETIS's measurement program has been written based on the assumption that the positron annihilation point is always located on the central surface equidistant of the two detectors that are deployed so that they face each other, it is possible to identify the annihilation point even from a single annihilation. That is, the system directly presents the measurement result in the form of images in real time, without requiring the image reconstruction process that is required in the case of PET equipment. On the other hand, PET measurement is based on a two-dimensional simultaneous measurement using detectors deployed in a circle, and what can be obtained from one event of PET measurement (one simultaneous measurement) is information on "chords" between detectors, which means that PET measurement identifies the chords on which the RIs decayed emitting positrons, but does not identify the exact points where the RIs existed. Therefore, it is not possible to know where in the disk-shaped two-dimensional space the RIs existed unless the intersection point of at least two events is determined. In addition, the premise for PET measurement is that the two chords identified from two events intersect. However, this does not always happen, because the directions of travel of annihilation gamma rays are probability events. Therefore, measurement must be repeated until an intersection point appears. This means that the noise component increases by that amount. On the other hand, PETIS is theoretically capable of identifying the point where a single RI existed.

Some researchers may think that obtaining accurate data is what is important in studies and question the value of obtaining data in the form of images in real time, but, as a member of the development team, I believe that PETIS has an irreplaceable value in the sense that it provides researchers with the excitement and joy of being able to see what is happening in front of them in the form of images in real time, which I believe causes supreme excitement and joy for researchers.

In PETIS-based measurement, the plant sample is placed in such a way that it is located exactly at the midpoint between the two two-dimensional position detection-type gamma ray detectors that are deployed so that they face each other. As an example, Fig. 16.4 presents an illustration showing how $^{11}CO_2$-based imaging measurement of the process of accumulation of photoassimilates in the ears of wheat is made. The annihilation gamma rays generated as a result of the annihilation travel in opposite directions and hit one of the BGO scintillators of the BGO scintillator arrays (comprised of BGO scintillators deployed in a grid pattern) installed in front of the detectors that are deployed so that they face each other. The incidence of the gamma rays causes the BGO crystal elements to emit light, and the light is transformed into electrons by the photomultiplier tubes installed behind the array. The electrons are then amplified through the electron avalanche phenomenon. The point of incidence is identified by measuring the distribution of the amplified electrons using the "x axis-y axis" mesh-type resistance chain installed as the last tier and calculating the position of the center by means of computation to calculate the center of gravity.

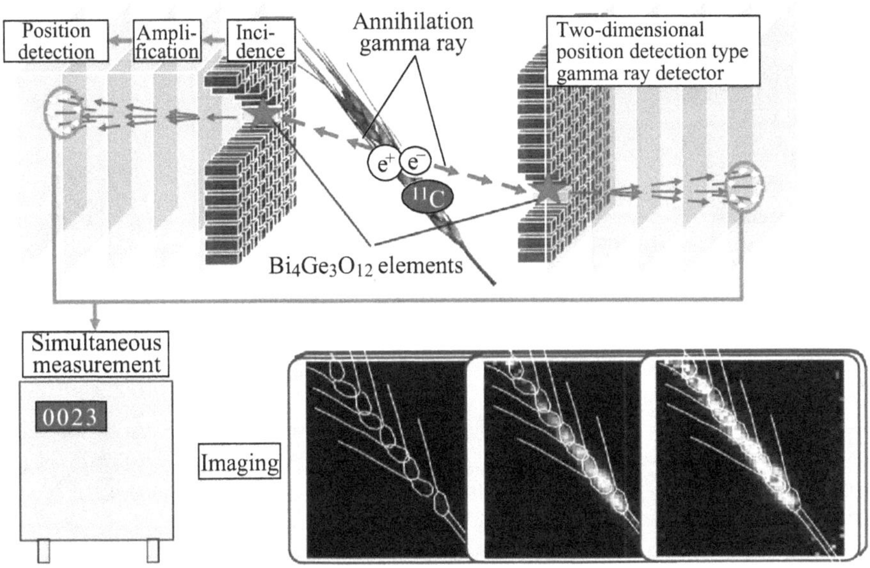

Fig. 16.4 Overview of PETIS-based positron imaging measurement

Development of Plant Positron Imaging Equipment

As described above, JAEA made an attempt to develop a two-dimensional imaging-based measurement technology dedicated for plant individuals of sizes that can be handled in a laboratory, based on the belief that such a technology is beneficial in that, unlike traditional imaging technologies such as autoradiography, it is not a technology for the acquisition of stationary distribution images which is inherently invasive and the belief that it can serve as an important tool for analyzing physiological functions of plants (especially those that involve movement of substances) like PET, which is a new image-based diagnostics technology being utilized in the field of medicine.

The first prototype equipment developed to achieve this was equipment with a measurement field of view of 4.8 cm (H) × 5.0 cm (W), which was even smaller than a trefoiled soybean. However, Kume et al. succeeded in absorbing the solution from the cut soybean and the rose stalk and observing how the solution was transported to soybean leaves and accumulated and transported to rose flowers (Fig. 16.5) [11, 12]. They named this two-dimensional imaging-based measuring equipment "PETIS" to distinguish it from PET equipment, which is equipment for three-dimensional tomographic imaging.

Because PETIS equipment is designed for two-dimensional imaging-based measurement only, it does not require the image reconstruction process required in PET equipment, which is equipment for three-dimensional imaging. For this reason, PETIS equipment has a higher S/N ratio than PET equipment. The PETIS

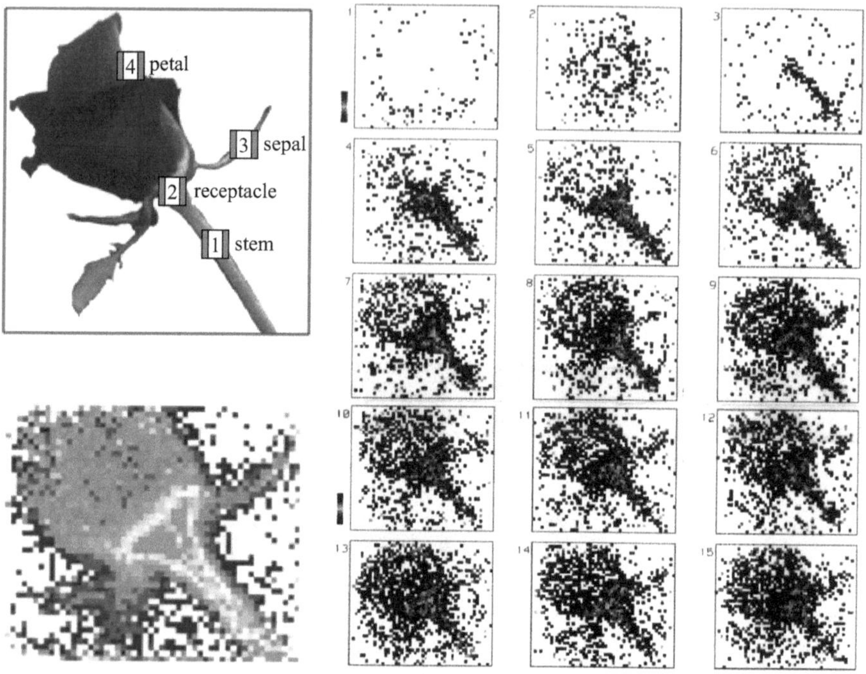

Fig. 16.5 PETIS images showing how ^{18}F that had been given to cut rose stems was transported to rose flowers. Upper left: the measured parts of rose with a 30 cm long stem. Lower left: an image showing the ^{18}F distribution 30 min after the dosing of ^{18}F. The 15 images on the right show the cumulative distribution of ^{18}F 2, 4, . . ., 28, and 30 min after the dosing of ^{18}F (The left image in the top row, the central image in the top row, . . ., the central image in the bottom row and the right image in the bottom row show the cumulative distribution of ^{18}F 2 min, 4 min, . . ., 28 min, and 30 min after the dosing of ^{18}F, respectively). The data collection time for each image was 2 min

equipment developed had a spatial resolution of 2.4 mm, which was higher than the resolution (3–4 mm) of PET equipment that was commercially available at the time.

A practical unit was developed based on the results of the test experiments conducted with the prototype measuring equipment, and it was named PETIS 1. PETIS 1 is comprised of two two-dimensional position detection type gamma ray detectors which are facing arranged, jigs to secure the detectors and the plant sample, the circuit section that processes signals from the detectors and make images. And a computer conducts a measurement, and collects and stores data. The detectors (which play the central role in PETIS) are capable of two-dimensionally detecting the point of incidence of the gamma ray. Each detector is comprised of a scintillator array comprised of BGO ($Bi_4Ge_3O_{12}$) scintillators with a cross section of 2 mm × 2 mm and a length of 20 mm deployed in a grid pattern at 2.2 mm intervals (an array with 22 columns and 23 rows) and a square two-dimensional position detection-type photomultiplier tube (Fig. 16.6). The white parallelepipeds shown in the photographs of Fig. 16.6 are a BGO scintillator

Fig. 16.6 Structure of PETIS's two-dimensional position detection-type detector. Photographs of the central part of a PETIS detector, which is comprised of a BGO scintillator array and a two-dimensional position detection-type photomultiplier tube, as seen diagonally from the front (photograph on the left) and from the side (photograph on the right)

array. Each of the grid meshes shown in the photograph on the left of Fig. 16.6 is the 2 mm × 2 mm front surface of a BGO crystal. Each BGO crystal has 2 mm × 20 mm side surfaces as shown in the photograph on the right of Fig. 16.6. More than 500 BGO crystallines are arranged neatly in a small 5 cm × 5 cm space. Because the photomultiplier tubes behind the BGO scintillator arrays have to collect signals from BGO scintillators in the form of light from them, it was necessary to employ a photomultiplier tube cross section that is larger than the cross section of the array, and this prevented the development team from achieving a compact detector base section. PETIS 3's detector base sections are much more compact because it employs metal package-type photomultiplier tubes developed later, which is much smaller than glass photomultiplier tubes.

PETIS 1 is a system comprised of a pair of detector sets with each detector set comprised of one to four detectors, each comprised of a BGO scintillator array (shown in Fig. 16.6) connected to a two-dimensional position detection-type photomultiplier tube, and with the two detector sets deployed so that they face each other. The maximum configuration (a pair of detector sets with each detector set comprised of four detectors (two detectors in the vertical direction and two detectors in the horizontal direction)) allows a maximum field of view of 15 cm × 15 cm to be achieved (Fig. 16.7), and the power supply system and computation circuit of PETIS 1 is designed to be capable of handling this. PETIS 1 in the maximum configuration is capable of obtaining images for the areas that are not covered by the detectors as well, by performing measurement between adjacent detectors simultaneously with the measurement between the detectors facing each other ("simultaneous cross measurement").

After the development of PETIS 1, continuous improvement of PETIS has been made. PETIS 1 is designed to be capable of being equipped with a maximum of four pairs of detectors, but PETIS 2 developed after PETIS 1 is designed to be capable of being equipped with a maximum of six pairs of detectors. The basic functions of PETIS 2 is the same as those of PETIS 1, but PETIS 2 can achieve a maximum

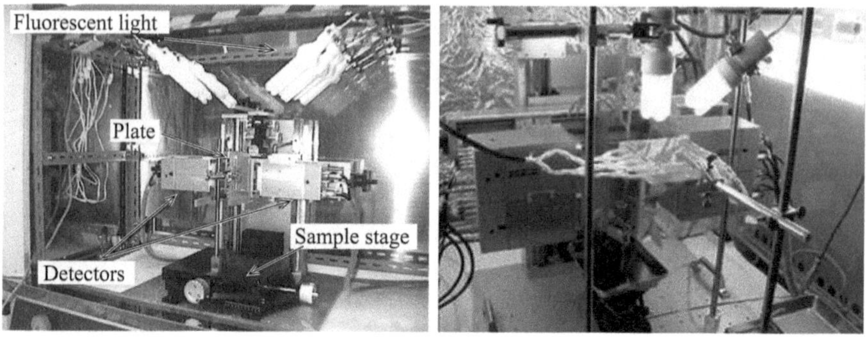

Fig. 16.7 PETIS 1 developed as a system for measuring functions of plants. The photograph on the left shows PETIS 1 configured to have a single pair of detectors (with a field of view of 5 cm × 5 cm), and the photograph on the right shows PETIS 1 configured to have two sets of two stacked detectors (with a field of view of 5 cm × 15 cm)

measurement field of view of 15 cm (W) × 25 cm (H), thanks to the larger number of detectors (Fig. 16.8). This is a field of view that is 15 times as large as that of PETIS 1 in the minimum configuration (one pair of detectors). Expanding the measurement field of view not only allows larger plants to be measured but also allows test plant individuals and control plant individuals to be measured simultaneously under the same conditions, thereby contributing to improving the accuracy and efficiency of experiments. An example is given in Fig. 16.9.

In PETIS 1 and PETIS 2, deterioration of the image around the detectors was unavoidable because the photomultiplier tubes used were glass photomultiplier tubes. However, PETIS 3 is capable of achieving far higher image quality, thanks to the various improvements introduced including the introduction of metal package-type photomultiplier tubes.

Each of the scintillator arrays employed in the detector base sections of the detectors of PETIS 3 is comprised of 2 mm × 2 mm × 20 mm BGO scintillators arranged in a grid pattern at intervals of 2.2 mm and is connected to a photomultiplier tube as in PETIS 1 and PETIS 2, except that the scintillator arrays of PETIS 3 are arrays with ten columns and ten rows. The PETIS system is so designed that signals are read out separately for each of the four detector blocks (two in the vertical direction and two in the horizontal direction), each comprised of a scintillator array and a photomultiplier tube. PETIS 3 is designed to be capable of being equipped with two stacks of three detector blocks with the two stacks arranged side by side and is so programmed that simultaneous diagonal cross measurement between detector blocks can be made. Because of this, PETIS 3 is capable of achieving a larger field of view (120.8 mm (W) × 186.8 mm (H)) without any no-sensing section between detector blocks. In addition, PETIS 3 has much higher measurement performance than PETIS 1 and PETIS 2, including the finest achievable spatial resolution of 1.6 mm.

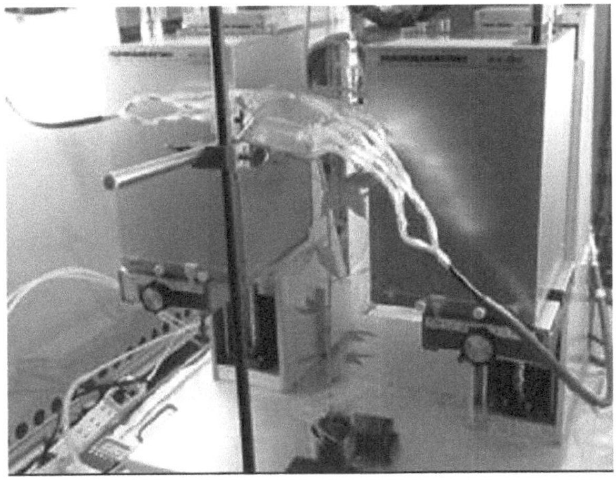

Fig. 16.8 PETIS 2 developed as a system that achieves a larger measurement field of view than PETIS 1 with the performance of PETIS 1 retained

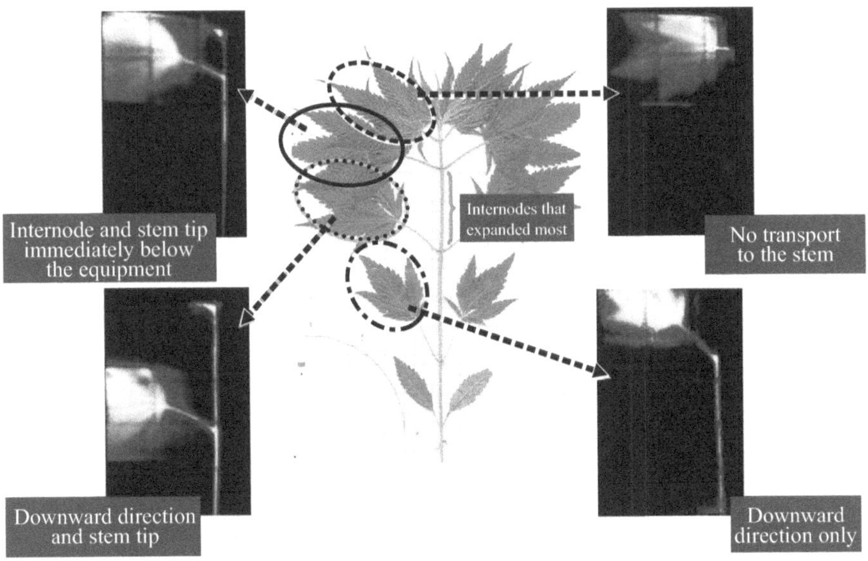

Fig. 16.9 PETIS 2 images showing how photoassimilates are transported in hemp. PETIS 2 images obtained in an experiment in which the experiment team had different leaves (encircled by ellipses drawn in different types of line) of a hemp plant individual absorb $^{11}CO_2$, which is a raw material for photosynthesis, and observed how and to where the photoassimilates were transported within the hemp. This was an excellent experiment conducted by Sakamoto et al. at JAEA which clearly showed the roles leaves at different stages of growth play in a plant individual. If it weren't for PETIS 2, which allows noninvasive measurement using a short half-life RI to be achieved, this experiment could not have been realized

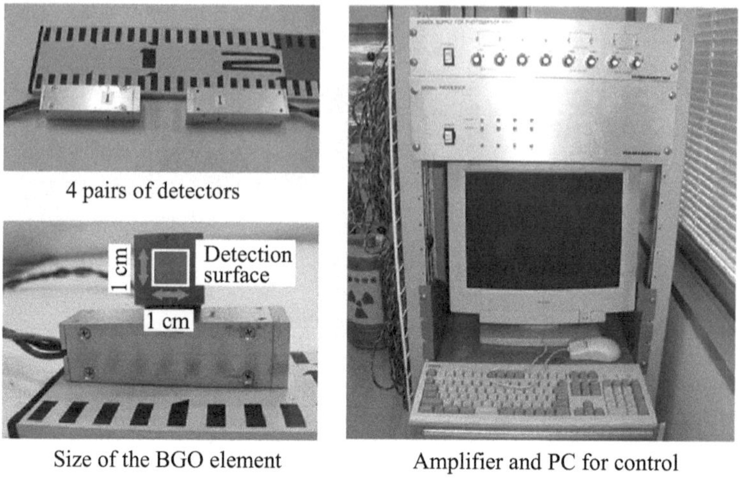

Fig. 16.10 Positron-based multipoint detection equipment. Four pairs of detectors (upper left) can be installed at different locations. Each detector contains a 1 cm × 1 cm × 2 cm BGO scintillator (lower left). The circuit system of this equipment is far smaller than that of PETIS because there is no need to display images (right)

Multipoint-Type Positron Imaging Measuring Equipment

PETIS was developed with the objective of visualizing in the form of images the behaviors in living plants of positron-emitting radionuclides as tracers, and its coverage area (measurement field of view) was gradually expanded. However, even though PETIS is measuring equipment for small plants, it is not always the case that the plant sample can be measured its entirety with a measurement field of view of the size of a B5 sheet of paper. In addition, there are cases where no image is necessary and knowing the change in the amount of RI that passed through a certain part suffices. For these reasons, positron-based multipoint detection equipment equipped with four pairs of probe-type detectors was developed which employs a simultaneous measurement technique like PETIS to improve the measurement accuracy but is not intended as equipment that obtains images (Fig. 16.10).

16.2.3 Positron-Emitting Radionuclides and Their Labeled Compounds

Utilization of Radioisotopes as a Tool for Plant Studies

Radioisotopes (RIs) are being utilized as a useful tool for biological studies and also as a tool for studies to investigate functions of plants. The most commonly used

Table 16.1 Radionuclides used for plant studies

Radionuclide	Decay type	Half-life	Energy (MeV)
^{3}H	β^-	12.3 years	0.0186
^{14}C	β^-	5730 years	0.156
^{32}P	β^-	14.26 days	1.709
^{35}S	β^-	87.5 days	0.167
^{125}I	EC	60.1 days	0.0355 (γ), 0.0275 (X)

radionuclides include tritium (^{3}H), carbon-14 (^{14}C), phosphorus-32 (^{32}P), sulfur-35 (^{35}S), and iodine-125 (^{125}I) (Table 16.1).

The half-lives of these radionuclides range from 14.3 days for ^{32}P to 5730 years for ^{14}C. A long half-life is necessary to synthesize compounds by labeling these radionuclides to substances that are present in living bodies or taken from outside into living bodies. Researchers do not always have to produce these radionuclides and their labeled compounds; they can choose to purchase them instead. A well-known example of utilization of these radionuclides in studies to elucidate functions of plants is the discovery of the Calvin-Benson cycle of photosynthesis through the use of ^{14}C [13]. ^{14}C is a popular RI that is also used for radiocarbon dating, but its utilization only started after nuclear reactor-based production of it became possible in 1945.

Most of the RIs shown in Table 16.1 are beta ray-emitting RIs (^{3}H, ^{14}C, ^{32}P, and ^{35}S), but ^{125}I is a gamma ray-emitting RI. The beta ray energy differs between RIs, but, with the exception of the beta ray emitted by ^{32}P (1.71 MeV), the beta rays emitted by these RIs have low energy (one hundred and several tens keV or less) and their penetrating power is weak. For this reason, usually a liquid scintillator or autoradiography is used as a means to detect these radionuclides. In particular, the energy of the beta ray emitted by ^{3}H is so low (18.6 keV, which is only about one-tenth of the energy of the beta rays emitted by ^{14}C, ^{35}S, etc.) that even the presence of a Mylar film with a thickness of 1 μm makes it necessary to extend the exposure time [14]. Thus measurement of beta ray-emitting radionuclides requires work such as grinding the living body sample to which the RI tracer has been given, extracting components, and preparing thin slices, which means that it is practically impossible to measure samples as samples in the same state as in living bodies.

To obtain information on the distribution and behavior of a radionuclide in a living body as the in vivo distribution and behavior, it is necessary to use a radionuclide that emits gamma ray with strong penetrating power. Gamma ray-emitting radionuclides include not only ones that directly emit gamma ray as they decay, such as ^{125}I, but also ones that indirectly emit gamma ray through the annihilation of positive electrons released as a result of the decaying. These radionuclides are called "positron-emitting radionuclides" and are being utilized in PET (positron emission tomography) -based imaging diagnosis of cancers and analyses of metabolic physiological functions in the field of medicine. Attempts to utilize them in molecular imaging studies are also being made [15–17].

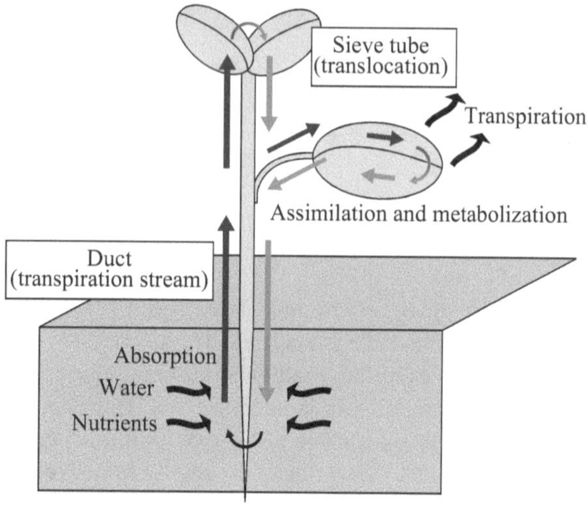

Fig. 16.11 Schematic diagram of the transport of water and nutrients within a plant individual. Plants absorb water and nutrients (such as nitrogen nutrients and metallic elements) using their roots and transport them to their aboveground part using the duct stream (or "transpiration stream"). They transport carbon nutrients they have assimilated through photosynthesis to the roots, leaves in the growth stage, fruiting bodies, etc., using the sieve tube stream

Utilization of Positron-Emitting Radionuclides in Plant Studies

Attempts to utilize positron-emitting radionuclides in studies of plants started many years ago. Before the production of the long half-life ^{14}C described above became possible, a study on photosynthesis using a positron-emitting radionuclide (^{11}C) produced using an accelerator was attempted at the end of the 1930s. The first researchers who utilized positron-emitting radionuclides were S. Ruben et al. They gave $^{11}CO_2$ they had produced using an accelerator (cyclotron) to barley, chlorella, etc., and studied the changes in the amount of the assimilated ^{11}C under various conditions [18–20]. After this study, plant studies using positron-emitting radionuclides such as ^{13}N and ^{18}F were attempted [21, 22]. These were utilization studies aimed at elucidating the metabolic paths and identifying the metabolic products, and it can be said that these studies focused on "the result of the movement" in terms of "from where to where a specific substance moved" (that is, they focused on comparison between the locations of the substance before and after the movement) like many other RI tracer-based plant studies conducted later, and that they did not include the concept of "the process." However, plants are living organisms like animals, and various substances are being dynamically transported in individuals of plants (Fig. 16.11). Utilization of positron-emitting radionuclides to investigate the mechanisms of such transport of carbon and nitrogen nutrients within individuals of plants includes time-series measurement of the amounts of radioactivity that passes through specific parts of a plant individual such as the stem and a leaf using radiation monitors such as scintillators. The first attempt to visualize time-series changes in the distributions of positron-emitting radionuclides in plant individuals in the form of images was made in 1988, by which time the development of PET for medical applications had entered the practical utilization stage. R.M.L. McKay et al. succeeded in which they cut soybean stems and had them absorb water solution

Table 16.2 Positron-emitting radionuclide tracers for studies of plants

Radionuclide	Half-life	Compounds
^{11}C	20 m	$^{11}CO_2$, ^{11}C-methionine, ^{11}C-formaldehyde
^{13}N	10 m	$^{13}NO_3^-$, $^{13}NH_4^+$, $^{13}N_2$
^{15}O	2.0 m	$H_2{}^{15}O$
^{18}F	110 m	$^{18}F^-$, ^{18}F-glucose, ^{18}F-proline
^{22}Na	2.6 y	$^{22}Na^+Cl^-$
^{48}V	16 d	$H_2{}^{48}VO_4^-$
^{52}Mn	5.6 d	$^{52}Mn^{2+}$
^{52}Fe	8.3 d	$^{52}Fe^{3+}$
^{62}Zn	9.3 h	$^{62}Zn^{2+}$
^{64}Cu	13 h	$^{64}Cu^{2+}$
^{105}Cd	56 m	$^{105}Cd^{2+}$
^{107}Cd	6.5 h	$^{107}Cd^{2+}$

of 18F they had produced using a Van de Graaff accelerator and obtained images showing the way the solution was transported from the stems [21].

Tracers for Plant Positron Imaging Measurement

Table 16.2 lists the positron-emitting radionuclides and their labeled compounds for which methods have been developed to produce them for use in plant positron imaging measurement or of which utilization in plant positron imaging measurement has been attempted.

In the case of production methods for positron-emitting radionuclide tracers for medical use, which have been developed together with PET, it is a matter of course that final products are positron-emitting radionuclide tracers dissolved in isotonic sodium chloride solution (a 0.9% NaCl water solution). However, in studies of plants, this NaCl (salt) can cause physiological stress in plants. Therefore, it was necessary to develop a new method to produce a final product (solution) that does not contain salt, even if the final product tracer is a radionuclide or its labeled compound had already been used in the field of medicine. Ishioka et al. were among the researchers who tackled the task of developing such a method from the onset. The development of plant positron imaging technologies tends to be regarded as the development of plant positron imaging "equipment," but it should not be forgotten that it would not have been possible to advance the development of plant positron imaging equipment without the researchers who made constant background efforts to develop and optimize many tracers for plant studies [22].

Positron-emitting radionuclides that can be utilized include not only the main nutrient elements for plants such as the carbon (^{11}C)) and nitrogen (^{13}N) described above but also essential trace elements such as manganese (^{52}Mn), iron (^{52}Fe), copper (^{64}Cu), and zinc (^{62}Zn). In addition, radionuclides of environmental pollutants such as cadmium (^{105}Cd, ^{107}Cd) and vanadium (^{48}V), which have been attracting attention in recent years in connection with the problem of the safety of food resources, can also be utilized. Therefore, suitable positron-emitting radionuclides can be selected from a wide range of positron-emitting radionuclides according to the research purpose [23].

In the production of positron-emitting radionuclides for PET, small accelerators (cyclotrons) capable of accelerating protons (H^+) to about 20 MeV are popularly used. These accelerators are usually designed to irradiate targets in the form of liquid or gas and are not designed to irradiate targets in the form of a solid. For this reason, the only radionuclides that can be produced are the so-called four PET radionuclides, namely, ^{11}C, ^{13}N, ^{15}O, and ^{18}F. (Author's note: Even with a small accelerator, it is possible to irradiate metal foil targets to produce positron-emitting radionuclides other than the four PET radionuclides if a port for irradiating targets in the form of a solid is provided. For example, the Faculty of Medicine of Gunma University has introduced a small accelerator capable of irradiating targets in the form of a solid.)

On the other hand, medium-sized accelerators capable of accelerating H^+ to about several tens to one hundred MeV have provisions for adjusting the irradiation environment more flexibly than in small accelerators including provisions for adding a port for irradiating targets in the form of a solid. This means that medium-sized accelerators allow the production of radionuclides that can cause nuclear reactions through irradiation of a metal foil target with a high-energy H^+ beam. In addition, medium-sized accelerators are capable of accelerating not only H^+ but also particles such as helium ions ($^4He^{2+}$), which allows the production of radionuclides utilizing nuclear reactions that are different from nuclear reactions achievable with small accelerators. In the case of the production of ^{18}F with a small accelerator, $H_2{}^{18}O$ is irradiated to utilize the ^{18}O (p, n) ^{18}F reaction. Therefore, it is necessary to purchase concentrated $H_2{}^{18}O$ as the target. On the other hand, the use of $^4He^{2+}$ generated with a medium-sized accelerator allows ^{18}F to be produced through the $^{16}O(\alpha(pn)^{18}F$ reaction. In the case of the method developed by Ishioka et al., $H_2{}^{16}O$, which most abundantly occurs in the natural world, can be utilized as the target [22].

The positron-emitting radionuclides listed in Table 16.2 have relatively short half-lives except for sodium-22 (^{22}Na) which has a long half-life of 2.6 years. In particular, the four PET radionuclides have very short half-lives (for example, the half-life of ^{18}F is as short as 110 min). Short half-lives are not suitable for synthesis of compound radiolabels requiring a complex reaction process or imaging measurement that is performed for a long period of time but allow the problem of individual differences (which always accompanies measurement of living organisms) to be avoided. For example, in the case of ^{11}C, whose half-life is 20 min, the amount of radioactivity will decay to one-eighth in 1 h and to one-sixty-fourth in 2 h. Therefore, in the case where the amount of radioactivity is a standard amount for imaging-based plant measurement, the amount of radioactivity in the plant will decay to the background level in several hours. This short decay time allows multiple time measurement to be performed on the same plant individual.

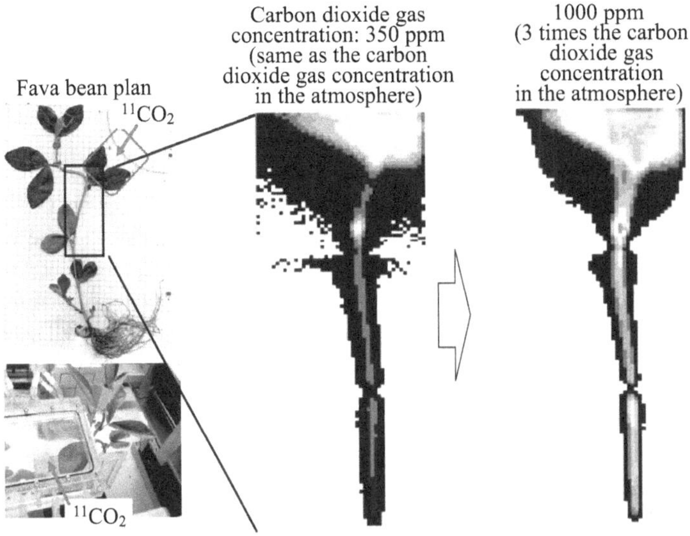

Fig. 16.12 Images from multiple time PETIS-based measurement of the transport of photosynthesis products in the same plant individual with the carbon dioxide concentration varied

Advantages of such short half-life radionuclides have been pointed out by J. Moorby et al [20]. Matsuhashi, Fujimaki, and their colleagues conducted imaging-based measurement of the transport of photosynthesis products multiple times in the same fava bean plant individual with the carbon dioxide concentrations in specific leaves varied (Fig. 16.12) [24].

Compounds labeled with radionuclide such as ^{18}FDG (^{18}F is the radiolabel glucose) and ^{11}C-methionine have been developed based on produced positron-emitting radionuclides, and these are being utilized in studies to investigate the relationship between sugar transport and signal transfer and studies to investigate the mechanism of the absorption of metallic elements from the soil.

16.2.4 Imaging-Based Measurement Technologies and Image Data Analysis Technologies

Effects of the Energy of Positrons

As mentioned above, positrons released from atomic nuclei have energy. However, the levels of energy possessed by individual positrons differ (these differences follow a certain distribution pattern) even between positrons emitted by the same positron-emitting radionuclide. One measure is the maximum positron energy. The maximum positron energy of a positron depends on the radionuclide that emits it. For example, the maximum positron energy of individual positrons emitted by ^{11}C, which is one of the radionuclides listed in Table 16.2, is 960 keV, whereas the maximum positron energy of individual positrons emitted by ^{13}N and the maximum positron energy of individual positrons emitted by ^{18}F are 1198 keV and 633 keV, respectively. The distance for which a positron travels around until it annihilates with an electron released by the atomic nucleus is proportional to the maximum positron energy. That is, the higher the maximum positron energy, the larger the distance is between the annihilation location and the location of the atom that released the positron. This distance as converted to the range in water is as long as 2.4 mm in the case of positrons from ^{18}F, 4.2 mm in the case of positrons from ^{11}C, and 5.4 mm in the case of positrons from ^{13}N. Matsuhashi et al. showed that these ranges due to positron energy deteriorate the quality of images obtained through imaging-based measurement. The higher the energy of a positron, the larger its effect is. In particular, in measurement of a plant with a small thickness, the range can be longer than the thickness of the plant sample, and some of the positrons may penetrate through the plant sample instead of annihilating in the sample. If this happens, the (apparent) number of measured annihilated positrons will be smaller than the number of positrons (the amount of RIs) that actually existed.

Measurement Environment of Imaging-Based Measurement

PETIS and PETIS-based measurement methods (that is, imaging-based measurement technologies) were developed through long trial-and-error-based development. In particular, in the development of measurement technologies for plants, intensive efforts were made to optimize the state measurement conditions because real-time measurement of behaviors of substances in living plants in the form of images was a type of measurement that had not been made in the past.

Positron imaging-based measurement requires both a facility for producing positron-emitting radionuclides and a facility that can handle unsealed radionuclides, regardless of whether the measurement target is human, animal, or plant. Positron imaging-based measurement also requires that these two types of facilities are located in the same building or adjacent to each other. The only facilities in Japan at which studies on positron imaging for plants can be conducted are TIARA of the JAEA Takasaki Advanced Radiation Research Institute, the National Institute of Radiological Sciences, and the PET Center of Hamamatsu Photonics K.K. JAEA's TIARA has positron imaging measurement equipment dedicated for plants.

PETIS is a main component technology for plant positron imaging, but measurement technologies for acquiring data of physiological significance and imaging data analysis technologies, as well as technologies to produce positron-emitting radionuclides and their radiolabel compounds, are also important component technologies. Plants sensitively react to changes in their environmental conditions such as light, temperature, and humidity and to vibrations due to wind, contact, etc. Therefore, to obtain good data in plant positron imaging-based measurement, it is

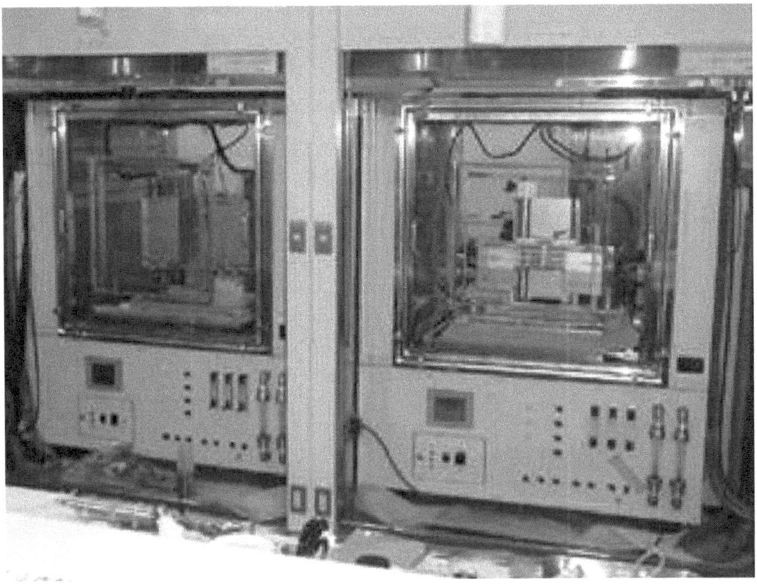

Fig. 16.13 JAEA's measurement environment (plant-growing vaults) for positron imaging-based measurement of plants. The vaults are capable of simulating the desired natural environment by controlling the temperature, humidity, and illumination of the inside space using a computer program and allow loads to be applied to specific parts of the plant through control of the concentrations and flow rates of atmospheric pollutants such as CO_2. As the measuring environments in the right and left vaults can be set separately (totally independent control), the test section and the control section can be measured in parallel

essential to make sure that the condition of the plant to measure does not change during the measurement. For this reason, JAEA has large plant-growing vaults (each having a 1 m (W) $\times$ 1 m (H) $\times$ 1 m (D) inside space) that control the illumination, temperature, and humidity in the inside space using a computer program and allow the operator to freely change the composition of the gas atmosphere in the inside space. By placing the plant sample and only the detector section of PETIS in one of these vaults, the plant sample can be measured in an environment that is completely independent of the outside environment (Fig. 16.13).

To conduct a study that uses a gaseous compound labeled with radionuclide such as $^{11}CO_2$, a completely closed system that can circulate the compound and supply it to only specific parts of the plant is required. In addition, to conduct a test to examine the responses of a plant to environmental condition changes, a system for controlling the composition of the gas in the inside space of the vault and the flow rate is required, because such a test involves mixing environmental pollution gases into the air supplied to the plant and making the carbon dioxide gas concentration higher or lower than the normal concentration. To make it possible to conduct such an experiment, a gas supply system was developed which is capable of pulse-

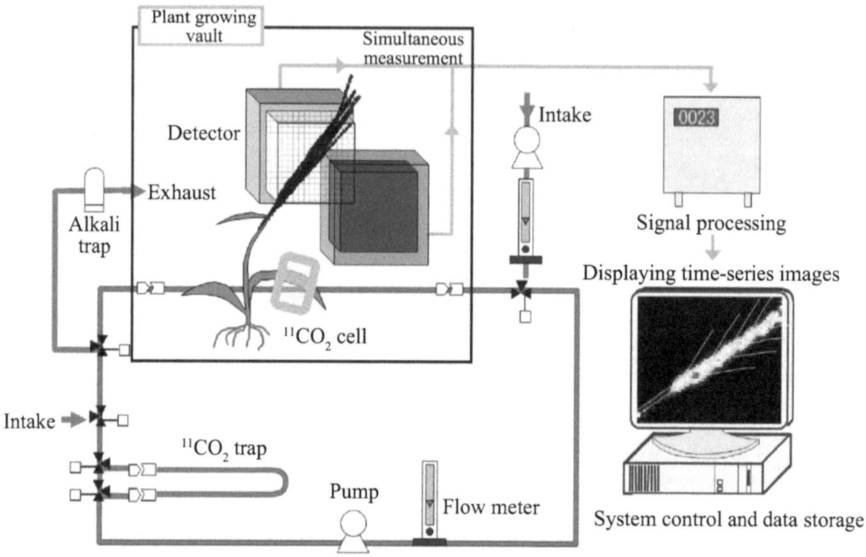

Fig. 16.14 Schematic diagram of the system for supplying gaseous positron-emitting radionuclide tracer

feeding $^{11}CO_2$ and exhausting it, replacing the air in the circuit with fresh air, and supplying gas at a constant flow rate (Fig. 16.14) [25, 26].

Visualization and Quantification of Functions of Plants Using Positron Imaging Technologies

In positron imaging-based measurement, images are constructed from time-series data that has been accumulated in the form of pixels (in PETIS, one pixel is 1.1 mm × 1.1 mm). Therefore, it is possible not only to make qualitative analyses such as comparison of images and comparison of the lengths of time tracers take to arrive at the specified area of interest but also to make quantitative analyses.

PETIS measurement data analysis methods that were used in the initial stage included methods that qualitatively compare the movement of tracers visualized in multiple measurement images or qualitatively compare the distributions of tracers in multiple measurement images and methods that calculate the lengths of time tracers take to arrive at the area of interest on the image from the time-activity curve (TAC). Then, as the quality of PETIS measurement technologies and image data from PETIS measurement became higher, the needs for "quantitative comparison" increased, and the development of methods for quantitatively analyzing PETIS image data started to be attempted.

In PETIS measurement, the movement of tracers in plant individuals, especially the transport of tracers in fibrovascular bundles, is measured. For this reason, studies were made to evaluate whether it is possible to introduce the transfer

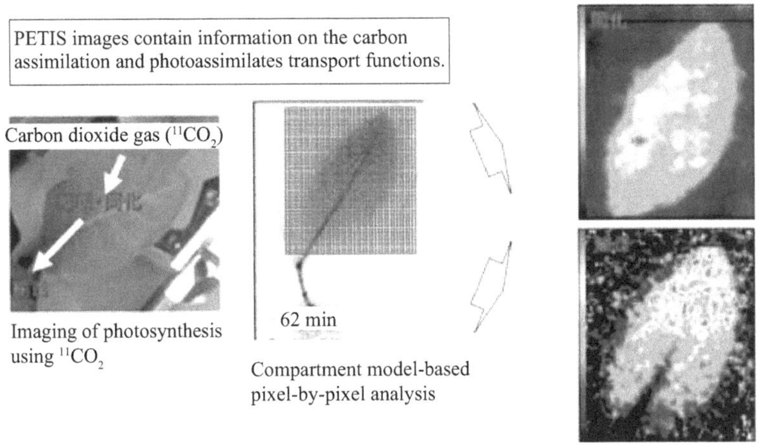

Fig. 16.15 Construction of images of functions based on compartment model-based analysis. The experiment focused on the behavior of carbon (^{11}C) and succeeded in separately extracting images of the distributions of intensity of the carbon assimilation function (upper right) and photoassimilates transport function (lower right) from a PETIS measurement image (center)

function method, which had track records in analyses of the transport of photosynthesis products in sieve tubes. Then, Fujimaki et al. developed a model system that simulates the behaviors of tracers in plants and constructed mathematical models and developed analysis methods. Matsuhashi, Fujimaki, and their colleagues conducted a study to analyze the speeds of transport of photosynthesis products in fava beans in a high-concentration carbon dioxide gas environment using these methods and succeeded in calculating the speeds of translocation of photoassimilates from PETIS image data [24].

The photosynthesis made by leaves of plants as seen in terms of the behavior of carbon can be divided into the "fixation" and "transport" functions [27–32]. Kawachi et al. succeeded in making simultaneous measurement of the amount of carbon fixed in leaves and the amounts of assimilation products transported, which had not been possible with the traditional methods, by expressing the behavior of carbon in leaves associated with photosynthesis with a compartment model (Fig. 16.15) [33].

Development of more advanced positron imaging-based measurement technologies and data analysis technologies will make it possible to construct images using the intensities of physiological functions of plants as indices, which is much better than merely obtaining tracer concentration distributions. In addition, realization of imaging-based measurement of functional differences on the molecule level can lead to realization of plant molecule imaging.

16.3 Utilization of Plant Positron Imaging Technologies in Studies of Plants

16.3.1 Behavior of Carbon in Plants

Plants acquire carbon nutrients through photosynthesis, which is their inherent function to acquire nutrients. Carbon is an element that forms the frameworks of sugars and amino acids. It also plays an important role as an energy source. Matsuhashi et al. visualized the way photoassimilates flow into and accumulate in the ears (the sink) of wheat in the process of maturing by PETIS measurement using $^{11}CO_2$ (Fig. 16.16). This experiment also showed that not only the light environment of the leaves (the source) but also the light environment of the ears (the sink) affected the transport of photoassimilates (Fig. 16.17).

Matsuhashi, Fujimaki, and their colleagues conducted an experiment in which they gave $^{11}CO_2$ to a fava bean plant individual with the carbon dioxide gas concentration varied and measured the transport of photoassimilates in the stem using PETIS, to investigate the effects of a high-concentration carbon dioxide gas environment on the transport of photoassimilates. They analyzed the obtained image data using the transfer function method and succeeded in calculating the speeds of transfer of photoassimilates at various parts of the stem and the proportions of photoassimilates that are unloaded from the sieve tube during the transport (Fig. 16.18). The high-concentration carbon dioxide gas environment increased the speeds of transfer of photoassimilates in both of the two plant individuals. The experiment showed that PETIS measurement is effective in achieving environmental response measurement that is not affected by differences between individuals.

Fujikake et al. discovered that reversible suppression of the activity of the root nodules of soybean plants in fixing nitrogen occurred when nitric acid was present and conducted an experiment in which they divided the roots of a soybean plant individual into two groups, applied different nitrogen nutrient treatment to the two groups, and measured the transfer of the $^{11}CO_2$ photosynthesis product using PETIS to investigate the effect of nitric acid on the supply of photosynthesis products to the root nodules [34]. The result of this experiment, together with the result of another measurement experiment conducted using $^{14}CO_2$, showed that the reduction in the amount of the photosynthesis product supplied to the root nodules due to the nitric acid treatment caused reversible suppression of the growth of the root nodules and the activity of the root nodules in fixing nitrogen.

$^{11}CO_2$ is used not only as a ground substance for the photosynthesis reaction but also as a starting substance for generating organic compounds, and ^{11}C-methionine is being used in studies. Nakanishi et al. conducted an experiment in which they dosed barley that was deficient in iron (which is an essential trace metal) with ^{11}C-methionine, which is a substance made by radiolabeling methionine (which is the precursor substance of mugineic acid, which is a chelate of iron that is excreted from the roots of barley deficient in iron), with ^{11}C, and observed the behavior of

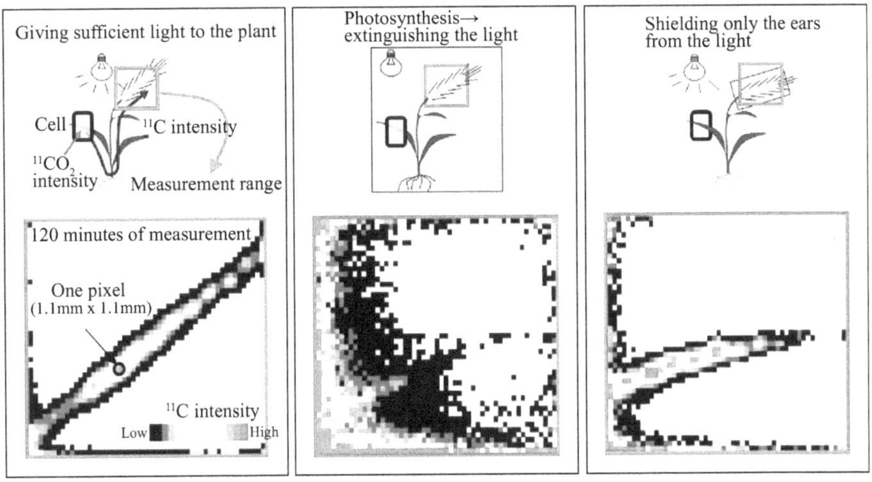

Fig. 16.16 Visualization and analysis of the way photoassimilates flow into and accumulate in the ears of wheat in the process of maturing [25, 26]. Images obtained in an experiment in which $^{11}CO_2$ was supplied for 10 min to a cell attached to a leaf of a wheat individual and the cumulative transfer of the ^{11}C compound generated by the assimilation through photosynthesis to the ears was measured for 2 h using PETIS. Left: result of the measurement performed with the light source kept on throughout the 10-min period for supplying $^{11}CO_2$ and the 2-h measurement period. Center: result of the measurement performed with the light source kept on for the 10-min period for supplying $^{11}CO_2$ and kept off (dark condition) throughout the subsequent 2-h measurement period. Right: result of the measurement performed with the ears shielded from the light with aluminum foil

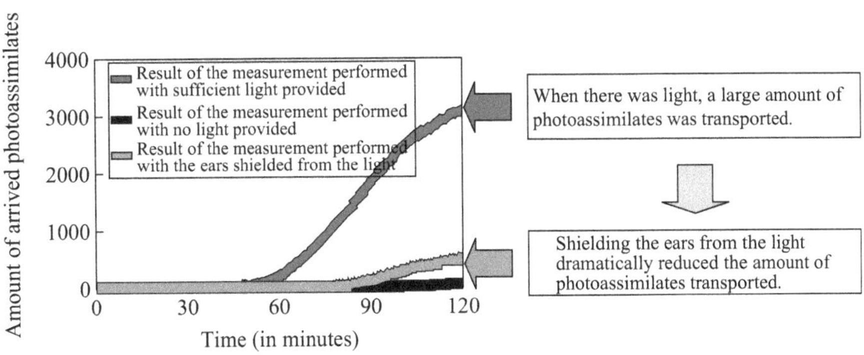

Fig. 16.17 Visualization and analysis of the way photoassimilates flow into and accumulate in the ears of wheat in the process of maturing. Profiles of the inflow of the ^{11}C photoassimilates into the ears and its accumulation therein obtained from the measurement performed in the three different light environments; this data suggests that it is not the light environment of the leaves (which is the part that makes photosynthesis) or the stem (which is the transport channel) but the light environment of the ears (which is the part that receives photoassimilates) that significantly affects the transfer of photoassimilates

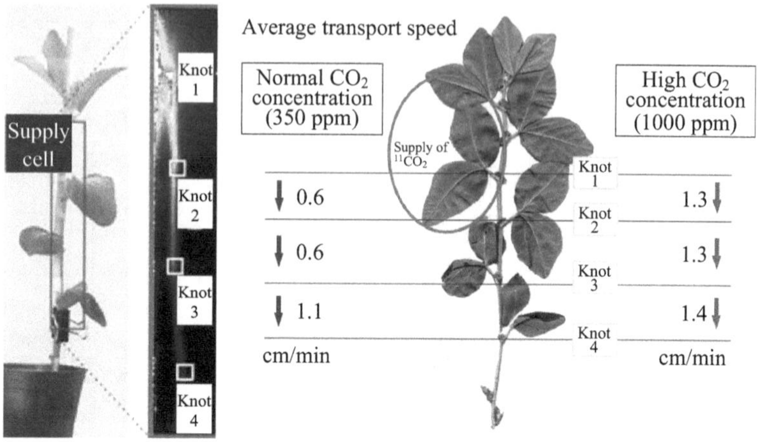

Fig. 16.18 Analysis of the effect of a high-concentration carbon dioxide gas environment on the transport of photosynthesis products. Left: the parts measured using PETIS and the ROIs set up on the image. Right: the speed of transfer of the photosynthesis product in the individual ROIs

the ^{11}C-methionine using PETIS, to identify the mugineic acid synthesis field. The experiment showed that, when ^{11}C-methionine was given to the barley from the tip of a leaf that had been cut for the experiment, ^{11}C was rapidly transferred to the roots and then to new leaves via the discrimination center (DC), but almost none of the ^{11}C was transferred to the matured leaves and roots [35]. Following this experiment, N. Bughio et al. conducted an experiment in which they found that a dramatic reduction in the portion of the ^{11}C-methionine absorbed through the roots of a barley plant individual which is transferred to the aboveground part of the barley individual only occurs when the barley individual is deficient in iron and, based on this and other findings, concluded that methionine, which is a substance used in the synthesis of mugineic acid, is generated in the roots (Fig. 16.19) [36].

As described above, ^{11}C, which is a positron-emitting radionuclide of carbon, has been utilized in the visualization of the transfer of photoassimilates in plants and analysis of metabolic physiological functions of plants and has been contributing to the acquisition of new findings. In addition, for the behavior of carbon in plants, the possibilities are not limited to simple visualization, as described in Sect. 16.2.3. That is, study results that can lead to realization of plant molecular imaging are being obtained, including technologies to calculate the detailed photoassimilates transfer speed during the transport and the detailed proportions of photoassimilates that are unloaded from the sieve tube during the transport through mathematical analysis method-based data processing and technologies to construct function images that show the intensities of component functions that comprise a series of physiological functions.

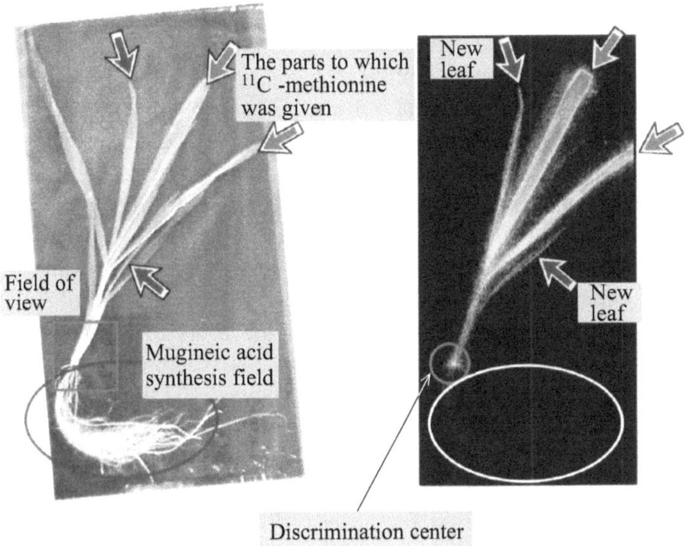

Fig. 16.19 Absorption of ^{11}C-methionine by an infant barley plant individual and the transfer of the absorbed ^{11}C-methionine in the individual

16.3.2 Behavior of Nitrogen in Plants

Nitrogen is an important nutrient for plants, as is carbon. Most plants absorb nitrate nitrogen (NO_3^-) or ammonium nitrogen (NH_4^+) through their roots, but some plants (such as leguminous plants) utilize the nitrogen gas (N_2) fixed by the root nodule bacteria that accrete with them in their roots. Hayashi et al. attempted PETIS-based measurement with ^{13}N, which is a positron-emitting radionuclide of nitrogen [37].

For the absorption and transfer of nitrogen nutrients by monocotyledons such as rice plants, PETIS-based measurement with $^{13}NO_3^-$ and $^{13}NH_4^+$ is being made. Hayashi et al. conducted an experiment in which they had a rice plant seedling absorb $^{13}NO_3^-$ (which is a nitrate ion tracer) through its roots and measured the transfer of the ^{13}N to the aboveground part of the seedling in the form of images using PETIS. The experiment showed that the ^{13}N reached the leaf sheaths 8 min after the start of the absorption. Similarly, Kiyomiya et al. conducted an experiment in which they had a rice plant seedling absorb $^{13}NH_4^+$ through its roots and measured the transfer of the ^{13}N in the form of images using PETIS [38]. The experiment showed that the ^{13}N took only 2 min to reach the DC after the absorption. In addition, Kiyomiya et al. compared the absorption and transfer of $^{13}NH_4^+$ with the absorption and transfer of $H_2{}^{15}O$, which is a radiolabel tracer for positron-emitting radionuclides of water, under various conditions. The result indicated that the transfer of ammonium ions from the roots to the DC depended on the transpiration stream in terms of passive dependency and on the ammonium ion transporter (or glutamine synthetase) in terms of active dependency [39].

Leguminous plants not only utilize nitrogen (N_2) in the atmosphere through symbiosis with root nodule bacteria but also are capable of utilizing nitrate ions. Sato et al. conducted an experiment in which they had a soybean plant individual absorb $^{13}NO_3^-$ through its roots and measured the transfer of the ^{13}N to the first post-cotyledon leaves using PETIS [40]. The ^{13}N signal was detected in the leaves 4 min after the dosing of $^{13}NO_3^-$, irrespective of whether or not root nodule bacteria were present in the root nodules, and increased linearly thereafter. Matsunami et al. conducted an experiment in which they gave $^{13}NO_3^-$ to only one of the roots of a soybean plant individual and to only one of the roots of a kidney bean plant individual, washed them immediately after, and observed the subsequent behavior of the ^{13}N using PETIS [41]. The experiment showed that there were areas where nitric acid accumulated intensively near the root tips. Ohtake et al., to investigate the effects of the states of nitrogen nutrients in soybean plants on the transfer of nitrogen to the seed pods, conducted an experiment in which they cut the stems of soybean plant individuals and had the stems absorb $^{13}NO_3$ from their bottom ends and measured the transfer of NO to the seed pods using PETIS [44]. $^{13}NO_3^-$ was transferred upward in the stem toward the step tip irrespective of the state of nitrogen nutrients, but, while clear images of seed pods started to be obtained 20 min after the dosing for individuals deficient in nitrogen, almost no transfer to seed pods was confirmed even 40 min after the dosing for individuals that had sufficient nitrogen, showing that, in soybean plants, the transfer of nitrogen to the seed pods is facilitated by a deficiency of nitrogen.

The fixation of atmospheric nitrogen by root nodule bacteria that accrete with leguminous plants is a very major topic for studies relating to nutrients for plants, and attempts to measure it in the form of images had been made since soon after the development of the first PETIS equipment, but these attempts had not succeeded because of various difficulties, including the fact that the amounts of atmospheric nitrogen absorbed by leguminous plant individuals are far smaller than the amount of $^{11}CO_2$ generated through photosynthesis (which is a popularly used substance for measurement); the need to supply the tracer to root nodules, which are normally located underground or underwater; the need to remove the impurities generated during the production of ^{13}N; and the short half-life of the radionuclide (10 min). No one had succeeded in measuring the fixation of nitrogen by root nodules in the form of images until 2009, when Ishii et al. succeeded in measuring the fixation of $^{13}N_2$ by root nodules in the form of images using PETIS [43].

16.3.3 Behavior of Water in Plants

Plants absorb water through their roots, transport it to their aboveground part through their ducts, and release it into the atmosphere through transpiration from their stomata. Plants absorb water together with nutrients from the soil and use the water together with carbon dioxide as the materials for their photosynthesis to transform the water and carbon dioxide into the hydrogen and oxygen contained in the photoassimilates (sugars). Because there had been no tracer suitable to observe the behavior of water in living bodies, it had not been clearly known how the water absorbed by living plants behaved in the plants.

Mori et al. conducted an experiment in which they had tomato and rice seedlings absorb water labeled with ^{15}O ($H_2{}^{15}O$) through their roots and observed the behavior of the absorbed $H_2{}^{15}O$ using PETIS [44]. This was the world's first successful visualization of the behavior of water in a living plant. The observation made by Mori et al. in this experiment showed that the $H_2{}^{15}O$ absorbed from the roots of the tomato seedlings was transferred to the stems in only 3 min and that the transfer to the aboveground part took place only when there was sufficient light. In addition, Mori et al. calculated the speed of transfer of water in tomato and rice seedlings at 1.9 and 0.4 cm per minute, respectively. Mori et al. also found that the amount of $H_2{}^{15}O$ transported was proportional to the amount of light, although the data has not been disclosed. Tsuji et al. also observed the transfer of water in tomato plants using PETIS and calculated the speed of transfer of water at 2.0–2.5 cm per minute, which matches well with the speed calculated by Mori et al. Tsuji et al. measured the transfer of $H_2{}^{15}O$ in tomato plants using PETIS 3 (which is PETIS equipment with enhanced performance including higher spatial resolution) and succeeded in obtaining clear moving images [45].

Kiyomiya et al. expanded the experiment of Mori et al. and investigated in detail the effects of the light environment of rice seedlings on the absorption of water by them and the transport of the absorbed water in them [46]. They found that (1) in the rice seedlings provided with sufficient light, the ^{15}O signal was detected in the DC 1 min after $H_2{}^{15}O$ was given to them from their roots, (2) giving light to rice seedlings that had been deprived of light caused accumulation of water in the DC to start in about 8 min, and (3) depriving rice seedlings of light immediately after their dosing with $H_2{}^{15}O$ caused the transfer of water to the DC to gradually weaken and ultimately stop 12 min after the deprivation of light. On the basis of these findings, they deduced that, in rice seedlings, the opening and closing of the stomata in response to light takes place in approx. 10 min. In addition, they found that sodium chloride and methylated mercury directly suppressed the absorption of water by rice seedlings through their roots and that methionine sulfoximine and abscisic acid were transported to laminas and weakened the transfer of water in rice seedlings by making them close their stomata. Kiyomiya et al. also investigated the relationship between the absorption of ammonium nitrogen ($^{13}NH_4{}^+$) by rice seedlings and the transfer of the absorbed ammonium nitrogen in the rice seedlings and the transpiration stream ($H_2{}^{15}O$) and discovered that, interestingly, depriving rice seedlings of light stopped the transfer of $H_2{}^{15}O$ completely, while 40% of the transfer of $^{13}NH_4{}^+$ was retained even after the deprivation of light. Tsukamoto et al. showed that the transfer of $H_2{}^{15}O$ from the roots of rice plants to the leaves and stems is facilitated by adding aminolevulinic acid (ALA) (a biologically active substance) and that ALA, etc., cause stomata to open to enhance the absorption and transfer of water [47].

Nakanishi et al. studied the mechanism of the absorption of water by soybean plants, etc., using $H_2{}^{15}O$ and ^{18}F water solution [48–50].

16.3.4 Behaviors of Sugars in Plants

Sugars are important energy sources for plants. In addition, they are used in the construction of plant cell walls and play a role in the transfer of signals. In an $^{11}CO_2$-based study of photosynthesis, the behavior of sucrose labeled with ^{11}C was measured. One of the positron-emitting radionuclide radiolabel compounds for sugars is ^{18}FDG (^{18}F-fluorodeoxyglucose; glucose labeled with fluorine-18), which is used in PET (positron emission tomography)-based diagnosis of cancers. Furuichi et al. developed a method to visualize calcium ion concentration increases in cells of *Arabidopsis thaliana* due to the reception of sugars by means of simultaneous measurement of the behavior of ^{18}FDG and the aequorin luminescence caused by calcium ion concentration increases in which the behavior of ^{18}FDG is measured with PETIS and the aequorin luminescence is measured with a high-sensitivity CCD camera [51]. Using this method, they succeeded in demonstrating, on the level of individuals, that the transfer of sugars and calcium ion concentration increases correlated.

It is expected that the transfer and distribution of a tracer labeled with ^{18}F in the form of ions in a water solution after it is taken in by plants will be very different from those of a tracer in the form of glucose labeled with ^{18}F, but this remains a mere assumption unless actual measurement is made. Figure 16.20 presents images showing the behaviors of $^{18}F^-$ and ^{18}FDG in hemp individuals after they were given to the individuals by having them absorb them through their roots. The images show that the behavior of the $^{18}F^-$ after absorption and the behavior of the ^{18}FDG after absorption were totally different.

16.3.5 Behaviors of Metallic Elements in Plants

Trace amounts of metallic elements such as iron, manganese, and zinc are essential for plants to grow. Because deficiencies of these essential metallic elements cause various nutritional disorders, elucidation of the mechanisms of the absorption and transfer of these essential metallic elements by plants is an important task for agricultural production. However, there still are many unknowns about the behaviors of these essential metallic elements in plants.

Gramineae plants excrete ferric oxide-chelating mugineic acids from their roots to solubilize ferric oxides in the soil into iron-mugineic acid complexes and absorb them via iron-mugineic acid complex transporters. Tsukamoto et al. conducted an experiment in which they measured with PETIS the absorption of iron by corn and barley individuals and the transfer of the absorbed iron in them using ^{52}Fe, which is

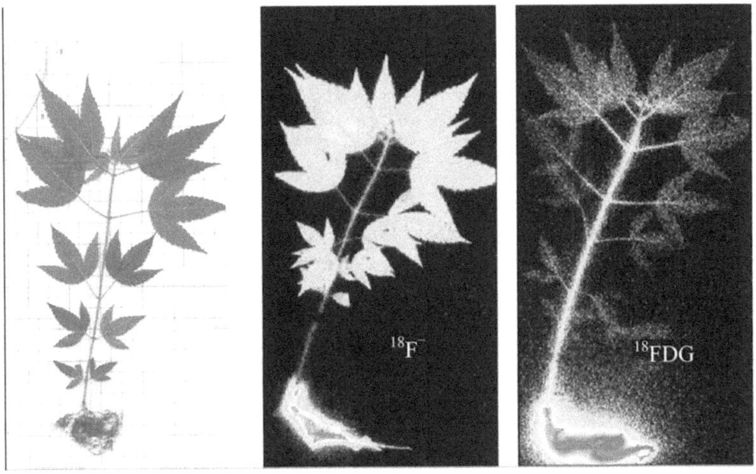

Fig. 16.20 Behaviors of $^{18}F^-$ and ^{18}FDG in hemp individuals after they were given to the individuals by having them absorb them through their roots

a positron-emitting radionuclide of iron [52]. This was the first experiment in the world that recorded the transfer of iron in living plants in the form of video. Tsukamoto et al. determined the amount of $^{52}Fe^{3+}$ deoxymugineic acid (DMA) absorbed by *ys1* corn plant mutants that do not have YS1, which is the iron-mugineic acid complex transporter for corn plants, and the amounts of ^{52}Fe transferred to the DC, the aboveground part, etc., and presented their analysis of these functions. In addition, they discovered that there was a new transport path to the newest leaves which had not been known, by conducting an experiment in which they blocked the duct stream, which is a stream for the transport of substances in plants, in a group of plant individuals and in an experiment in which they blocked the sieve tube stream, which is another stream for the transport of substances in plants, in another group of plant individuals (Figs. 16.21 and 16.22) [53–57].

Tsukamoto et al. conducted an experiment in which they measured the transfer of manganese in barley individuals with PETIS using ^{52}Mn, which is a positron-emitting radionuclide of manganese [58]. The experiment showed that the amount of manganese transferred from the roots to the leaf sheaths in barley individuals deficient in manganese was much larger than that in barley individuals that had a sufficient amount of manganese, indicating that a deficiency of manganese facilitates the absorption of manganese from the soil and the introduction of the absorbed manganese into the vascular bundle. The experiment also showed that the transport of manganese from the roots to the base section of the leaf sheaths was suppressed in barley individuals that had an excessive amount of manganese. In addition, Tsukamoto et al. discovered that the transport of manganese to the youngest leaves in barley individuals that had a sufficient amount of manganese was not suppressed even when the transpiration stream was suppressed by deprivation of light. This

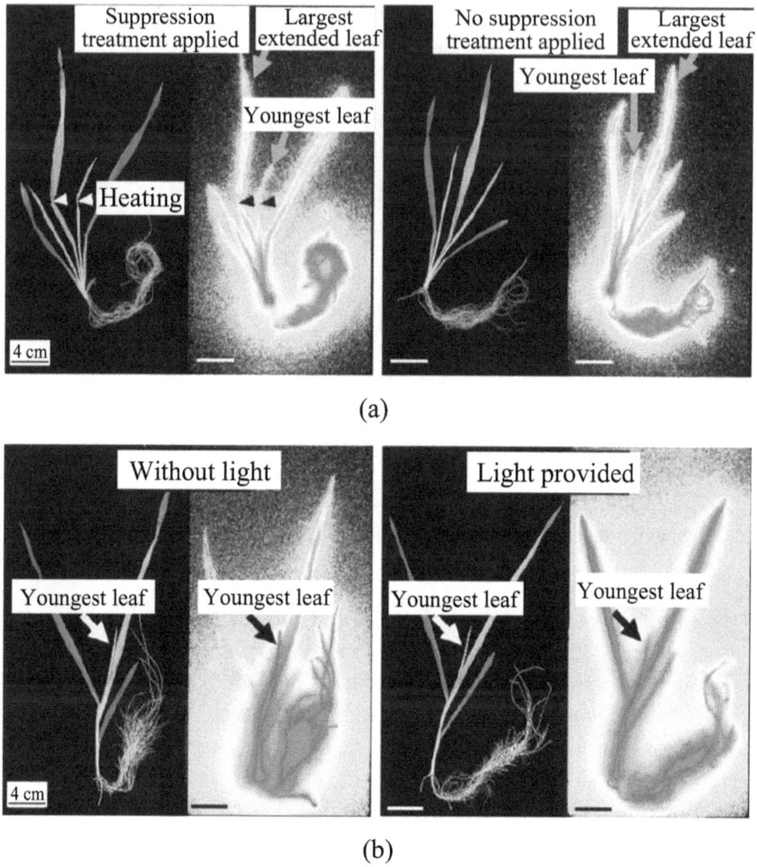

Fig. 16.21 Result of an analysis of the mechanism of the transport of iron to young leaves of barley plant individuals. (**a**) Images showing the transfer of iron to new leaves of barley plant individuals whose translocation stream has been suppressed. Left: an image of the transfer of iron to new leaves of a barley plant individual whose translocation stream has been blocked at the location marked with the white arrow by means of steam treatment. Right: an image of the transfer of iron to new leaves of a control barley plant individual. (**b**) Images showing the transfer of iron to new leaves of barley plant individuals whose transpiration stream has been suppressed. Left: an image of the transfer of iron to new leaves of a barley plant individual whose transpiration stream has been blocked by depriving it of light. Right: an image of the transfer of iron to new leaves of a control barley plant individual

finding implies a possibility that the transport of manganese in barley to the youngest leaves may not use the duct, as in the case of iron.

Suzuki et al. conducted an experiment in which they dosed a group of barley individuals deficient in zinc with ^{52}Fe and ^{62}Zn (a positron-emitting radionuclide of zinc) in the form of inorganic ions and another group with ^{52}Fe and ^{62}Zn in the form of chelate compounds with mugineic acids (MAs) (^{52}Fe-MAs and ^{62}Zn-MAs) and measured the behavior of ^{62}Zn using PETIS [59]. The experiment showed that

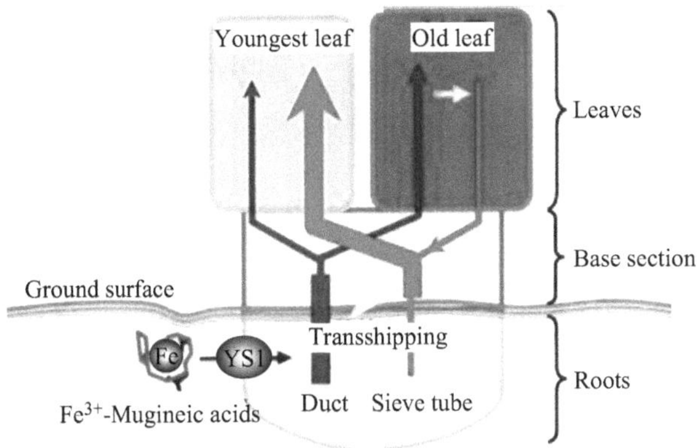

Fig. 16.22 Schematic diagram of the newly discovered iron transport path to the youngest leaves of barley. Tsukamoto et al. presented a theory that there was a new path through which the ^{52}Fe that has been absorbed through the roots and put on the duct stream is transshipped to the sieve tube in the roots or base section and directly transported to the youngest leaves, based on the result of an analysis of the mechanism of the transport of iron (^{52}Fe) in barley plant seedling individuals shown in Fig. 16.21. The path for transporting the iron absorbed through the roots that had been known by that time was from the roots to old leaves (extended leaves), where the iron is transshipped to the sieve tube, to the base section, and then to the youngest leaves

^{62}Zn-MAs were most easily absorbed by barley. On the basis of the results of the experiment, Suzuki et al. deduced that the biosynthesis and excretion of MAs induced by a deficiency of zinc would not be induced by a deficiency of iron.

In recent years, Watanabe et al. succeeded in developing a method for producing ^{64}Cu, which is a positron-emitting radionuclide of copper [60]. The study is a novel approach which is very meaningful and noteworthy in the sense that it shows the way future development of RIs and studies for their utilization should be, because, unlike other researchers who had conducted studies on new RIs, they considered ^{64}Cu as a starting substance for utilization in life sciences as a whole without being bound by a belief that ^{64}Cu should be utilized for medical purposes or plant study-related purposes, induced ^{64}Cu into chemotypes and its compounds suitable for the respective study purposes, presented their analysis on the usefulness of ^{64}Cu, and conducted experiments that covered all aspects of the development of new RIs from irradiation methods for new RIs to separation and refinement to RI-labeled compound synthesis and their utilization.

They had soybean seedlings absorb ^{64}Cu ions through their roots and measured the absorption of ^{64}Cu ions by the seedlings and the transport of the absorbed ^{64}Cu ions in them. PETIS images of the transport of ^{64}Cu ions showed that the transport of ^{64}Cu ions progressed to the extent where the shape of the stem was slightly visible about 12 h after the start of the dosing and thereafter ^{64}Cu ions started flowing into the stem tip part and the post-cotyledon young leaves immediately below it and started accumulating there (Fig. 16.23). Transport of ^{64}Cu ions to the post-cotyledon leaves belonging to the first knot below the knot to which the post-cotyledon young leaves immediately below the stem tip part belonged did take

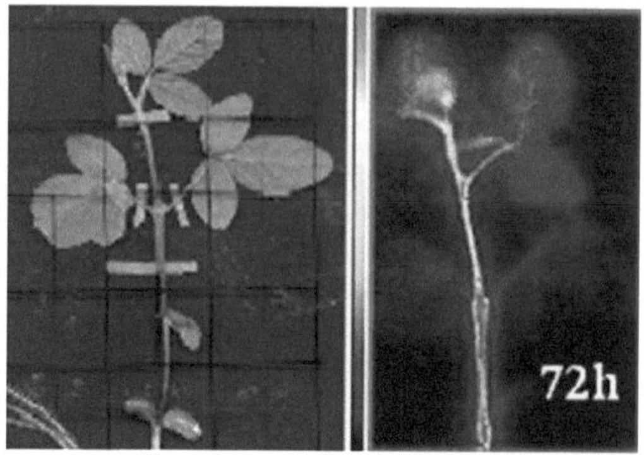

Fig. 16.23 Absorption of copper by soybean seedlings and transport of the absorbed copper in them. PETIS measurement image (right) showing the distribution of ^{64}Cu ions in a soybean seedling (left) 72 h after the seedling was dosed with ^{64}Cu ions by having it absorb ^{64}Cu ions through its roots

place, but the accumulation of ^{64}Cu ions in them progressed only to the extent where the leaves were barely visible. These facts showed that there was a clear difference between the amount of ^{64}Cu ions transported to the post-cotyledon young leaves immediately below the stem tip part and the amount of ^{64}Cu ions transported to the post-cotyledon leaves belonging to the first knot below the knot to which the post-cotyledon young leaves immediately below the stem tip part belonged. Considering the fact that copper is involved in various catalytic reactions in living bodies, the author thinks that this is very precious data that clearly indicates that the demand for copper in a plant varies between individual tissues of the plant in proportion to the degree of growth by showing that copper was preferentially transported to tissues that were growing at a fast rate, which had higher demand for copper, while only the required minimum amount of copper was transported to tissues that have already grown to a sufficient degree.

16.3.6 Behaviors of Harmful Pollutants in Plants

Reducing or controlling the absorption of harmful environmental pollutants by plants is important for securing food safety. To develop technologies to achieve this, it is necessary to obtain accurate knowledge about how harmful pollutants are absorbed by plants and how the absorbed harmful pollutants are transferred in the plants.

Vanadium (V) is a pollutant that is generated as a result of the use of fossil fuels. L.X. Tham et al. conducted an experiment in which they used ^{48}V, which is a positron-emitting radionuclide of vanadium, to show that irradiation decomposition products of chitosan, which has bioactivities such as antimicrobial activity, relieve the growth-suppressing stress caused by vanadium in rice plants [61]. Furukawa et al. made an analysis of the effects of the absorption of vanadium on cowpeas. The result of the analysis suggested that vanadium taken in by plants hinders their absorption of water through their roots [62].

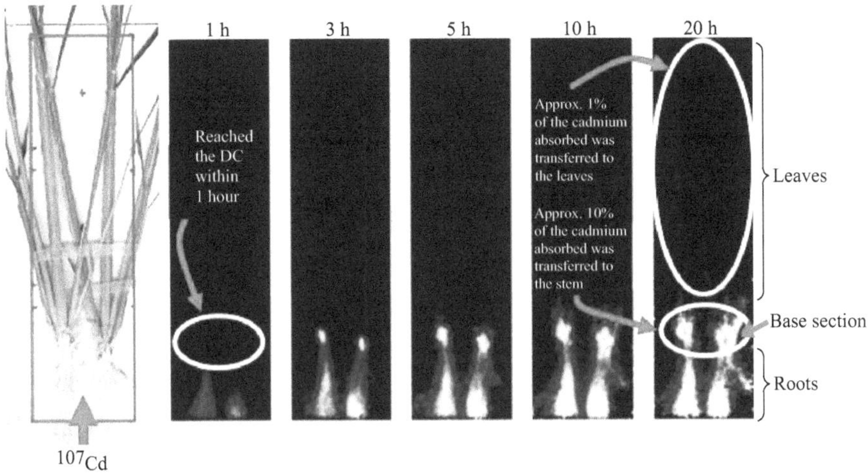

Fig. 16.24 Visualization of the absorption of cadmium by a young rice plant individual and the transfer of the absorbed cadmium in the individual [6]. The moving images show that the cadmium absorbed through the roots reached the DC within 1 h after the absorption and a significant portion of the absorbed cadmium accumulated in the base section, while only a very small portion of the absorbed cadmium was transferred to the laminas

Cadmium is the causal substance for the Itai-itai disease. In Japan, the maximum allowable cadmium concentration for rice is specified in the Food Sanitation Act. Currently, international standard maximum allowable cadmium concentration values for a wide range of foods are being set by the Codex Commission, and the development of technologies to prevent the pollution of agricultural produce by cadmium is being viewed as an urgent task. Cadmium has two positron-emitting radionuclides: ^{105}Cd and ^{107}Cd. ^{105}Cd is a positron-emitting radionuclide with a short half-life of 56 min, but the rate of the positron-emitting β^+ decay of ^{105}Cd is high at 26%. ^{107}Cd is a positron-emitting radionuclide with a long half-life of 6.5 h, but the rate of the positron-emitting β^+ decay of ^{107}Cd is very low at 0.23%. Therefore, it is necessary to use these two positron-emitting radionuclides appropriately according to the measurement purpose. Studies have already been conducted which attempted to make positron imaging-based measurement using the long half-life ^{107}Cd. These studies successfully recorded the absorption of cadmium by young rice plant individuals and the transfer of the absorbed cadmium in the young rice plant individuals in the form of video [63, 64] (Fig. 16.24). PETIS-based measurement of the way cadmium is absorbed by plants is being utilized not only for securing food safety but also in studies aimed at developing environmental cleanup technologies (phytoremediation technologies) that utilize plants.

16.4 Future Prospects for Plant Positron Imaging

This chapter outlined the features of plant positron imaging-based measurement methods that have been developed as a means to investigate functions of living plants and presented examples of practical measurement that has been made using the methods. Because this chapter is not an academic paper but a part of a textbook, the author has added some of his personal opinions as a person who used to be a member of the PETIS development team, while taking care not to be too subjective in doing so.

As mentioned in this chapter, PETIS is evolving from equipment that merely shows the way substances (tracers) are transferred in plants in the form of images to a powerful tool for extracting information on plant functions hidden in images and elucidating the meanings of such functions. Utilization of RIs in the field of plant studies to deepen the understanding of plants as living organisms is presently far behind the utilization of RIs in the field of medicine. However, this means that there are more possibilities for obtaining new findings by appropriately utilizing various RI tracers. Achieving this is by no means an easy task, but the author hopes that this chapter will help people who are willing to pursue such possibilities.

Exercises

1. Complete the sentences by selecting the appropriate words for [a] to [k] from the word list below referencing the text of this chapter:

Positron imaging is a technology for measuring, in the form of images and in a [a] manner, compounds that are transferred within living bodies without damaging them using, as [b], positron-emitting radionuclides that emit [c] (positrons) when they decay. In the field of medicine, positron imaging is being utilized in the form of [d] (positron emission tomography) for image-based diagnosis of cancers.

When a positron-emitting radionuclide β^+ decays, a [e] in the atomic nucleus changes into a [f], and a positron is emitted as the differential electric charge between the [e] and [f]. For example, decaying of ^{11}C (carbon-11), which is a positron-emitting radionuclide, turns it into ^{11}B (boron-11), whose [g] is the same as that of ^{11}C and whose atomic number is one less than that of ^{11}C. Positrons are called [c] because they are antiparticles of electrons which have [h] electric charges. A positron emitted from an atomic nucleus has [i]. It will fly around in random directions and ultimately [j] together with an electron. This annihilation generates a pair of 511 keV gamma rays (annihilation gamma rays). One of these gamma rays corresponds to the [g] of the positron and the other to the [g] of the electron. These two annihilation gamma rays travel in opposite directions (i.e., in directions 180° apart from each other). In positron imaging measurement, the pair of annihilation gamma rays is measured using the [k] method.

Word List

1. Positive electrons
2. Coincident counting
3. Tracers
4. Mass
5. Annihilates
6. Noninvasive
7. energy
8. PET
9. Positive
10. Proton
11. Neutron

2. Refer the text of this chapter, mark the correct ones out of the following five statements ([1] to [5]) with a ○ mark and the incorrect ones with a × mark.

[1] Short positron-emitting radionuclides are of no advantage for studies of plants, because most of them have a short half-life and their radioactivity decays in a short period of time.

[2] The only positron-emitting radionuclides that can be used for studies of plants are the so-called four PET radionuclides: ^{11}C (carbon-11), ^{13}N (nitrogen-13), ^{15}O (oxygen-15), and ^{18}F (fluorine-18).

[3] Because positron-emitting radionuclides emit gamma rays with strong penetrating power, they can be utilized as tracers to measure the transport of nutrients in plants in the form of images.

[4] Images that can be obtained with positron imaging are not shape images like ones from X-ray CTbut function images that reflect functions of living bodies.

[5] The measurement principles of PET equipment, which is used for cancer diagnosis, etc., and the measurement principles of PETIS (plant positron imaging equipment), which is used for studies of plants, are the same.

References

1. Nohara K (2009) History of PET device development. Houshasen 35(1)
2. Basic knowledge on PET "Positron Emission Tomography Course" (Introduction). http://www.med.osaka-u.ac.jp/pub/tracer/pet_text.htm
3. Mutoh A (2001) History and future prospect of cyclotron and PET. Jpn Soc Radiol Technol 57(10)
4. Murayama H, Now and future of the optical world –Positron CT-http://annex.jsap.or.jp/OSJ/50th_cd/main/mokuji/index.htm
5. Senda M, Mr. PET's nuclear medicine classroom. http://www.asca-co.com/nuclear/
6. Ishioka NS (2009) Nihon no RI kenkyuu kaihatu no genjou. Genshiryoku eye 55(7) (in Japanese); Ishioka N (2009) Current status of research and development of RIs in Japan. Atomic Eye 55(7) (in Japanese)

7. Ishioka NS, Fujimaki S (2009) Seimei kagaku bunya de yakudatu pojitoron housyutu kakusyu o mochiita hishinsyuuteki keisoku gijyutu. Houshasen to sangyo 124 (in Japanese); Ishioka N, Fujimaki S (2009) Positron-emitting radionuclide-based non-invasive measurement technologies that are useful for life sciences. Radiat Ind 124 (in Japanese)
8. PET to gan kenshin (2005) Nihon houshasen gijyutu gakkai zassi 57(10) (in Japanese); PET and Cancer Diagnostics (2005) J Jpn Soc Radiol Technol 57(10) (in Japanese)
9. Matsuhashi S et al (2000) Shokubutu kenkyuu no tameno pojitoron imejinguhou no kaihatu. Houshasenkagaku 70 (in Japanese); Matsuhashi S et al (2000) Development of positron imaging methods for studies of plants. Radiochemistry 70 (in Japanese)
10. Uchida H et al (1998) Syokubutu keisoku you pojitoron imejingu souchi. Houshasen to sangyo 80 (in Japanese); Uchida H et al (1998) Positron imaging equipment for measuring plants. Radiat Ind 80 (in Japanese)
11. Uchida H et al (2004) A compact planar positron imaging system. Nucl Instr Meth Phys Res A 516:564
12. Kume T et al (1997) Uptake and transport of positron-emitting tracer (^{18}F) in plants. Appl Radiat Isot 48(8):1035
13. Calvin M (1962) The path of carbon in photosynthesis. Science 135:879–889
14. Motoji N et al (1995) Studies on the quantitative autoradiography. II. Radioluminography for quantitative autoradiography of ^{3}H. Biol Pharm Bull 18(1):94
15. Kusakabe K (2007) PET shindan no susume. Genshiryoku eye 53(9) (in Japanese); Kusakabe K (2007) Rreasons why you should use PET-based diagnostics. Atomic Eye 53(9) (in Japanese)
16. Fukuda H et al (2007) PET ni yoru gan shindan no genjou to kongo no hatten. Genshiryoku eye 53 (9) (in Japanese); Fukuda H et al (2007) Current status and expected future evolution of PET-based cancer diagnostics. Atomic Eye 53(9) (in Japanese)
17. Suzuki M et al (2004) Positron Emission Tomography (PET) hou ni yoru seitai no bunshi imejingu kenkyu to souyaku iryou eno ouyou. Seibutu buturi 44(6) (in Japanese); Suzuki M et al (2004) Positron Emission Tomography (PET) Imaging-based Molecule-level studies of living bodies and applications for pet imaging in the fields of pharmaceutical development and medicine. Biophysics 44(6) (in Japanese)
18. Ruben S et al (1939) Photosynthesis with radio-carbon. Science 90:570
19. Ruben S et al (1939) Radioactive carbon in the study of photosynthesis. J Am Chem Soc 61:661
20. Moorby J et al (1963) The translocation of 11C-labelled photosynthate in the soybean. J Exp Bot 14:210
21. McKay RML et al (1988) The use of positron emission tomography for studies of long-distance transport in plants: uptake and transport of ^{18}F. Plant Cell Environ 11:851
22. Ishioka NS et al (1999) Production of positron emitters and application of their labeled compounds to plant studies. J Radioanal Nucl Chem 239(2):417
23. Matsuhashi S et al (2007) Pojitoron imejingu ni yoru syokubutu kinou no kashika. Denki hyouron 515 (in Japanese); Matsuhashi S et al (2007) Positron imaging-based visualization of functions of plants. Rev Electr-Relat Stud 515 (in Japanese)
24. Matsuhashi S, Fujimaki S et al (2005) Quantitative modeling of photoassimilate flow in an intact plant using the positron emitting tracer imaging system (PETIS). Soil Sci Plant Nutr 51(3):41
25. Matsuhashi S et al (2000) Pojitoron imejingu souchi (PETIS) to imejingu pureto (IP) ni yoru syokubutu kenkyu no tame no pojitoron imejingu. Radioisotopes 49(11) (in Japanese); Matsuhashi S et al (2000) Positron imaging-based plant studies that use positron imaging equipment (PETIS) and imaging plates (IPs). Radioisotopes 49(11) (in Japanese)
26. Matsuhashi S et al (2006) A new visualization technique for the study of accumulation of photoassimilates in wheat grains using [^{11}C]-CO_2. Appl Radiat Isot 64:435
27. Troughton JH et al (1977) Relations between light level, sucrose concentration, and translocation of carbon 11 in *Zea mays* leaves. Plant Physiol 59:808

28. Jahnke S et al (1981) Translocation profiles of ^{11}C-assimilates in the petiole of *Marsilea quadrifolia* L. Planta 153:56
29. Roeb GW, Fuhr F (1990) Use of the short-lived ^{11}C radioisotope in phytophysiological research. Plant research and development, vol 32, Ed. Institute for Scientivid Co-Operation in Conjunction with the Federal Research Center for Forestry and Forest Products and Numerous Members of German Universities:55–70
30. Minchin PEH, Troughton JH (1980) Quantitative interpretation of phloem translocation data. Ann Rev Plant Physiol 31:191
31. Keutgen N et al (2002) Transfer function analysis of positron-emitting tracer imaging system (PETIS) data. Appl Radiat Isot 57:225
32. Fujimaki S et al (2004) Four elemental profiles of time-activity curves obtained from PETIS experiments, JAERI-Review, pp 2004–2025
33. Kawachi N et al (2006) Kinetic analysis of carbon-11-labeled carbon dioxide for studying photosynthesis in a leaf using positron emitting tracer imaging system. IEEE Trans Nucl Sci 53(5):2991
34. Fujikake H et al (2003) Quick and reversible inhibition of soybean root nodule growth by nitrate involves a decrease in sucrose supply to nodules. J Exp Bot 54:1379
35. Nakanishi H et al (1999) Visualizing real time [^{11}C] methionine translocation in Fe-sufficient and Fe-deficient barley using a positron emitting tracer imaging system (PETIS). J Exp Bot 50:637
36. Bughio N et al (2001) Real-time [^{11}C] methionine translocation in barley in relation to mugineic acid phytosiderophore biosynthesis. Planta 213:708
37. Hayashi H et al (1997) Detection and characterization of nitrogen circulation through the sieve tubes and xylem vessels of rice plant, Ando T et al (Eds.), Plant nutrition – for sustainable food production and environment, Dordrecht, Kluwer Academic Publishers, pp. 141–145
38. Kiyomiya S et al (2001) Real time visualization of ^{13}N-translocation in rice under different environmental conditions using positron emitting tracer imaging system. Plant Physiol 125:1743
39. Kawachi T et al (2002) Effects of anion channel blockers on xylem nitrate transport in barley seedlings. Soil Sci Plant Nutr 48:271
40. Sato T et al (1999) Analysis of nitrate absorption and transport in non-nodulated and nodulated soybean plants with $^{13}NO_3^-$ and $^{15}NO_3$. Radioisotopes 48:450
41. Matsunami H et al (1999) ^{13}N-nitrate uptake sites and rhizobium-infected region in a single root of common bean and soybean. Soil Sci Plant Nutr 45:955
42. Ohtake N et al (2001) Rapid N transport to pods and seeds in N-deficient soybean plants. J Exp Bot 52:277
43. Ishii S et al (2009) Real-time imaging of nitrogen fixation in an intact soybean plant with nodules using ^{13}N-labeled nitrogen gas. Soil Sci Plant Nutr 55:660
44. Mori S et al (2000) Visualization of ^{15}O-water flow in tomato and rice in the light and dark using a positron-emitting tracer imaging system (PETIS). Soil Sci Plant Nutr 46:975
45. Tsuji A et al (2001) Uptake of ^{18}F-water, ^{15}O-H_2O and $^{13}NO_3^-$ in tomato plant. JAERI-Review:2001–2039
46. Kiyomiya S et al (2001) Light activates $H_2{}^{15}O$ flow in rice: detailed monitoring using a positron-emitting tracer imaging system (PETIS). Physiol Plant 113:359
47. Tsukamoto T et al (2004) $H_2{}^{15}O$ translocation in rice was enhanced by 10um 5-aminolevulinic acid as monitored by positron emitting tracer imaging system (PETIS). Soil Sci Plant Nutr 50:1085
48. Nakanishi TM et al (2001) Circadian rhythm in ^{15}O-labeled water uptake manner of soybean plant by PETIS. Radioisotopes 50
49. Nakanishi TM et al (2001) Comparison of ^{15}O-labeled and ^{18}F-labeled water uptake in a soybean plant by PETIS (positron emitting tracer imaging system). Radioisotopes 50
50. Nakanishi TM et al (2001) ^{18}F used as tracer to study water uptake and transport imaging of a cowpea plant. J Radioanal Nucl Chem 249

51. Furuichi T et al (2003) Real time-monitoring for translocation and perception of signal molecules in Arabidopsis thaliana mature plant. JAERI-Review:2003–2033
52. Tsukamoto T et al (2002) Absorption and transfer of the iron (^{52}Fe) in the iron absorption maize mutant '*ys1*'. JAERI-Review, pp:2002–2035
53. Tsukamoto T et al (2003) Phloem transport of ^{52}Fe from the discrimination center to immature sink was suggested by ^{52}Fe translocation in barley plants under dark condition. JAERI-Review, pp 2003–2033
54. Tsukamoto T et al (2004) Fe is translocated directly from DC or/and roots to the youngest leaf via phloem in graminaceous plants. JAERI-Review, pp 2004–2025
55. Ishimaru Y et al (2006) Rice plants take up iron as an Fe^{3+}-phytosiderophore and as Fe^{2+}. Plant J 45:335
56. Ishimaru Y et al (2007) Mutational reconstructed ferric chelate reductase confers enhanced tolerance in rice to iron deficiency in calcareous soil. PNAS 104:7373
57. Tsukamoto T et al (2009) ^{52}Fe Translocation in barley as monitored by a positron-emitting tracer imaging system (PETIS): evidence for the direct translocation of Fe from roots to young leaves via phloem. Plant Cell Physiol 50(1):48
58. Tsukamoto T et al (2006) ^{52}Mn translocation in barley monitored using a positron-emitting tracer imaging system. Soil Sci Plant Nutr 52:717
59. Suzuki M et al (2006) Biosynthesis and secretion of mugineic acid family phytosiderophores in zinc-deficient barley. Plant J 48:85
60. Watanabe S et al (2009) Production of no-carrier-added ^{64}Cu and application to molecular imaging by PET and PETIS as a biomedical tracer. J Radioanal Nucl Chem 280:199
61. Tham LX (2001) Effect of radiation-degraded chitosan on plant stressed with vanadium. Radiat Phys Chem 61:171–175
62. Furukawa J et al (2001) Vanadium uptake and an effect of vanadium treatment on ^{18}F-labeled water movement in a cowpea plant by positron emitting tracer imaging system (PETIS). J Radioanal Nucl Chem 249:495
63. Matsuhashi S (2006) Shokubutu pojitoron imejingu gijyutu ni yoru syokubutu kinou no kaiseki – Seicyou cyu no ine ni okeru kadomiumu doutai no kashika. Genshiryoku eye 52(8) (in Japanese); Matsuhashi S (2006) Plant positron imaging technology-based analysis of functions of plants – visualization of the behavior of cadmium in rice plants in the process of growing. Atomic Eye 52(8) (in Japanese)
64. Matsuhashi S (2007) Shokubutu no kadomiumu kyusyu yusou – Syokubutu pojitoron imejingu ni yoru kashika –, Isotope NEWS, No. 634 (in Japanese); Matsuhashi S (2007) Plant positron imaging technology-based visualization of the absorption of cadmium by plants and of the intra-plant transport of the absorbed cadmium. Isotope NEWS, No. 634 (in Japanese)

Chapter 17
Medical Utilization of Radiation

Shinpei Matsuhashi and Noriko S. Ishioka

More than 115 years have passed since the discovery of X-ray by W.C. Röntgen in 1895. During this period, radiation has been utilized for diagnosis and therapy of diseases and in elucidating mechanisms and workings of humans and life. Nowadays, radiation plays a role as a technology supporting medicine and medical care. In particular, radiation is playing an important role in the diagnosis and therapy of cancers, which is the number one cause of death of Japanese citizens. This chapter explains how radiation is being utilized and is playing an indispensable role at the actual scene of medical treatment and in the medical research and development that supports it, with a focus on the diagnosis and therapy of cancers.

17.1 History of Utilization of Radiation in Medicine

Medical utilization of radiation started in 1985 when W.C. Röntgen of Germany discovered an unknown ray with strong substance penetration power (X-ray), while he was conducting a study of vacuum discharge using a Crookes tube (a three-pole vacuum glass tube that has a cathode inside). Röntgen showed that X-ray allowed bones inside the human body to be seen (Fig. 17.1). This was an epoch-making achievement everybody had wanted to make, but no one had been able to. This was a discovery that was sufficient to stimulate people's intellectual curiosity (seeing -> knowing), which is an inherent desire of humans. As all of you know, Röntgen became the first Nobel laureate in physics in 1901.

S. Matsuhashi (✉) · N.S. Ishioka
National Institutes for Quantum and Radiological Science and Technology, Chiba, Japan
e-mail: matsuhashi.shimpei@qst.go.jp

H. Kudo (ed.), *Radiation Applications*, An Advanced Course in Nuclear Engineering 7, https://doi.org/10.1007/978-981-10-7350-2_17

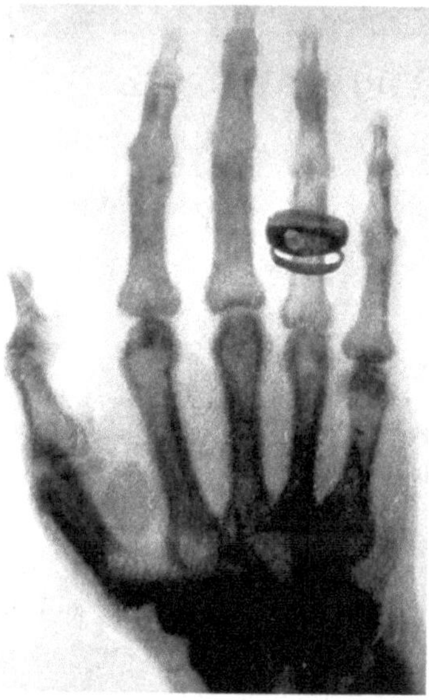

Fig. 17.1 Mrs. Röntgen's hand photographed by Röntgen using X-ray

Röntgen named this ray with unknown properties "X-ray," but many people called it "the Röntgen ray" to pay respect to the discovery. However, it is said that Röntgen himself did not like this name.

The first people who responded to the discovery of this X-ray which allowed the inside of substances to be seen from the outside were people of the medical community. On January 6, 1896, which was only 1 week after Röntgen's report on December 28, 1895, of his discovery, the Berlin Society of Internal Medicine had a discussion on the prospect of medical utilization of X-ray. The medical community expected that X-ray could be utilized in treating various diseases and made trials and errors to confirm the effects of X-ray. As a result, the first successful utilization of X-ray in skin cancer therapy was made in 1899. In the area of X-ray generation technologies, the Coolidge tube was developed in 1913. This further accelerated the medical utilization of X-ray. Even today, almost 100 years after the development of Coolidge tubes, the majority of X-ray generators for medical applications are modifications of Coolidge tubing. X-ray generation equipment such as the Coolidge tube and the Crookes tube used by Röntgen led to subsequent development of radiation generation equipment that uses accelerators, etc.

One year after the discovery of X-ray by Röntgen, A.H. Becquerel discovered that uranium compounds, which had been known as fluorescent substances, were capable of emitting radiation (the Becquerel ray) themselves. M. Curie was interested in this study by Becquerel and, together with her husband (Pierre Curie), succeeded in separating polonium and radium from uranium compounds in 1898. In this experiment, M. Curie used the terms "radiation" and "radioactivity" for the first time in the world. Radium separation was a very cumbersome task, and M. and P. Curie had to process several tons of uranium extraction residues over a period of 3.5 years to obtain only 0.1 g of radium. M. and P. Curie received the Nobel Prize in

Fig. 17.2 Certificates of the Nobel Prizes awarded to M. Curie

Physics in 1903, and M. Curie received the Nobel Prize in Chemistry in 1911 (Fig. 17.2).

P. Curie showed that radium had effects on living bodies which were similar to those of X-ray. This led to the utilization of radium as an irradiation source for diseases for which therapeutic effects can be obtained with X-ray. This was the dawn of the utilization of radioisotopes as radiation sources. Because radium did not require any equipment unlike X-ray, attempts were made to treat uterine cancer patients by using radium that has been put in a glass tube. However, radium caused burns in the parts of the body which came in contact with it. The cause of these burns was alpha and beta rays. The use of radium as a radiation source for medical treatment became popular after the development in 1907 by Dominici of a method wherein radium is covered with lead and the development of the Dominici tube that allows only the gamma ray, which is the radiation necessary for the treatment, to be obtained. This is the origin of the present sealed radiation sources.

Alpha and beta rays were discovered by E. Rutherford in 1898, which is the year in which M. and P. Curie discovered radium and polonium, and gamma ray was discovered in 1900 by P.V. Villard.

Medical utilization of radiation in Japan started in 1896 when Muraoka, who had been stimulated by the news of the discovery of X-ray by Röntgen, succeeded in generating X-ray after trials and errors. After that, the Shimadzu Corporation succeeded in producing the first medical X-ray equipment in Japan in 1909 and delivered it to the Kohnodai Army Hygiene Hospital (Fig. 17.3). With regard to introduction of medical radiation equipment at medical facilities, X-ray equipment

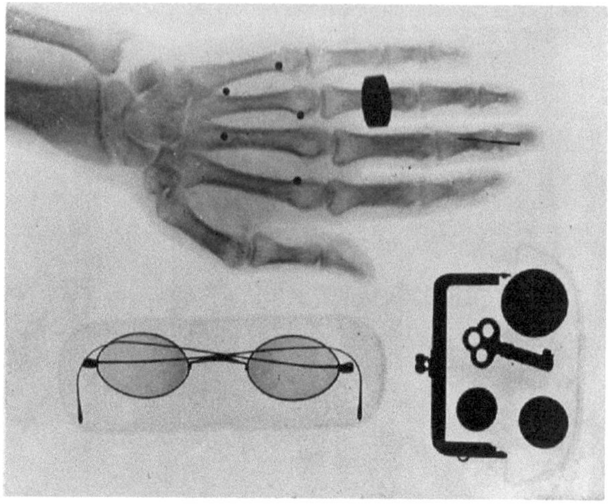

Fig. 17.3 One of the first X-ray photographs taken in Japan (Photo by courtesy of the Shimadzu Corporation)

for medical treatment was introduced at the Department of Dermatology of Tokyo University in 1911, and a radium treatment room was set up at the Aoyama Brain Hospital in 1912. Radiation-based treatment steadily became popular in Japan as in other advanced countries, and, according to the result of a survey, there were a total of 91 pieces of deep therapy equipment in Japan (including both domestically produced and imported ones) as of 1926, which was 30 years after the successful generation of X-ray by Muraoka.

In the early years of medical utilization of radiation, X-ray, which had been discovered by Röntgen, and gamma ray from radium, which had been discovered by M. and P. Curie, were used. Then, around 1930, new radiation generation equipment and artificial RIs started to be developed. A particle accelerator was developed in 1929, and Van de Graaff ultrahigh-voltage generation equipment was developed in 1931. A cyclotron was developed in 1932, and an artificial radionuclide was produced using a cyclotron in 1934. In 1936, studies on neutron-based therapy using cyclotrons were started. With regard to RI-based treatment, leukemia therapy using ^{24}Na and ^{32}P was started in 1936, and treatment of Basedow disease using ^{130}I was started in 1941. Shipment of nuclear reactor RIs from the United States was started in 1946, and utilization of RIs progressed further. In 1950, a linac (linear accelerator) for medical treatment was developed.

In addition to the utilization of radiation for treatment of diseases, utilization of radiation for diagnosis was also started. A patent for tomography was applied in 1921, and tomography equipment was commercialized in 1934. Utilization of RIs for purposes other than therapy was also started. In 1935, a tracer experiment was conducted using ^{32}P, and new developments started to take place in the utilization of radioisotopes in the field of life sciences.

As described above, technologies and methods for medical utilization of radiation were originally intended as technologies and methods dedicated for treatment of diseases, but their range has gradually been expanded to cover other purposes such as utilization of radiation for diagnosis, and this facilitated new developments in medical utilization of radiation.

17.2 Present Status of Medical Utilization of Radiation in Japan

17.2.1 Statistics on Radiation-Based Medical Treatment

Statistical data on the current status of medical utilization of radiation includes "Statistics on the Use of Radiation in Japan" and "Statistics on the Distribution of Radioisotopes in Japan" that are published yearly by the Japan Radioisotope Association. For example, "Statistics on the Use of Radiation in Japan 2008" shows that there were a total of 1433 pieces of radiation generation equipment (equipment that requires legal permission for use) in Japan as of the end of March 2008. It also shows how many of them were owned by medical institutions, educational institutions, research organizations, and private-sector companies, the types of pieces of generation equipment installed in Japan, and the numbers of individual types of pieces of generation equipment installed in Japan. According to "Statistics on the Use of Radiation in Japan 2008," 1039 pieces of radiation generation equipment, which represented 73% of the national total, were installed at medical institutions as of the end of March 2008, and 86% and 13% of them were linacs (linear accelerators) and cyclotrons, respectively. "Statistics on the Use of Radiation in Japan 2008" also shows that the number of medical institutions that owned radiation generation equipment as of the end of March 2008 was 826 and that the average number of pieces of radiation generation equipment owned per institution was approx. 1.3. "Statistics on the Use of Radiation in Japan 2008" also shows that the number of medical institutions that were handling sealed or unsealed radiation sources (which require legal permission) as of the end of March 2008 was 492, which represented only about 60% of the total number of medical institutions that owned radiation generation equipment. "Statistics on the Use of Radiation in Japan" also provides an overview of the utilization of sealed and unsealed RIs, but "Statistics on the Distribution of Radioisotopes in Japan" provides more detailed statistical data. For example, "Statistics on the Distribution of Radioisotopes in Japan 2009" shows that Cobalt-60 (Co-60) represented about 90% of the sealed RIs supplied to medical institutions in FY 2008 and that it was used as the radiation source for gamma knives for remote irradiation treatment. With regard to unsealed RIs, which are supplied as radiopharmaceuticals (for in vivo use), Technetium-99m (Tc-99m) and its generators were the very distant first place, representing more than 80% of the total. The second and third places were Iodine-123 (I-123) and Iodine-131 (I-131) (which are used for diagnosis and therapy of thyroid gland diseases), respectively, but they together represented only about 7% of the total. Fluorine-18 (F-18), which is used for such purposes as PET diagnosis of cancers, represented only about 4% of the total. Careful examination of the data presented in "Statistics on the Distribution of Radioisotopes in Japan" reveals that the amount of Strontium-89 (Sr-89) supplied was not counted in "Statistics on the Distribution of Radioisotopes in Japan 2006" and the preceding "Statistics on the Distribution of Radioisotopes in Japan," and the amount of Yttrium-90 (Y-90) supplied was not counted in "Statistics on the Distribution of Radioisotopes in Japan" and the preceding "Statistics on the Distribution of Radioisotopes in Japan." The data on these two radionuclides presented in "Statistics on the Distribution of Radioisotopes in Japan" shows that they were approved as the second and third radionuclides for medical treatment long after the approval of Iodine-131 (I-131), which had been the only approved unsealed radionuclide for medical treatment in Japan. In the sense that "Statistics on the Distribution of Radioisotopes in Japan" provides information on such a trend in medical utilization of radioisotopes, it can be said that the statistical data presented in "Statistics on the Distribution of Radioisotopes in Japan" is important data.

As an example of a material that summarizes the situation of radiation use in a viewpoint closer to the site of medical treatment, there is "The Present State of Nuclear Medicine Practice in Japan". For instance, "The Present State of Nuclear Medicine Practice in Japan 2008" shows that, in 2007, a little less than 33,000 examinations were conducted per month which used PET, which had been approved as an effective means of cancer diagnosis and designated as a means of medical examination covered by the National Health Insurance system. In addition, "The Present State of Nuclear Medicine Practice in Japan" contains information compiled from the standpoint of the actual clinical scene, such as the types of nuclear medicine equipment installed, the numbers of pieces of individual types of nuclear medicine equipment installed, and the trends in diagnosis and therapy. This information is useful to understand the current status of nuclear medicine.

17.2.2 Statistics on the Economic Effects of Medical Utilization of Radiation

Aside from the abovementioned publications, there are publications that provide data about how much radiation-based therapy and diagnosis are contributing to the economy. One of these is the report of the "Research on the Economic Scale of the Utilization of Radiation" conducted by the Cabinet Office in FY 2007. According to this report, the economic scale of the utilization of radiation in medicine and medical treatment in FY 2005 was approx. 2 trillion yen (approx. 1.57 trillion yen for medical therapy and diagnosis and approx. 0.43 trillion yen for irradiation equipment (medical equipment)), which represented 0.4% of the GDP in FY 2005 (approx. 502 trillion yen). It can be said that the economic scale of the utilization of radiation in medicine and medical treatment is quite large considering the fact that it is approx. one-fiftieth of the economic scale of all service industries combined and the fact that the gross product of agriculture, forestry, and fishery (primary industries) combined is 1.5% of the GDP.

Medical utilization of radiation can be divided into utilization of radiation in diagnosis and utilization of radiation in therapy. Each of these is outlined in the following sections.

17.3 Radiation-Based Diagnosis

In present-day medicine, image-based diagnosis is an indispensable source of information for determining the state of the disease or injury prior to treating it. Image-based diagnosis includes plain radiography, X-ray contrast examination, X-ray CT, nuclear medicine scanning (PET, SPECT, etc.), magnetism-based MRI (nuclear magnetic resonance imaging), ultrasonic (echo) examination, and funduscopy. Because these diagnosis technologies are used alone or in combination depending on the purpose and application, MRI, ultrasonic examination, funduscopy, and other image-based diagnosis technologies that do not use radiation are treated in the same way as image-based diagnosis technologies that use radiation.

This section explains about plain radiography, X-ray contrast examination and X-ray CT, which use a radiation source placed outside the body of the patient, and nuclear medicine examination, which uses a radiation source that has been put in the body of the patient.

17.3.1 Plain Radiography

Principle of X-Ray Generation and X-Ray Tube

Basically, the principle of X-ray generation has not changed since the time of Röntgen. That is, a high voltage in the range between several tens kV and 100 and several tens kV is applied to a vacuum tube that has a filament cathode and a metal (tungsten, molybdenum) target anode, to make the filament generate thermions. The thermions generated move toward the anode while accelerating and hit the anode target to generate X-ray. The X-ray generated consists of bremsstrahlung X-ray and characteristic X-ray, but most of the X-ray generated is bremsstrahlung X-ray, and the proportion of characteristic X-ray is less than 10%. In the case of an X-ray tube for diagnosis, the rate of conversion of the kinetic energy of electrons into X-ray is only 0.5% or less, and the remainder of the energy becomes the thermal energy of the anode target. It can be said that the history of the development of X-ray tubes over the past 100 years or so is the history of the development and improvement of technologies to make the anode in the vacuum

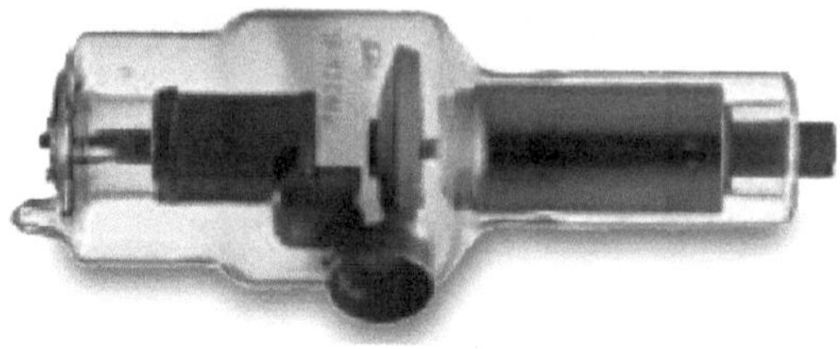

Fig. 17.4 "Be Window Rotor Anode" for mammography produced by Toshiba Electron Tubes and Devices Co., Ltd. (From the "Toshiba X-ray Tube 1915" brochure of Toshiba Electron Tubes and Devices Co., Ltd.)

tube rotate at a high speed to prevent this heat generation to make the service lives of X-ray tubes longer. An example of an X-ray tube is shown in Fig. 17.4.

Plain Radiography

Plain radiography is the radiation image-based diagnosis technology with the longest history. It is a technology that is familiar to most people, because it is used for diagnosis of bone fractures, chest X-ray examination made in mass checkups, and diagnosis of decayed teeth at dentists. It is also called plain X-ray photography or Röntgen photography.

Radiography utilizes X-ray transmittance differences between individual components of the human body such as bones, muscles, fat, and air. The subject is sandwiched between two components of the X-ray generation equipment, and a film is exposed to the X-ray projected from the X-ray generation equipment. The parts with low X-ray transmittance such as bones are shown in white, and the parts with high X-ray transmittance such as air are shown in black.

Radiography is being widely utilized for diagnosis of bone lesions such as bone fractures and osteoporosis; dental diagnosis; diagnosis of lung lesions such as lung cancer, pneumonia, tuberculosis, pleural fluid, and pneumothorax; and diagnosis of kidney stones, ureteral stones, and intestinal obstruction (Fig. 17.5).

In radiography, the method has been used for a long time wherein images (so-called Röntgen photographs) are acquired by directly exposing films to X-ray. In this method, the only data that can be obtained from each time of imaging is a single film, and the quality of the data for diagnosis depends on the skill of the radiographer. In addition, this method is cumbersome in that it is necessary to develop the films after the imaging and the developed films must be stored as they are.

Nowadays, "computed radiography" is becoming popular as a result of advances in computer technologies including image digitization technologies. In the broad sense, the term "computed radiography" refers to radiography that uses computer-based image processing. However, in the strict sense, it refers to radiography that uses "imaging plates" (IPs) coated with X-ray-sensitive material instead of X-ray

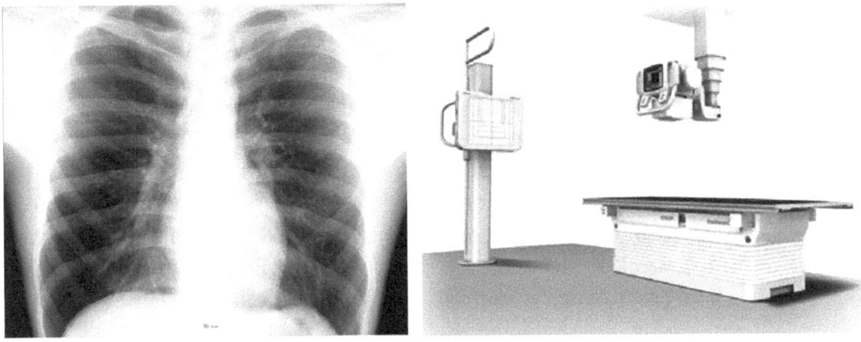

Fig. 17.5 Chest X-ray photograph of an adult male (left) and ordinary radiography equipment (right) (The photograph in the right was taken from the website of the Hitachi Medical Corporation with permission)

films. "Flat panel detectors" (FPDs) that use a semiconductor detector with a large area are often discriminated from this. Digital images are advantageous in that the tint, contrast, etc. can be changed after photographing and that it is possible to construct images that can be used for diagnosis even in the case where the contrast of the subject part is high. Although computed radiography has these advantages, films are still much more frequently used as the media at the actual scene of diagnosis, as you can easily imagine from the fact that most doctors are still using films when they explain to their patients the states of their diseases or injuries.

17.3.2 X-Ray Contrast Examination

You may have heard that there is a stomach examination (which is often made as part of a medical checkup) in which the subject is asked to "drink barium." This is one type of contrast medium-based X-ray contrast examination. X-ray contrast examination is used because it is difficult to obtain clear images of the alimentary canal including the stomach and intestines with plain radiography. X-ray contrast examination is an examination in which contrast media with different X-ray absorption rates are given to the subject prior to the imaging so that high-contrast images can be obtained. In X-ray contrast examination, two types of contrast media are used: the negative contrast medium whose X-ray absorption rate is lower than the surrounding tissue (such as air) and the positive contrast medium such as barium sulfate. Double-contrast method-based examination that uses barium and blister gas or air in combination is playing an important role in detecting small lesions such as stomach cancers in their early stages by virtue of its high imaging accuracy that allows detailed images of the internal surface of the wall of the alimentary canal to be obtained. X-ray contrast examination can be used for the alimentary canal (from the pharynx to the rectum), blood vessels, uterus, oviduct, and pancreas.

X-ray contrast examination uses X-ray television equipment. X-ray television equipment is equipment that shows the part under examination in real time. In X-ray contrast examination, it is possible to obtain images of the subject while changing the posture of the subject and pressing the body surface of the subject so that best possible images can be obtained.

Advances in image processing technologies in the area of the digitization of imaging have been contributing to improving the accuracy of diagnosis by maximizing the amount of information that can be extracted from the photographed images and improving the accuracy of image-based diagnosis. For example, in angiography, it is possible to render only the blood vessels in real time through subtraction processing in which the image before the injection of the contrast medium into the blood vessels (the mask image) is subtracted from the image after the injection of the contrast medium into the blood vessels (the live image). Its formal name is "digital subtraction angiography (DSA)," but the abbreviation DSA sometimes refers to digital imaging equipment capable of performing such subtraction processing. Subtraction processing is an image processing technology that is being utilized not only in contrast medium-based examination but also in diagnosis and examination that use other image-based diagnosis equipment such as plain radiography, X-ray CT, and MRI.

17.3.3 X-Ray CT

Computed Tomography (CT)

CT equipment is diagnosis equipment that is being popularly used at the actual scene of diagnosis as is plain radiography (Röntgen photography). Some of you may have an experience of being recommended "a CT" by a doctor. In X-ray CT, the patient is sandwiched between an X-ray tube (the X-ray source) and a detector that are deployed so that they face each other, and images are reconstructed by computer-processing the attenuation coefficient of the X-ray obtained from rotation scanning.

Average CT imaging equipment consists of an X-ray tube (the X-ray source) and a detector which are deployed so that they face each other and are contained in a donut-shaped gantry. In the actual imaging, the rotating X-ray source and detector contained in the gantry are moved along the body axis of the subject lying flat on the bed, and the subject is irradiated with X-ray from all directions. The X-ray projected at the human body is received by the X-ray detection equipment and recorded as data after its absorption by the human body and the resultant decaying. The data obtained shows from what direction X-ray is absorbed by what part of the body and to what extent. Then, images are constructed by Fourier-transforming the obtained data using a computer.

As described above, the concept of tomography was already presented in the 1920s. However, it took about 50 years after that for the first commercial CT scanner to be developed. G.N. Hounsfield and A.M. Cormack each developed their own equipment and received the Nobel Prize in Physiology or Medicine in 1979. It is said that the first prototype equipment developed in 1971 was equipment that performs 160 parallel reading scanning in 1-degree increments and that the equipment required more than 5 min to complete 180-degree scanning and the image construction required 2.5 h even with a mainframe computer. The first X-ray CT equipment was used to obtain tomographic images of the brain. The equipment took approx. 4 min to acquire two pieces of tomographic data and approx. 7 min to construct one tomographic image, which are longer than the time required by present-day equipment.

X-ray CT equipment was originally developed as equipment that constructs images of thin slices of the human body by layering multiple tomographic images (two-dimensional images). However, as a result of subsequent dramatic improvement in processing capacities of computers and development of equipment capable of being equipped with larger numbers of detectors, it has become possible to directly construct three-dimensional stereoscopic images with shorter measurement time. In particular, helical CT equipment has dramatically reduced the examination time and has been greatly helping reduce the burden on the patient in an X-ray CT examination. For example, traditional CT equipment was designed to acquire one image of a slice of the human body at a time. For this reason, it was necessary to ask the subject to stop breathing 20 times to obtain 20 layers of images. Because approx. 30 s was required to acquire one image of a slice of the human body (including the time required to move the CT equipment (or the table on which the subject is lying)), obtaining 20 layers of images took approx. 10 min. On the other hand, the time for which the subject must stop breathing in a helical CT equipment-based examination is only 20 s. In addition, helical CT equipment is capable of reconstructing an image of a slice at any desired location and allows a high-accuracy three-dimensional display image to be obtained from one-time data collection.

In X-ray CT, the X-ray attenuation coefficient values of air, water, and hard bones are defined to be -1000 HU (Hounsfield unit), 0 HU, and $+1000$ HU, respectively, with the X-ray attenuation coefficient value of water used as the reference X-ray attenuation coefficient value, so that the X-ray absorption can be indicated on a 2000-division scale. This is called the CT number.

CT methods include non-enhanced CT, contrast-enhanced CT, and dynamic CT.

Non-enhanced CT is a tomography method for the case where sufficient observation is possible without using a contrast medium such as examinations of cerebral hemorrhage, lung shape, and morphological abnormalities of bones or the case where it is desirable not to use a contrast medium.

Contrast-enhanced CT is a tomography method in which iodinated contrast medium with a high X-ray absorption rate is injected into the blood vessels. The contrast medium is transported to all parts of the body in bloodstreams and also diffuses to outside the blood vessels through capillaries, thereby making the organs look dyed. Because the blood vessels through which the contrast medium is transported and tissues with high bloodstream flow rates are drawn with stronger tint, images with clear contrast can be obtained. In addition, contrast-enhanced CT is suitable for observing tumor tissues, because in many cases they have higher bloodstream flow rates than the surrounding normal tissues. Other contrast medium-based special CT methods include the "dynamic contrast CT method" in which contrast medium is rapidly injected into the veins and the changes in the behaviors of the blood vessels over time are observed, the "CT angiography method" which is a blood vessel CT method for diagnosis of arterial diseases such as aneurysms, and the "interventional radiology CT (IVR-CT) method" in which contrast medium is directly introduced into the arteries or veins during a catheter examination so that only specific organs can be imaged with strong contrast.

X-ray CT is sometimes regarded as a tomography technology that is similar to MRI, but it is necessary to bear in mind that X-ray CT examinations and MRI examinations have totally different clinical meanings. Roughly speaking, CT is in many cases useful for examining emergency diseases such as bone and lung diseases and hemorrhage, while MRI is in many cases useful for examinations that require soft tissue resolution such as examination of brain tumors and the uterus, ovaries, and muscle diseases.

17.3.4 Nuclear Medicine Examination

Nuclear medicine is a field of medicine in which radioisotopes (RIs) are used for diagnosis (examination) or therapy. Usually, the term "nuclear medicine" refers to the nuclear medicine that uses unsealed RIs. Nuclear medicine is categorized into "in vivo nuclear medicine," "in vitro nuclear medicine," and "radionuclide therapy" according to the purpose and method of utilization.

In vivo nuclear medicine examination is an examination in which an RI is given to the patient and the changes in the distribution and behavior of the RI over time are analyzed. Scintigraphy-, SPECT-, and PET-based in vivo nuclear medicine imaging comprises most nuclear medicine. In vitro nuclear medicine examination is an examination in which trace substances such as hormones in a sample taken out of the body of the patient are measured by radioimmunoassay, etc. Radionuclide therapy is a therapy in which a drug that selectively accumulates in cancers is labeled with an RI and given to the patient to treat the patient using the radiation emitted from the RI that has accumulated in the cancers.

Table 17.1 exhaustively lists the RIs that are currently being used in Japan for diagnosis and treatment in radiology as a whole (as opposed to only in the field of nuclear medicine).

Radionuclides that are being used for in vivo diagnosis include Technetium-99m (Tc-99m), Iodine-123 (I-123), Thallium-201 (Tl-201), and Fluorine-18 (F-18). Radionuclides for in vitro examinations include Iodine-125 (I-125) and Iron-59 (Fe-59).

Table 17.1 Radionuclides that are currently being used in Japan for clinical purposes

Purpose	Method		Radionuclides used
Diagnosis	In vivo examination	SPECT-based measurement	Ga-67, Kr-81m, Tc-99m, In-111, I-123, I-125, I-131, Xe-133, Tl-201
		PET-based measurement	C-11, N-13, O-15, F-18
	In vitro examination		H-3, Fe-59, I-125, I-131
Therapy	Irradiation from outside the body		Co-60
	Irradiation after introduction into the body	Insertion of a sealed radiation source	Co-60, Cs-137, I-125, Ir-192, Au-198
		Dosing with a drug labeled with an RI	Sr-89, Y-90, I-131

In an in vivo nuclear medicine examination, the patient is dosed with a radiopharmaceutical (drug labeled with an RI) that reflects the state of the metabolism of organs of the patient or the state of a specific function of organs of the patient, and the state is measured in the form of images using such equipment as PET and SPECT. In in vivo examinations, radiopharmaceuticals labeled with gamma ray-emitting RIs are used, but the equipment used for the measurement differs depending on the mode of decaying of the RI. In the case of RIs that β^- decay and emit a single gamma ray as a result of the decaying, such as Tc-99m, SPECT is used for the measurement. In the case of radionuclides that β^+ decay and emit two gamma rays through the annihilation resulting from the decaying, such as F-18, PET is used for the measurement (the abbreviations "SPECT" and "PET" refer to the methods for image-based measurement, but they are also used as terms that refer to the equipment for the measurement. For example, equipment for SPECT measurement may be called "SPECT equipment" or simply "SPECT").

SPECT and Scintigraphy

SPECT (single-photon emission computed tomography) and scintigraphy are image-based diagnostic methods that use single-photon emitters such as Tc-99m. The RIs used for the imaging are radionuclides that emit only gamma ray or characteristic X-ray without emitting alpha or beta ray, such as Tc-99m and I-123. The energy of the gamma ray emitted by these radionuclides is approx. 100–400 keV. Their half-lives are sufficiently short to prevent them from remaining longer than necessary after the images are obtained.

The measuring equipment used in SPECT and scintigraphy is also called "gamma cameras," because they detect gamma rays and turn them into images. SPECT gamma cameras include single-detector (single-head) SPECT gamma cameras, dual-detector (dual-head) SPECT gamma cameras (these are also called "opposed-detector- or opposed-head-type SPECT gamma cameras"), triple-detector (triple-head) SPECT gamma cameras, and SPECT gamma cameras dedicated for head imaging. Gamma cameras consist of a scintillator that will emit light when it receives radiation and a two-dimensional position detection-type photomultiplier deployed behind the scintillator which is capable of electronically amplifying the light emitted by the scintillator and acquiring information on the location of incidence of the radiation using a mesh-like resistor chain. For the scintillator, a NaI crystal is normally used. A collimator made of titanium, etc. is deployed in front of the scintillator to prevent oblique incidence of radiation. Because the gamma rays emitted by radionuclides for SPECT have different levels of energy, it is necessary to optimize the settings on the detector side including the collimator settings. In addition, it is desirable to bring the gamma camera as close as possible to the patient during SPECT measurement, because the sensitivity and spatial resolution decrease as the distance from the radiation source (the patient dosed with the diagnostic RI) increases.

PET

PET (positron emission tomography) examination is a cancer diagnosis technology that is increasingly attracting attention and uses glucose labeled with F-18 (fluorodeoxyglucose (^{18}F-FDG)). Unlike SPECT, PET uses positron-emitting radionuclides that β^+ decay as radiation sources. As the name implies, positron-emitting radionuclides emit positrons (positive electrons (antiparticles of electrons that have a positive charge and the same mass as electrons)) from their nuclei as they decay. Each of the released positrons gradually loses energy and ultimately annihilates with an electron. This annihilation generates a pair of gamma rays each having energy of 511 keV, which corresponds to the mass of the electron/positron. The two gamma rays travel in opposite directions (i.e., in directions 180° apart from each other). Unlike SPECT, PET does not require a physical collimator because it simultaneously measures the pair of annihilation gamma rays generated by the annihilation using gamma ray detectors that are deployed so that they face each other. In PET-based measurement, it is not necessary to change the settings on the detector side according to the radionuclide that labels the drug, because it detects the annihilation gamma rays. In terms of measurement performance, the detection efficiency of PET is two to three orders of magnitude higher than that of SPECT, and the spatial resolution of PET is about two times as high as that of SPECT.

Early PET was equipment for obtaining images of slices of the human body through imaging. However, 3D-PET, which performs image-based real-time three-dimensional measurement instead of constructing three-dimensional images through image processing after the measurement, has already been developed thanks to technological advances in recent years.

Unlike X-ray CT, which renders the form, PET and SPECT measure the strengths of metabolic functions, etc. of the subject in the form of images. For instance, F-18-labeled FDG is an analogue of glucose and is metabolized as pseudo-glucose by the glycolysis system once it is taken into the human body. However, because FDG has F-18 built into its structure unlike glucose, the metabolism stops

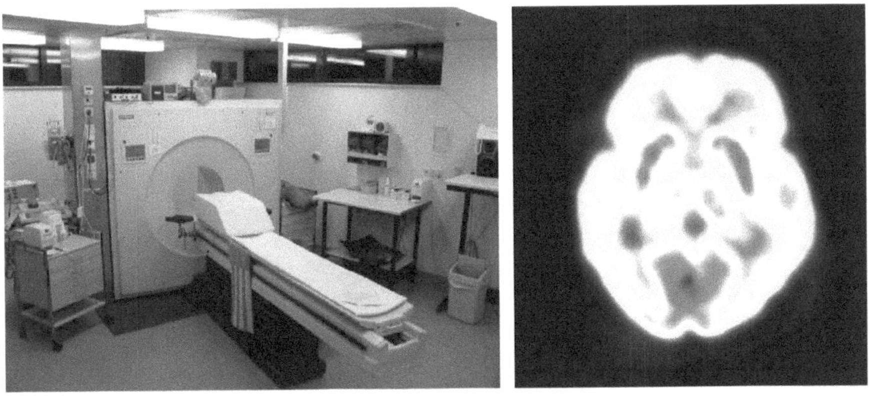

Fig. 17.6 PET equipment (left) and an example of brain PET image (right)

at a certain stage of the metabolism by the glycolysis system, and it accumulates at the location where the metabolism stopped. FDG-PET is a technology for the acquisition of images based on the strength (or activity) of the glycolysis function. Cancer tissues make cell divisions more actively than normal tissues and consequently require more sugar as the energy source. For this reason, cancer tissues have heavier accumulation of FDG than normal tissues and consequently are drawn as darker areas in PET images. However, there are cancers that are difficult to detect even with PET examination such as certain stomach and lung cancers, cancers that normally do not have significant accumulation of FDG such as liver cancers, and ureter system cancers such as kidney cancers, bladder cancers, and prostate gland cancers. Nevertheless, ^{18}F-FDG is the grand champion radiopharmaceutical for cancer diagnosis at present, and no compound that can replace it has been developed so far. A photograph of PET equipment and an example PET image are given in Fig. 17.6.

One of the features of the RIs that are currently being used as PET radionuclides is their short half-lives. For example, the half-lives of the so-called four PET radionuclides, namely, C-11, N-13, O-15, and F-18, are 20, 10, 2, and 110 min, respectively (all on the order of minutes). For this reason, using these radionuclides requires an accelerator to produce them, and facilities having PET had been forced to have a small accelerator. However, for F-18-labeled FDG, a "delivery service" became available in 2005 wherein pharmaceutical companies produce and ship it. This made it possible for facilities that have PET but do not have an accelerator to make diagnosis. According to the "The Present State of Nuclear Medicine Practice in Japan: A Report of the 6th Nationwide Survey in 2007," approx. 40% of the facilities that are conducting PET examinations do not have a cyclotron and are using FDG delivered from pharmaceutical companies. However, the number of PET examinations in which delivery drug was used represented only about 20% of all PET examinations made. In the remaining 80%, drugs produced by the facilities that made PET examinations were used.

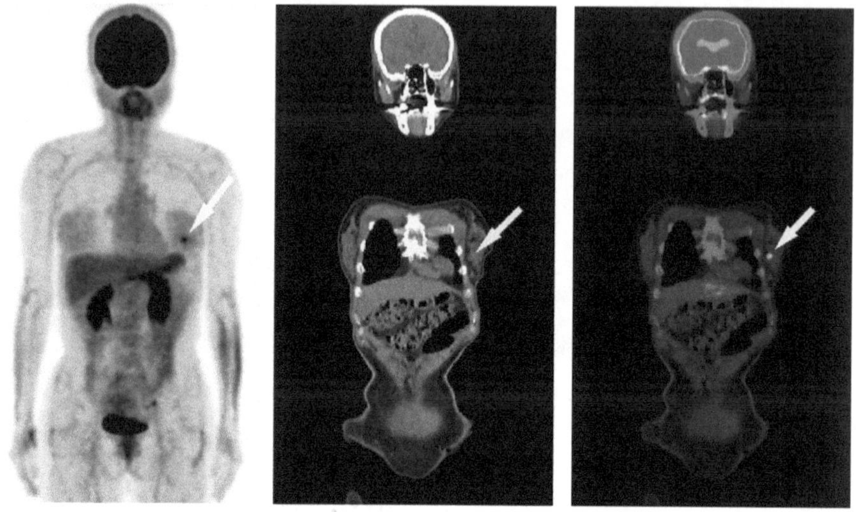

Fig. 17.7 An image from a PET-CT examination
Constructing an image (right) by superimposing an X-ray CT shape image (center) on a PET cancer focus detection image (left) allows the location and size of the cancer focus to be determined more accurately. The arrows indicate the breast cancer focus (The images were taken from the website of the Nagano PET and Image-based Diagnostics Center of the JA Nagano Health and Welfare Federation of Agricultural Cooperatives with permission)

Combining Function Images with Form Images

PET allows images with higher spatial resolution than SPECT to be obtained. However, the highest achievable spatial resolution is 1–2 mm even with equipment with the highest spatial resolution. In addition, the images obtained do not necessarily reflect the shape, etc. of the organs because they are images that use metabolic functions, etc. of the subject's body as the indices. For this reason, image-based diagnosis technologies have been developed which combine function images obtained with PET with form images obtained with other measuring equipment. The typical example is PET-CT. As the name implies, PET-CT is equipment that combines PET with X-ray CT. It allows both PET images and CT images to be obtained in a single examination (Fig. 17.7). Superimposing a CT image on a PET image is effective in not only determining the presence/absence of cancer tissues but also determining the locations, sizes, and states of the cancer tissues. According to the "2008 Report of the Research of the Status of Nuclear Medicine in Japan," approx. 300 pieces of PET-CT equipment had been introduced in Japan.

In addition to the abovementioned technology to combine PET images with CT images, attempts are being made to develop technologies to combine PET images with MRI images. Because MRI does not require X-ray irradiation unlike CT, it is expected that PET-MRI will reduce the exposure of patients during examinations. However, the task of evaluating the clinical advantages of PET-MRI over PET-CT (which is a prior technology) remains to be addressed.

17.4 Radiation-Based Therapy

Since the discovery of radiation, attempts have been made to utilize it for treating various diseases. However, radiation therapies that are being used today are practically limited to those for treating cancers with the exception of radioiodine-based radiation therapies for treating thyroid diseases. This section explains technologies to utilize radiation for treating diseases, with a focus on technologies to utilize radiation for cancer therapy.

Cancer therapy methods include the surgical therapy that physically removes cancer tissues by surgeries, the anticancer pharmaceutical-based chemotherapy, and the radiation therapy that reduces or kills cancer tissues by irradiating them with radiation.

Radiation-based therapy can be divided into the method in which the patient is irradiated using a radiation source located outside the body of the patient and the method in which a radiation source is introduced into the body of the patient first, and then the patient is irradiated using the introduced radiation source. In either method, cancer cells are irradiated with radiation to kill them. Radiation therapy is advantageous in that the physical and mental burdens on the patient are smaller because it does not significantly deform the body unlike surgeries that cut off tissues and eliminates the need to address problems relating to the physical strength of the patient. On the other hand, there are tasks that need to be addressed about radiation therapy, including the fact that there are certain types of cancers for which its treatment effect cannot be expected and the fact that normal tissues surrounding the cancer tissue are also irradiated with radiation.

Table 17.2 compares the features of the surgical therapy, chemotherapy, and radiation therapy.

It should be noted that the feature comparison presented in the table is a comparison based on overviews of the three therapeutic methods as seen from a general viewpoint, and the actual features, advantages, and/or shortcomings may differ from those presented in the table (i.e., they may be better or worse than the features, advantages, and/or shortcomings presented in the table) depending on the case and/or future advances made in relation to the three therapeutic methods. In addition, it is necessary to avoid interpreting the advantages of these therapeutic methods too favorably. At the actual scene of medical treatment, too, it is not uncommon to use two or all of the three therapeutic methods in combination to reduce the burdens on the patient and enhance the treatment effect by making the most of the advantages of the respective therapeutic methods and making up for the shortcomings of each therapeutic method with the other therapeutic method (s) according to the state of the patient and the cancer (combined modality

Table 17.2 Types of cancer therapy and their features

	Surgical therapy	Chemotherapy	Radiation therapy
Application	Early- to middle-stage cancers	Distant metastasis-type cancers (mainly those at stage IV) and leukemia	Early-stage cancers and inoperable locally advanced cancers
	Cancers that have developed only in specific parts of the body	Cancers that have spread throughout the body	Cancers that have developed only in specific parts of the body
Advantages	Good possibility of complete cure	Prevention of the aggravation of cancers	Functional and morphological losses suffered by the body of the patient are fewer and smaller
		Life of the patient may be prolonged	Adverse effects on the other parts of the body of the patient are fewer and smaller
			For early-stage cancers, the treatment track record is equivalent to that of the surgical therapy
Shortcomings	Functional and morphological losses suffered by the body of the patient are large	Adverse effects on the other parts of the body of the patient are strong (there are strong side effects)	For locally advanced cancers, the possibility of complete cure is smaller than the surgical therapy
	Depending on the location of the cancer and the condition of the patient (age, complications, etc.), it may not be possible to apply the surgical therapy	Low possibility of complete cure	More expensive than the surgical therapy and chemotherapy

treatment). Examples include dosing the patient with an anticancer pharmaceutical combined with the irradiation of the patient with radiation and irradiation of the cancer during an operation.

17.4.1 Irradiation of the Patient with Radiation from Outside the Body of the Patient

The main radioactive rays used to treat cancers by irradiating patients with radiation from outside their bodies are X-ray and gamma ray, which are photon beams

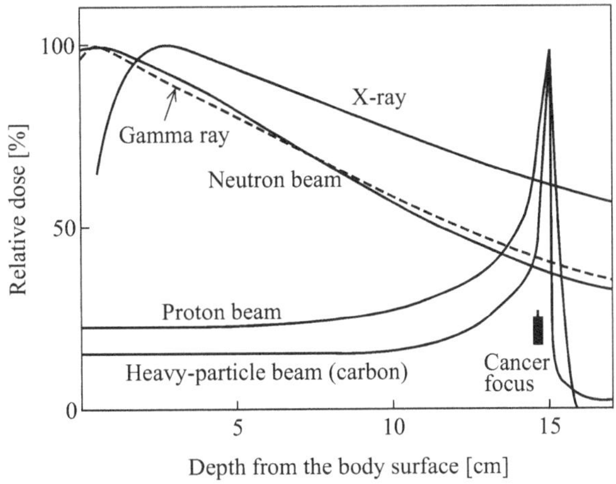

Fig. 17.8 Profiles of energy of various radioactive rays projected onto the body from outside the body (The graph was taken from the "Heavy-particle beam-based Treatment – HIMAC" brochure of the National Institute of Radiological Sciences with permission)

(electromagnetic waves), and proton beams and heavy-particle beams, which are particle beams. Figure 17.8 presents a graph showing the changes in the levels of energy of various radioactive rays projected onto the body from outside the body with the depth from the body surface. It can be seen from the graph that there are radioactive rays that have their energy peaks at shallow depths from the body surface and impart increasingly lower energy to the body as the depth increases, such as X-ray and gamma ray, and radioactive rays that advance from the body surface to a certain threshold depth without imparting high energy to the body and suddenly impart high energy upon reaching the threshold depth, such as the proton beam and heavy-particle beams. You may deduce from the graph that only particle beams are effective in treating cancers and X-ray and gamma ray are of no use. However, there are more than 800 medical institutions in Japan which have introduced linacs that generate X-ray and/or Co-60 gamma ray irradiation equipment, whereas there are only six particle beam-based treatment facilities in Japan. This means that the number of patients who can be treated with X-ray or gamma ray is more than 100 times greater than the number of patients who can be treated with particle beams, if you disregard the sizes and efficiencies of the facilities. Therefore, it can be said that all radioactive rays have both advantages and shortcomings. So, it is necessary to evaluate radioactive rays based on a comprehensive analysis that considers all advantages and shortcomings of individual radioactive rays instead of judging the value of a radioactive ray on the basis of a single advantage or shortcoming.

Photon-Based Treatment

X-Ray

Linear Accelerator (Linac)-Based Treatment

Linear accelerators are the most commonly used radiation therapy equipment that accelerates electrons to approx. 20 MeV using high frequencies. The electron beam ejected from an electron gun is accelerated in an acceleration tube to a speed that is close to the speed of light, and the accelerated electron beam is bent with a bending magnet to the direction of the body of the patient. In the case where the cancerous part of the body is located at a shallow depth from the body surface, the cancerous part is directly irradiated with the electron beam. In the case where the cancerous part is located at a deep depth, the accelerated electron beam is projected onto a target made of a heavy metal such as tungsten to generate X-ray, and the cancerous part is irradiated with the X-ray. A linac with a standard structure has a gantry that looks like an overturned tray and is attached to the tip of an L-shaped arm and a square window that serves as the radiation ejection port. The arm is so structured that it can be rotated about an axis that looks like an umbilicus. The patient lies on a bed, and the positions of the bed and gantry are adjusted so that the cancerous part can be accurately irradiated with radiation.

(i) Ordinary irradiation treatment that uses linac: In irradiation treatment that uses linac, each of the directions in which the radiation is projected is called a "gate." If the irradiation treatment involves irradiation from two directions, it is called "two-gate irradiation treatment." Accordingly, the term "opposed two-gate irradiation" refers to irradiation from two directions that are apart from each other by 180°. Currently, opposed two-gate irradiation is the mainstream irradiation mode in radiation therapy. Irradiation from four directions, that is, from the front, rear, right, and left of the cancerous part, is called four-gate irradiation or "box irradiation." In this mode of irradiation, the four projected beams geometrically look like two lines that perpendicularly intersect at the cancerous part. It can be said that four-gate irradiation is an effective method for treating parts surrounded by important organs because there is no risk that the target part may be highly irradiated and exposed only in one direction.

(ii) Conformal radiotherapy: Conformal radiotherapy (CRT) is a radiation therapy method that aims at irradiating the cancerous part selectively and intensively through multi-gate irradiation while minimizing unnecessary exposure. The idea of CRT was presented by Shinji Takahashi as early as 1960. Takahashi's idea of CRT is that making an aperture of a shape that fits the shape of the cancer and irradiating the body of the patient from various directions while making the aperture move around the body allow selective intensive irradiation of the cancer to be achieved.

Early CRT used equipment equipped with a collimator for adjusting the aperture attached to the gantry rotation mechanism. CRT then evolved into irradiation methods that use a multi-leaf collimator (MCL) and hollow-cut irradiation methods. Thanks to subsequent advances in computer technologies, it has become possible to make the MCL aperture width smaller and adjust the irradiation field shape more precisely.

Conformal radiotherapy has evolved into various irradiation methods including the linac-based fixed multi-gate irradiation method, the irregular irradiation field-type irradiation method, and three-dimensional conformal radiotherapy.

(iii) Three-dimensional conformal radiotherapy: Conformal radiotherapy originated from Takahashi's "attempt to obtain a high-dose region that fits the cancer focus" as described above. CRT started as a "coplanar radiotherapy" wherein the surface on which the linac radiation source can be rotated and moved is in the same plane that the cancer focus lies in. Then, during the three-dimensional irradiation realization process, CRT was affected by the gamma knife technology, and the "stereotactic multiple arc radiotherapy" (SMAR) wherein the linac radiation source is not in the same plane that the cancer focus lies in (noncoplanar irradiation), that is, the cancer focus is irradiated from various directions, was adopted in linacs. This led to the evolution of CRT into three-dimensional conformal radiotherapy (3D-CRT). 3D-CRT is being improved to achieve more selective and intensive irradiation of the cancerous part, and this is bringing conformal radiotherapy closer to the originally intended treatment technology.

(iv) Stereotactic radiation therapy: Stereotactic irradiation is defined as a technology for narrow beam-based selective and intensive irradiation of cancers which meets the following conditions:

1. The system must be capable of keeping the center of irradiation within a fixed tolerance in the body of the patient or a coordinate system that is connected to the body of the patient.
2. A stereotactic operation frame or detachable fixing equipment must be used.
3. The precision of the irradiation equipment regarding the center of irradiation must be 1 mm or higher.
4. The center of irradiation must be kept within the fixed tolerance specified above throughout the treatment.

Irradiation that meets the conditions listed above is called "stereotactic irradiation (STI)." Stereotactic irradiation in which irradiation is performed only once is called "stereotactic radiosurgery (SRS)," and stereotactic irradiation in which irradiation is performed a multiple number of times as split irradiation is called "stereotactic radiation therapy (SRT)."

The realization of stereotactic radiation therapy has made it possible to irradiate cancer focuses with radiation at higher doses than doses that can be used in traditional irradiation methods, thereby allowing much higher cancer control rates to be achieved. The dose that can be used in one-time SRT irradiation is 3–35 Gy, which is higher than doses that can be used in traditional methods. If the dose of 3 to 4 times of 15 Gy SRT irradiation is converted to biological effective dose (BED) to compare it with the ordinary split irradiation dose, the BED value will be 100 Gy or higher. It is considered that the biological effect of the 15 Gy × (3–4) SRT irradiation is about 2 times as strong as that of ordinary 60 Gy split irradiation, because cancers do not re-propagate because of the short irradiation time.

(v) Intensity-modulated radiation therapy (IMRT): Intensity-modulated radiation therapy is an irradiation-based therapy in which the "clinical target volume (CTV)" (the cancer focus) is irradiated with radiation at a sufficient dose and the dose received by the normal tissues surrounding the cancer focus is so regulated that it remains at a safe level by optimizing the distribution of the dose given to the CTV through precise control of the beam strength in the irradiation field for each of the small segments (Fig. 17.9). Normally, dose intensity profiles for five to nine gates are calculated in IMRT. In this calculation, the dose distribution in the CTV is designed first, and then the irradiation conditions are calculated so that they match the dose distribution (this is called "inversed planning" or "inductive planning"), unlike the traditional approach wherein the irradiation method is determined first and then the dose distribution is calculated to verify that the irradiation method satisfies the conditions for the treatment (forward problem). IMRT is an external irradiation treatment method that is very effective in treating prostate cancers, etc.

In IMRT, it is necessary to maintain precise control of the point of irradiation, because the irradiation plan is designed to minimize the dose received by the normal tissues surrounding the target tissues that are to be irradiated at a high

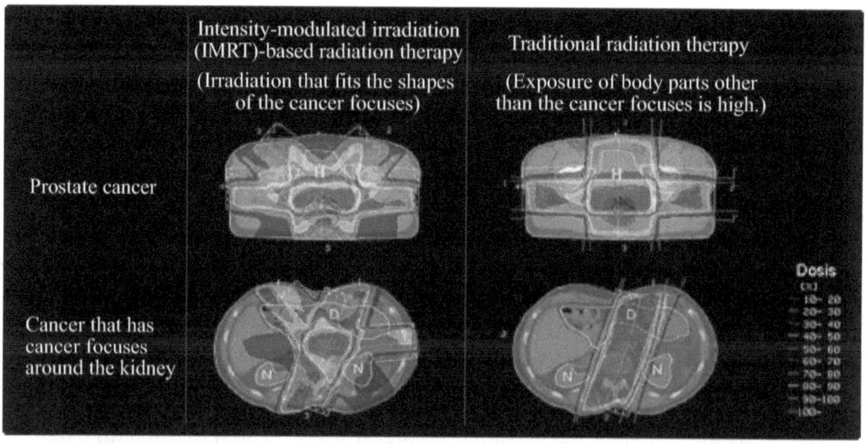

Fig. 17.9 IMRT-based dose construction that takes the shapes of the cancer focuses into account (The image was taken from the web site of Clinic C4 with permission)

dose. In addition, IMRT requires complicated design work to optimize the irradiation conditions.

Advanced Irradiation Treatment Equipment

(i) Image-guided radiation therapy: In the traditional radiation therapies, the location of the cancer focus is confirmed in advance by an image-based examination such as an examination that uses CT, and then the cancer focus is irradiated with radiation with the radiation beam accurately trained on the cancer focus. However, it is difficult to have the patient take the posture during the irradiation which is exactly same as his/her posture at the time of the imaging. In addition, it is inevitable with the traditional radiation therapies that normal tissues vulnerable to radiation are affected depending on the location and/or shape of the cancer focus. Image-guided radiotherapy (IGRT) is a technology that was developed to solve these problems by allowing accurate irradiation treatment to be performed with the locational errors (deviations) of the body of the patient compensated for in real time based on image information. IGRT is an important technology that supports advanced, complex, and precise irradiation treatment such as IMRT and is playing an indispensable role in present-day radiation therapies.

(ii) CyberKnife: CyberKnife is radiation therapy equipment developed in the United States. It consists of an ultra-compact linac weighing 135 kg attached to a 6-axle-control-type Japanese industrial robot and is capable of intensively irradiating any given point of cancerous parts with a weak X-ray from any desired position and angle. This equipment is so designed that the exact location of the cancerous part is memorized by a computer and the cancerous part is irradiated with the X-ray precisely trained on the cancerous part using the six-dimensional information on the cancerous part obtained from CT. Locational deviations of the body of the patient during irradiation due to the movement of the patient are detected with two fluoroscopy cameras that monitor the movement of the patient, and the detected deviations are compensated for. CyberKnife allows uniform distribution of irradiation to be achieved even for lesions with complex shapes and insufficient irradiation of the lesion part and unnecessary irradiation of healthy tissues to be avoided. CyberKnife-based treatment requires having the patient wear a mesh mask to confirm the location of the lesion. However, unlike in the case of gamma knife-based treatment, there is no need to fix the head of the patient with a metal fixing apparatus. In Japan, the use of CyberKnife is limited to treatment of tumors in the head section such as brain tumors and tumors in the domains of otolaryngology and oral surgery. In the United States, however, it is being used to treat lung and pancreas cancers.

(iii) Elekta Synergy: Elekta Synergy is a linac for IGRT whose gantry section is equipped with equipment capable of performing three-dimensional CT imaging. This equipment has an X-ray tube (keV voltage) and a flat panel-type detector that are deployed perpendicular to the direction of the treatment

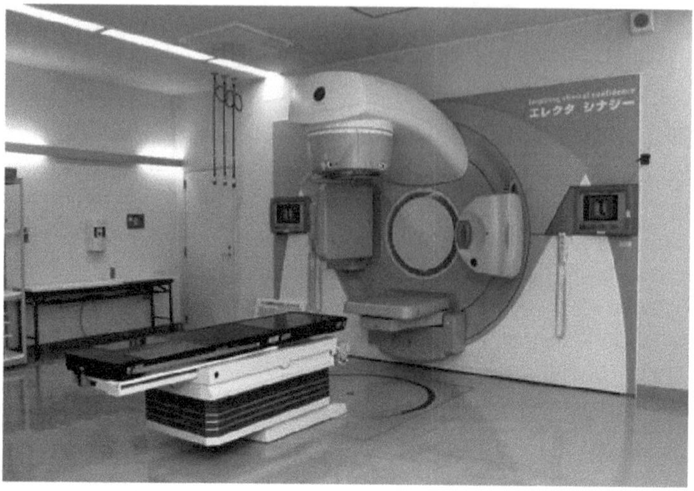

Fig. 17.10 Elekta Synergy (linac with X-ray CT) (The photograph was taken from the web site of the Department of Radiology of The University of Tokyo Hospital with permission)

beam. The equipment is capable of performing not only single-image imaging and continuous (X-ray fluoroscopic) imaging but also three-dimensional CT imaging. The three-dimensional CT imaging is achieved by rotating the gantry 360° with the patient lying on the treatment bed in the same way as in the actual treatment and fixed to the bed. The three-dimensional images obtained are superimposed on the reference CT images used to prepare the treatment plan to calculate the errors. By adjusting the position of the bed using the values obtained, precise irradiation of the irradiation point specified in the treatment plan can be achieved. Elekta Synergy is treatment equipment that can achieve high-accuracy irradiation such as IMRT and SRT irradiation (Fig. 17.10).

(iv) Novalis: Novalis is equipment developed for stereotactic radiation therapy. It is capable of generating 6 MeV X-ray using a linac and can be used for intensity-modulated radiation therapy as well. The locational deviations of the body of the patient are precisely calculated using two positioning mechanisms (infrared camera and X-ray imaging from two directions), and the deviations are automatically corrected for by moving the bed. Novalis can also be used for image-guided radiotherapy (IGRT). The radiation ejection port of Novalis is equipped with a collimator with a width of 3 mm called micro-multi-leaf collimator, which can be operated to adjust the beam according to the shape of the cancer. The name "Novalis" is not an abbreviation of a phrase describing the features of the treatment equipment but a Latin word that stands for "the new Frontier."

Novalis has been introduced at ten facilities in Japan.

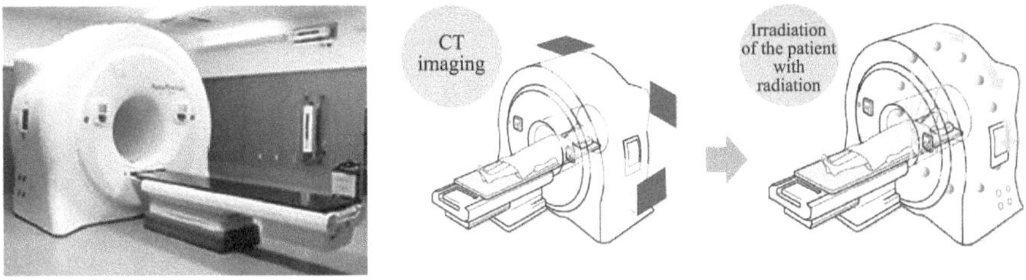

Fig. 17.11 Tomotherapy equipment equipped with X-ray CT equipment and capable of not only performing pinpoint irradiation but also irradiating two or more lesions at different locations simultaneously (left) and conceptual diagrams of the flow of tomotherapy from CT imaging to irradiation treatment (center and right) (The photograph and diagrams were taken from the web site of the Uji Takeda Hospital with permission)

(v) Tomotherapy: Tomotherapy is equipment that was developed in 2003 by combining computed tomography equipment with irradiation equipment. The name "tomotherapy" is a coinage made by combining the word "tomography" with the word "therapy." This equipment is capable of irradiating the patient while performing CT imaging. In tomotherapy, the irradiation equipment is continuously rotated through the use of the helical CT technology, and, at the same time, the irradiation table on which the patient is lying is moved under computer control (Fig. 17.11). Tomotherapy equipment has 1 radiation ejection port equipped with 64 tungsten baffle plates. The baffle plates can be opened and closed under computer control to adjust the irradiation area and dose to achieve intensity-modulated radiation therapy (IMRT).

This equipment is capable of not only performing pinpoint irradiation but also irradiating lesions with complex shapes and irradiating two or more tumors at different locations simultaneously. In addition, the equipment can be utilized for treating prostate cancers and multiple lung cancers, which have traditionally been considered as difficult cancers to irradiate. Tomotherapy-based treatment is covered by the National Health Insurance, but only 11 medical institutions had introduced tomotherapy equipment as of June 2008, because the introduction of tomotherapy equipment requires an initial cost of approx. 500 million yen.

Microtrons (Circular Accelerators)

Figure 17.12 shows an illustration of a microtron. Microtrons are equipment that accelerates electrons using microwaves and irradiates the cancerous part directly with the high-energy electron beam generated or irradiates the cancer tissues with the X-ray generated. In microtrons, the spread of the beam due to the beam deflection in the gantry is smaller because the variation in the beam energy is smaller. Equipment that uses electron beam, X-ray, and gamma ray has already become popular, but most of them are equipment with energy of 20 MeV or less. However, increasing the energy would reduce the dose received by the skin surface and increase the portion of the beam which reaches deeper parts of the body. For

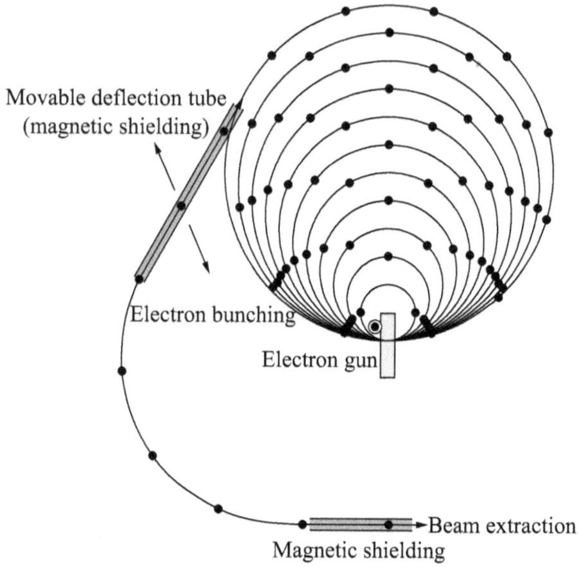

Fig. 17.12 Schematic diagram of the beam acceleration and transport by microtron (The diagram was taken from the "Explanation" section of the "Atomin" web site (a web site that provides energy and atomic energy education support information) with permission)

this reason, techniques to achieve higher energy with smaller equipment are being devised including repeated acceleration of electrons with linear accelerators and microtrons.

Gamma Ray

Cobalt-60-Based Rotating Treatment Equipment

Figure 17.13 shows a concept diagram of the equipment. This is well-known radiation treatment equipment equipped with a Cobalt-60 radiation source enclosed in a sealed container and a thick lead shield to block the gamma ray. The irradiation equipment is designed to rotate around the patient lying on the bed so that the cancerous part (the target) is irradiated with radiation at the maximum possible dose and the dose received by the normal tissues surrounding the cancerous part is minimized through dose distribution.

Gamma Knife

Figure 17.14 shows photographs of a gamma knife system. Gamma knife is a gamma ray-based radiation therapy for lesions in brains developed in 1968 by Professor Lexell at the Neurosurgery Department of the Karolinska University Hospital. It was named gamma knife because the lesion after treatment looks as if it has been cut off with a knife. Gamma knife is equipped with 201 Co-60 radiation sources deployed in concentric half circles in a semispherical structure. Gamma knife is so designed that the gamma rays from all Co radiation sources converge at a

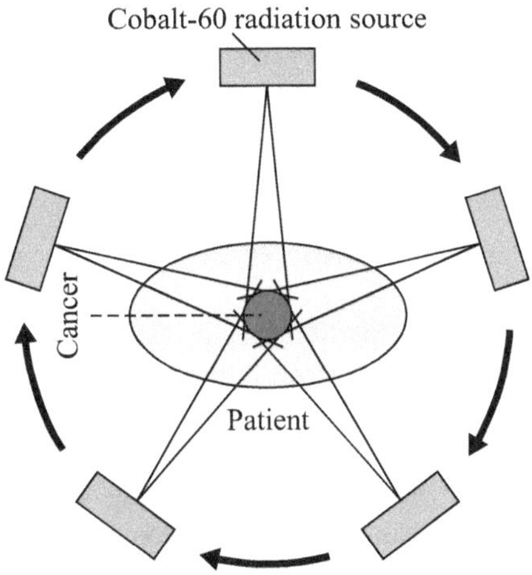

Fig. 17.13 Conceptual diagram of irradiation with Cobalt-60-based rotating treatment equipment (The diagram was taken from the "Explanation" section of the "Atomin" web site (a web site that provides energy and atomic energy education support information) with permission)

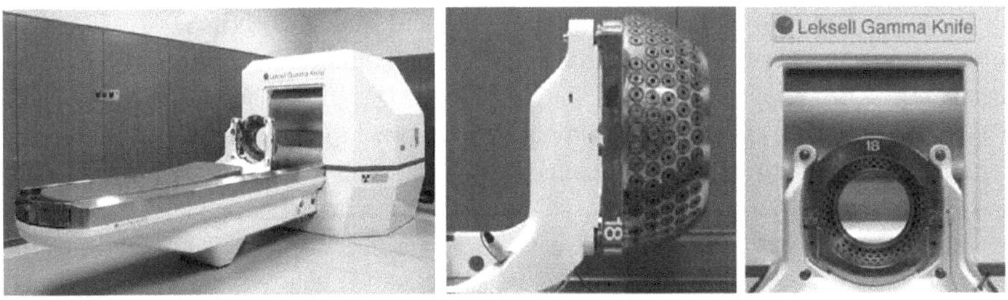

Fig. 17.14 Photograph showing an entire gamma knife (left) and photographs of the helmet equipped with Cobalt-60 radiation sources deployed in concentric half circles as seen from the side (center) and front (right) (The photographs were taken from the web site of the Apoplexy and Cranial Nerves Center of Aizawa Hospital with permission)

single point so that only the lesion part is irradiated at high energy. The advantage of gamma knife is that the adverse effects on the surrounding normal tissues can be reduced and the posttreatment disorders such as dermatitis, hair loss, and suppression of functions of the marrow can be prevented. This is because the positional accuracy of irradiation by gamma knife is very high (0.5 mm or higher) and the gamma rays from the individual radiation sources are weak. It has been proven that gamma knife is very effective in treating small benign tumors (about 3 cm or smaller) such as acoustic tumors, meningioma, and hypophyseal tumors and malignant tumors such as metastatic brain tumors that have metastasized to brains from other parts of the body. In addition, it has been proven that gamma knife is very effective in treating cerebral arteriovenous malformation, which can cause subarachnoid hemorrhage. However, it is necessary to fix the head of the patient securely with a metal fixing apparatus so that it does not move.

Gamma knife is irradiation equipment developed to treat brain tumors and similar tumors and has been used for that purpose. However, gamma knife has been attracting attention in recent years as a prospective clinical means of treating orbital malignant melanomas and age-related macular degeneration, for which no definitive treatment method currently exists.

As of July 2007, a total of 57 gamma knife units were in operation in Japan.

Particle Beam-Based Treatment

The term "particle beams" refers to accelerated particles that are heavier than electrons. However, in the context of radiation-based treatment, the term "particle beams" refers to accelerated ions of elements heavier than hydrogen.

The history of particle beam-based treatment began in 1946 with the presentation of the idea of accelerating protons to use them for treatment purposes by R.R. Wilson. The first proton beam-based treatment was performed in 1954 to suppress the functions of the pituitary gland of a breast cancer patient. In Japan, the history of particle beam-based treatment began in 1979 with the commencement of operation at the National Institute of Radiological Sciences (NIRS) of a 70 MeV proton beam-based treatment facility for superficial tumors. In 1994, heavy-particle beam-based treatment was started at NIRS. In 2003, NIRS's heavy-particle beam-based treatment was designated as a highly advanced medical treatment method.

At present, using a particle beam for cancer treatment requires a large-scale accelerator facility. However, only seven large-scale accelerator facilities were in operation in Japan as of October 2009. Of these seven facilities, six are facilities for proton beam-based treatment, and two are facilities for heavy-particle beam-based treatment (the Hyogo Ion Beam Medical Center has a facility capable of using both proton beam and heavy-particle beam). In addition, the Gunma University Heavy Ion Medical Center (Fig. 17.15) began in 2010 treatment at its heavy-particle beam-based treatment facility, and there are plans to construct heavy-particle beam-based treatment facilities in Nagano, Aichi, Fukui, Osaka, Saga, and Kagoshima prefectures.

Particle beam-based treatment facilities require a synchrotron accelerator to accelerate protons or heavy particles (carbon ions). It is not possible at present to construct a small-scale particle beam-based treatment facility because of the size of the synchrotron accelerator. Since it is evident that particle beam-based treatment is effective treatment, it is important to popularize particle beam-based treatment facilities. To do so, it is essential to reduce the cost of realizing and maintaining small-scale particle beam-based treatment facilities.

In particle beam-based treatment of cancers, proton beams (hydrogen ion beams) and heavy-particle beams (beams of ions heavier than helium) are being used. The advantage of particle beam-based treatment is that it allows the beam to be so controlled that the target cancer lesion is intensively attacked by the energy that is simultaneously released by all particles taking advantage of the phenomenon wherein the energy released by an accelerated particle reaches its peak just before it stops its movement (See Fig. 17.8). For this reason, particle beam-based treatment is effective in treating such cancers as pancreas cancers that have been difficult to

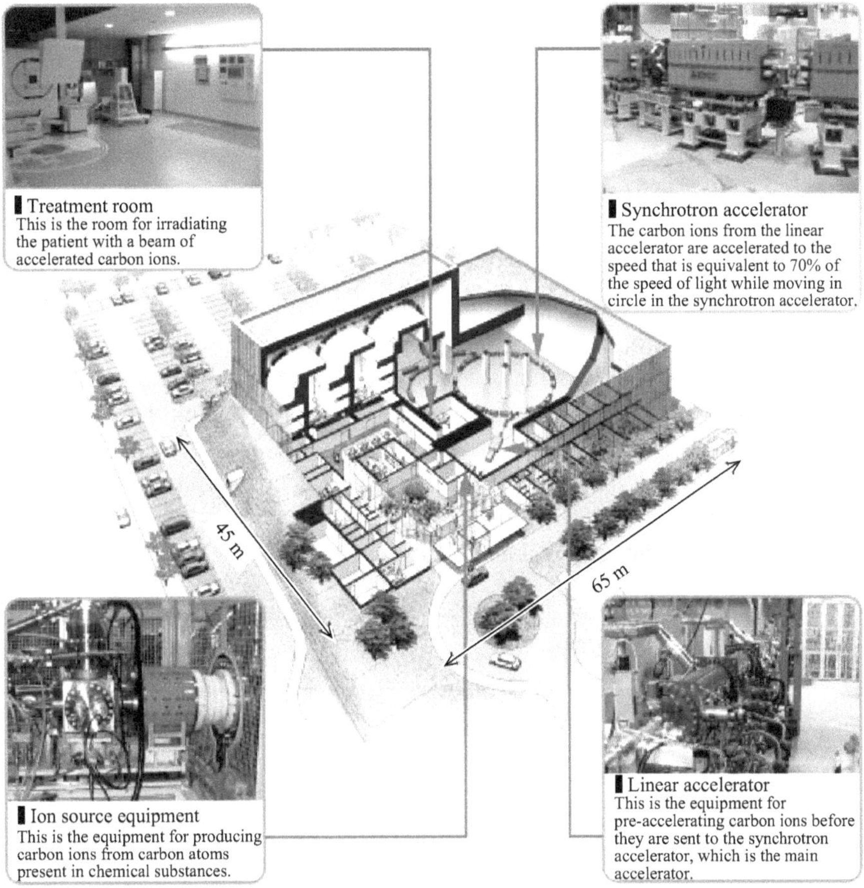

Fig. 17.15 Heavy-particle beam-based treatment facility of the Gunma University Heavy Ion Medical Center

treat with traditional radiation therapy methods such as X-ray and gamma ray-based radiation therapy methods.

Because the scattering of a carbon heavy-particle beam in a substance is weaker than the scattering of a proton beam in the same substance, it is considered that higher contrast can be achieved between the dose received by the cancer tissues and the dose received by the surrounding normal tissues. In addition, the biological effects (i.e., the effects on cells) of heavy-particle beams are stronger than those of proton beams and other radioactive rays at the same physical dose. Heavy-particle beams are also effective in treating tumors with low oxygen concentration. However, the dose levels of heavy-particle beams become a little higher than the dose levels of proton beams at depths deeper than the Bragg peak depth, because heavy-particle beams involve a spallation reaction.

17.4.2 Internal Irradiation

Internal irradiation-based therapeutic methods include the sealed small radiation source-based therapeutic method and the unsealed RI-based therapeutic method. In the former therapeutic method, a radionuclide such as radium or iridium is processed into sealed needle-shaped, tubular or granular radiation sources and directly put into the cancer to treat it. In the latter therapeutic method, the patient is dosed with a radioisotope or a radiopharmaceutical by means of oral administration or intravenous injection, and the radiation emitted by the RI that has accumulated in the cancer is used to treat it.

Another method is neutron capture therapy, which makes the tumor tissues take in an element in advance, irradiates the element with a neutron beam to cause a nuclear reaction, and treats the tumor by irradiating it with the generated radioisotope.

Sealed Radiation Source-Based Treatment Method

This is a treatment method that is usually effective for treating small cancers. In this method, an RI such as Ra-226, Ir-192, or I-125 is encapsulated in small tubular, needle-shaped, wiry or granular metal capsules, the capsules are put into the body of the patient, and the capsules are used to irradiate the cancer with radiation from inside the body to treat it.

Radium-226 (Ra-226), which had been discovered by the M. and P. Curie at the end of the 1800s and whose suitability for use as a radiation source had been confirmed through subsequent studies, had been used for many years as the RI for this treatment method. In the mid-twentieth century, it became possible to artificially produce RIs using nuclear reactors, etc. and Cobalt-60 (Co-60), Cesium-137 (Cs-137), etc. started to be used for the sealed small radiation source-based treatment method. Then, it became possible to use Gold-198 (Au-198), Iridium-192 (Ir-192), and Iodine-125 (I-125). Nowadays, Ra-226, Co-60, and Cs-137 are very rarely used.

The sealed radiation source-based treatment method is divided into the "implant irradiation method" for cancers in the oral cavity, tongue, breast, and prostate gland, in which the radiation source is directly inserted into the cancer tissues and tissues surrounding them, and the "intracavitary irradiation method" for esophageal cancers, cervical cancers, and lung cancers, in which a tube is attached at an appropriate location in a cavity such as the esophagus, uterus cavity, or windpipe in advance and then the radiation source is inserted using the tube.

A typical example of therapy by the former method is I-125-based treatment of prostate cancers. The radiation source used in this treatment is very small pieces (each with a length of approx. 5 mm and a diameter of approx. 1 mm) of I-125 that are chemically bonded to the surfaces of short lengths of thread made of silver and put in very small capsules made of pure titanium. Each radiation source capsule is called "a seed." About 60–90 seeds are used in one time of treatment. The I-125 seeds inserted into the body of the patient permanently remain in the body. However, because the tissue half-value thickness of the radiation emitted by I-125 is small (2 cm), the effects on the surrounding normal tissues are small. In addition, because the half-life of I-125 is 59.4 days, I-125 ceases to serve as a radiation source after the passing of 1 year after its insertion. Ir-192 is also used as an RI that temporarily serves as a radiation source in the body of the patient.

In the latter method, equipment called an "after-loading system" is used. This equipment consists of a very small radiation source (with a length of approx. 5 mm and a diameter of approx. 1 mm) attached to the tip of a wire. The equipment is inserted into lumens such as the esophagus, bronchus, and uterus or stuck into the lesion itself to treat the lesion. The radiation source used is Co-60 or Ir-192. Presently, Ir-192 is used in most cases. The treatment is achieved by irradiating the lesion for a period of 1 day to 1 week or by performing irradiation of the lesion several times for several minutes, depending on the strength of the radiation source.

Unsealed RI-Based Treatment Method

This treatment method is also called the radionuclide therapy. In Japan, the term "radionuclide therapy" had referred for a long time to Iodine-131 (I-131)-based treatment of thyroid cancers that have metastasized to the lungs, etc. and hyperthyroidism. However, strontium chloride labeled with Strontium-89 (Sr-89) and an anti-CD20 antibody labeled with Yttrium-90 (Y-90) were approved in 2007 and 2008, respectively.

(a) I-131: I-131 is a radioisotope that has a half-life of 8.0 days and emits gamma ray (364 keV) and beta ray (606 keV). The intra-tissue range of the beta ray emitted by I-131 is only 2 mm. Once I-131 gets into the body, it is selectively taken in by the thyroid tissue. Because of these characteristics, I-131 is being utilized to treat thyroid cancers and hyperthyroidism and being used in small amounts for diagnosis of these diseases. In Japan, I-131 became available in 1952. It was initially used as a diagnostic tracer and started to be used for therapy in the 1960s. It is orally administered to the patient in the form of a sodium iodide capsule containing I-131. G.H.W. Bush, who was the 41st president of the United States, helped demonstrate the effectiveness of I-131; he suffered from the Basedow disease during the Gulf War but was healed of the disease as a result of I-131-based therapy.

(b) Sr-89: Sr-89 (commercial name = Metastron) is an alkali earth metal that is similar to calcium. It behaves like calcium in the human body and accumulates in the parts whose osteogenesis activity is high. For this reason, Sr-89 intensively accumulates in bone metastasis lesions with increased osteogenesis activity and alleviates pain. Sr-89 is a radionuclide that emits beta ray only. The maximum energy is 1.49 MeV, but the average intra-tissue range is only 2.4 mm (the maximum range is approx. 8 mm), and the half-life is rather long

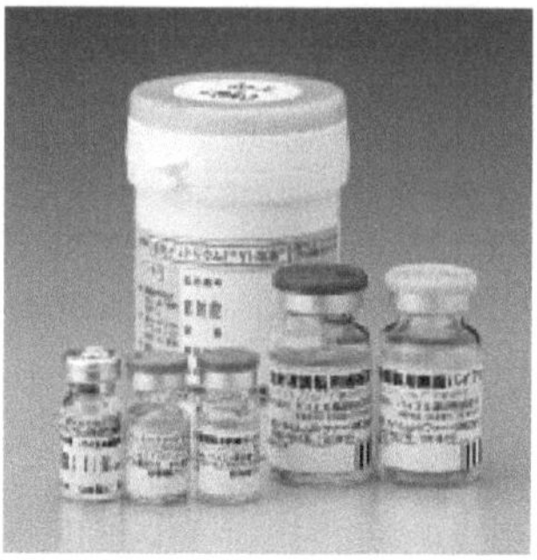

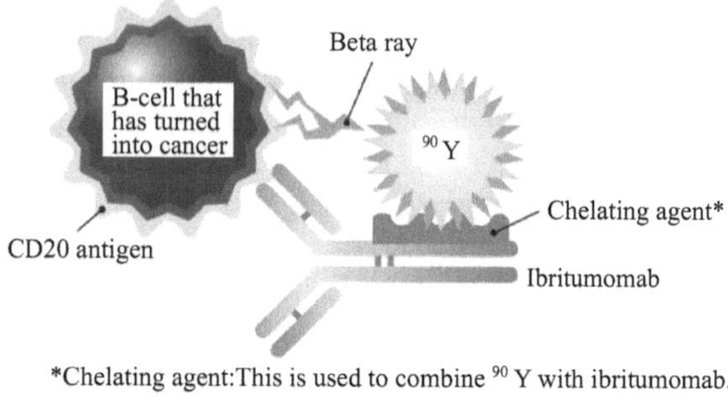

Fig. 17.16 Photograph showing a Zevalin intravenous injection set that uses an antibody labeled with Yttrium-90 (left) and a conceptual diagram of the structure of Zevalin, the bonding to a B cell that has turned into cancer through the antigen-antibody reaction, and the beta ray irradiation (right) (The photograph and diagram were taken from the web site of Bayer Yakuhin, Ltd. with permission)

(50.5 days). For this reason, it is possible to continuously irradiate the near field of the area in which Sr-89 exists, and the pain alleviation effect of one-time administration lasts for 3–6 months.

(c) Y-90: An anti-CD20 antibody labeled with Y-90 (commercial name = Zevalin, generic name = Y-90 ibritumomab tiuxetan) is being used as a pharmaceutical for radioimmunotherapy for non-Hodgkin's lymphoma of the B-lymphocyte type (B-cell malignant lymphoma). The advantage of Zevalin is that, by radiolabeling Rituxan (a monoclonal antibody for CD20 antigen; generic name = rituximab), which has been certified as a therapeutic pharmaceutical that utilizes the antigen-antibody reaction-based immune activity, with Y-90 using a chelating agent, not only the lymphoma cells that Rituxan has recognized and combined with but also the surrounding cancer cells can be treated (Fig. 17.16).

In the monoclonal antibody-based therapy, the cancer cells with which Rituxan has combined are killed by inducing apoptosis or programmed cell death in them. CD20 antigen is the most common cell surface antibody that occurs in B-cell malignant lymphomas at an occurrence rate of 95%, but it does not occur in human cells other than the B cells. Therefore, unnecessary combination of the antibody with cells other than the target cells can be avoided. In addition, rituximab, which is the monoclonal antibody for the CD20 antigen, is a chimeric antibody of humans and mice and remains on the cell surfaces after acting on the CD20 antigen. On the other hand, CD20 will not be isolated into the blood even if this specific antibody reacts. It is being utilized as a target in radioimmunotherapy as well.

Y-90 is a radioisotope with a relatively short half-life of 64 h. Because this is about the same as the biological half-life of ibritumomab, which is a mouse-type monoclonal antibody, irradiation of tissues other than the target cells with radiation can be minimized. In addition, Y-90 emits high-energy (2.28 MeV) pure beta ray only. The average intra-tissue range is 5.3 mm (maximum range = approx. 11 mm).

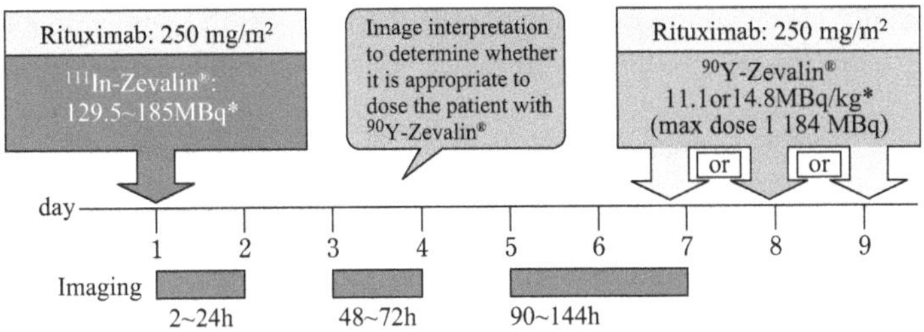

Fig. 17.17 Flow of treatment using Zevalin (an antibody pharmaceutical labeled with Y-90) (The diagram was taken from the web site of Bayer Yakuhin, Ltd. with permission)

This range allows the target cells to be effectively irradiated with radiation. It also allows the adjacent cancer cells to be killed.

In the actual treatment, the suitability of treatment using Zevalin labeled with Y-90 is assessed first, rather than unconditionally dosing the patient with Zevalin labeled with Y-90. This is achieved by dosing the patient with Zevalin labeled with Indium-111 (In-111), performing gamma camera-based imaging 48–72 h after the dosing and evaluating the distribution of the In-111 in the body of the patient and whether In-111 has been taken in by the bone marrow based on the scintigrams obtained. Treatment using Zevalin labeled with Y-90 is performed only in the case where it has been confirmed by the assessment that the distribution of the In-111 in the body of the patient is appropriate (Fig. 17.17).

Boron Neutron Capture Therapy (BNCT)

The boron neutron capture therapy is a treatment method that is based on neutron irradiation from outside the body of the patient. However, this is not a treatment method that utilizes the irradiation effect of neutrons themselves. In BNCT, a boron compound is fed to the cancer cells of the patient first, and then the cancer cells are irradiated with a neutron beam to cause a nuclear reaction that transforms the boron compound into lithium. The irradiation effect of the helium (^{4}He (alpha ray)) generated during this transformation process is utilized for the treatment. Therefore, it can be said that BNCT is essentially a technology for treating cancers by irradiating them with radiation from inside the body of the patient.

Because the alpha ray source (boron) is localized in the cancer cells, the alpha ray generated causes the irradiation effect only in the cancer cells and does not affect the surrounding normal cells. Because the range of alpha ray in the human body is only about 10 μm, BNCT is an excellent method for irradiating cancers with radiation at the cell level. In addition, because the boron compound itself will not generate radiation until it is irradiated with neutrons, unnecessary exposure of the patient and the treatment staff can be avoided. However, irradiation of normal tissues located closer to the radiation source than the lesion is unavoidable during neutron irradiation in BNCT as in other external irradiation-based treatment methods.

Although BNCT is such an excellent radiation treatment method, technical problems remain. One is the tumor selectivity of key boron compounds for treatment with BNCT, and the other is the development of a new neutron source that does not rely on nuclear reactors. The currently used boron compound is a compound that is taken in by not only cancer cells but also normal cells. Therefore, the development of a boron compound with high selectivity which is only taken in by cancer cells is essential. In addition, a neutron source is essential to perform BNCT. However, because there are only two nuclear reactors at present that can be used as neutron sources for BNCT (one at JAEA and the other at Kyoto University), only a limited number of patients can be treated at present. Attempts are being made to develop accelerators that can serve as new nonreactor-type neutron sources, but they are still in the research stage and have not reached the stage where they can be used for BNCT. BNCT can become a more effective cancer treatment method when these challenges are resolved.

17.5 Other Types of Utilization of Radiation in the Field of Medicine

17.5.1 Disinfection of Medical Supplies, Etc.

Nowadays, medical supplies are developed and produced as individually packaged disposable medical supplies for one-off use. For example, you should notice when you get an injection at a hospital that the syringe is made of plastic and not of glass that used to be used in the past. For the disinfection of disposable medical supplies and plastic pharmaceutical containers and packaging, ethylene oxide gas (EOG) has frequently been used. However, EOG-based disinfection is being replaced by radiation-based disinfection as a result of environmental problems in recent years, the need to secure occupational health and safety, and the need to comply with the validation guidelines. Radiation-based disinfection treatment is an effective means of disinfecting seal-packaged plastic medical supplies, etc. because it leaves no residual substance after the treatment unlike EOG-based disinfection treatment and does not heat the medical supplies, etc. unlike the high-pressure vapor-based disinfection treatment. Examples of medical supplies that are being disinfected by radiation-based disinfection include disposable needles and syringes, disposable blood collection devices, blood transfusion devices, surgical gloves, nylon sutures, sticking plasters with gauze, petri dishes, culture bottles, centrifuge tubes, and non-woven fabric products such as gowns, drapes, and cover cloths.

In addition, radiation-based disinfection treatment is being used to disinfect experiment animal feeds (especially feeds for "specific pathogen-free (SPF)" experiment animals), because radiation-based disinfection treatment satisfies the strict requirements relating to feeding facilities and environments.

17.5.2 Tracers for In Vitro Examinations and Studies

In vitro examinations are high-accuracy and high-sensitivity examinations that detect trace substances present in samples of blood, urine, etc. of the patient such as trace hormones and biologically active substances, which are difficult to detect with ordinary laboratory-level examinations, by making them react with radioactive pharmaceuticals in test tubes. The RIs used for in vitro examinations are tritium (H-3), Iron-59 (Fe-59), Iodine-125 (I-125), and Iodine-131 (I-131).

Radioimmunoassay, which is one of the in vitro examination methods, is an immunological test method developed as a method for measuring the amounts of trace antigens using RIs. The enzyme-linked immunosorbent assay (EIA or ELISA) method, which is being popularly used, is also based on the same principle. Radioimmunoassay is a method that was developed in the 1950s by R.S. Yalow and S.A. Berson. It was an epoch-making method at the time because it allowed even very trace amounts of hormones to be quantified. R.S. Yalow received the Nobel Prize in Physiology or Medicine in 1977 for his development of a radioimmunoassay method for insulin.

After that, in vitro measurement methods have been languishing as a result of the development of other measurement methods and for other reasons. Isotope distribution statistics show that the amount of RIs used for in vitro measurement has been decreasing year by year. In 2008, only Fe-59 and I-125 were supplied as RIs for in vitro measurement. The amount of these two radioisotopes supplied in 2008 was only one-ten thousandth of the amount of RIs supplied for in vivo measurement.

Development of a new pharmaceutical requires not only verification of the new pharmaceutical regarding its effects as drugs but also verification of the new pharmaceutical regarding its safety and other aspects. For this reason, RIs such as tritium (H-3), Carbon-14 (C-14), and Iodine-125 (I-125) are being used as tracers. These RIs are being used as a means of radiopharmaceuticals so that their distribution and accumulation in experiment animals can be confirmed by autoradiography. In addition, it is possible to know the rate at which a drug is taken in by the human body and the rate at which a drug is excreted from the human body by measuring the amounts of RIs present in organs taken out of bodies of experiment animals and excreta of experiment animals. Thus, the use of RIs allows the behaviors of substances that are candidates for new drugs in living bodies to be examined in detail.

RIs are also being utilized in biotechnology and genetic engineering studies as a means to determine the locations of specific DNA in chromosomes by combining small amounts of DNA with RIs and examining the sequences of the bases that comprise the DNA using the RIs as markers.

17.6 New Attempts to Utilize Radiation: From Radiation to Quantum Beams

One hundred and fifteen years have passed since the discovery of X-ray by Röntgen. This chapter has exhaustively outlined how various types of radiation have started to be used during this period in the field of medicinal science and medical treatment and how technologies for medical utilization of radiation are serving as important technologies that are indispensable for our daily lives today. Everyone knows that radiation is very effective in diagnosis and therapy of various types of diseases and injuries, but one thing not many people know is the fact that continuous efforts are being made at the actual scene of research and development to realize higher quality diagnosis and therapy and further improve the quality of life of patients. To conclude this chapter, this section presents examples of ongoing ambitious research and development efforts aimed at making innovations in radiation-based medical treatment.

17.6.1 Quantum Beam Technologies

This chapter has discussed how various types of radiation have been and are being utilized in the field of medicine. It goes without saying that radiation generation technologies form the basis of the utilization of radiation. Results of efforts that have been made since the discovery of radiation by Röntgen to develop radiation technologies including advanced ones have been reflected in today's radiation therapies. Therefore, we would like to explain here about quantum beam technologies, which are important not only in the sense that they are core technologies for the present radiation-based medicine but also in the sense that they are advanced technologies that will extend the frontiers of radiation therapies in the future.

The radioactive rays that fall under the definition of "radiation" under the "Act on the Definitions of Nuclear Fuel Substances, Nuclear Raw Material Substances, Nuclear Reactors and Radiation" are as follows:

1. Alpha rays, deuteron beams, proton beams, beta rays, and other heavily charged particle beams
2. Neutron beams
3. Characteristic X-ray (generated by electron capture) and gamma rays
4. X-ray and electron beams (1 MeV or higher)

Quantum beam technologies are advanced scientific technologies for improving the quality of radioactive rays and achieving advanced utilization of radioactive rays. Quantum beams are useful for "production," "observation," and "therapy," and these functions of quantum beams are being effectively utilized in the field of medicine as well. For example, particle beam-based cancer treatment is a technology that incorporates the best of quantum beam accelerator technologies, and RI-based diagnosis and therapy are technologies that combine various radiation-based state-of-the-art measurement and image analysis technologies. Thus, advances in quantum beam-related research and development can provide highly advanced technologies that are closely connected to our daily life and make impossible things possible.

17.6.2 Research and Development of RIs for More Accurate and Precise Diagnosis and Therapy

As described above, radiopharmaceuticals are effective means of diagnosis and therapy. Currently, studies are being conducted which are aimed at developing RIs for achieving more accurate and precise diagnosis and higher therapeutic effects and radiopharmaceuticals. For example, delivering RIs to only the cancer tissues in which an antigen has appeared through the use of an antibody that has very high affinity with the cancer tissues (as is done with the Y-90 Zevalin described in this chapter) enhances the effects of drugs. The authors call this the "RI-drug delivery system" (RI-DDS) to distinguish it from systems that use simple labeled compounds with low biological affinity (Fig. 17.18).

Developing an effective RI-DDS requires research and development to develop an RI and a DDS that are suitable for the RI-DDS as well as studies to make the RI and DDS (i.e., the components) work as a system. With regard to the RI, it is necessary to choose an RI that is suited to the characteristics of the molecules used

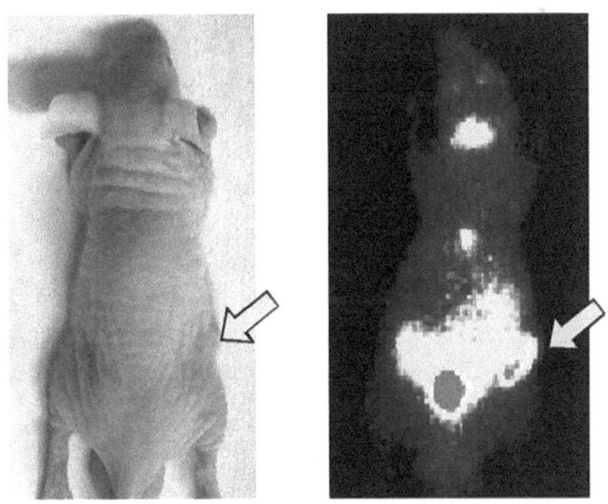

Fig. 17.18 Research and development aimed at constructing a new RI-DDS. RI-DDS that uses Br-76 as a new positron-emitting radionuclide for PET diagnostics
(Photos by courtesy of Noriko Ishioka at JAEA)

for the DDS. Consider a situation in which an antibody capable of selectively binding to a specific cancer cell is obtained and the antibody is used to develop RI-DDS for PET diagnosis. Because it normally takes about 1 day for an antibody to accumulate in cells that have an antigen, even with F-18, which is the radionuclide with the longest half-life among the PET radionuclides that have been put into practical use, the radioactivity level will have decayed by the time of the measurement to about one-ten thousandth of the level at the time of the dosing, which means that it will be impossible to make diagnosis. For this reason, studies on methods for producing positron-emitting radionuclides with longer half-lives than F-18, including Copper-64 (Cu-64) (half-life: 12.7 h) and Bromine-76 (Br-76) (half-life: 16.1 h), as well as studies to develop compounds radiolabeled with these radionuclides, are being conducted.

It goes without saying that research and development to develop new RIs requires establishing a basic production method. To realize the popular use of new RIs at the clinical level, it is essential to construct and install facilities and equipment for supplying sufficient amounts of RIs with suitable quality as drugs for diagnosis or therapy and to develop production processes for the new RIs. For example, some newly developed RIs have not yet advanced to the practical use stage because there are problems that have yet to be solved regarding production processes for them, despite the fact that laboratory experiments have shown that they have excellent characteristics.

By supply source, radiopharmaceutical can be divided into those imported from overseas, those domestically produced by Japanese manufacturers, and those produced at medical institutions where the actual clinical use is made. Examples of RIs whose supply relies on imports from overseas include Tecnetium-99m (Tc-99m), which is the RI that is being most frequently used in nuclear medicine diagnosis, and Yttrium-90 (Y-90), which is being used in radioimmunotherapy. In particular, if a serious problem occurs at a Tc-99m production facility, the supply of Tc-99m in Japan will stop, and treatment plans will be seriously affected.

A realistic way to supply new RIs is to utilize the 130 or so small accelerators installed in medical institutions in Japan for in-house production of radiopharmaceuticals for PET for examinations for cancers. Medical institutions that have small accelerators naturally have facilities and equipment for nuclear medicine diagnosis such as PET. In-house production would make it easier to utilize even radionuclides with relatively short half-lives. In addition, such in-house production would involve the production of the only necessary amounts of the necessary RIs, because there would be no need to distribute produced RIs. Because these small accelerators are capable of accelerating hydrogen ions (H^+) using energy of about 20 MeV, it is possible, for example, to produce the Cu-64 and Br-76 described above. Therefore, it is expected that studies to develop RI production methods for utilizing these accelerators and develop drugs labeled with produced RIs will play an important role in the future.

17.6.3 Laser-Accelerated Proton Beam-Based Treatment

Particle beam-based treatment is a very effective new method of radiation therapy for cancers. This requires a large facility equipped with an accelerator, as mentioned above. To popularize particle beam-based treatment as a method of radiation therapy for cancers, accelerators should ideally be downsized at least to a level where they can be installed in a treatment room of an ordinary hospital, if not to a size equivalent to the size of linacs, which is the mainstream equipment at present. However, it is difficult to downsize conventional accelerators without compromising their ability to produce particle beams with sufficient energy for treatment, because of the limitations associated with the principle of particle acceleration they employ.

For this reason, studies are being conducted to develop equipment that is based on a completely different principle to generate particle beams. It is known that a proton beam can be generated by irradiating a target such as a metal or macromolecule with high-intensity laser. The studies aim at utilizing this principle for radiation therapy. A recent study proposed a theoretical model that generates a high-energy ion beam by making use of the powerful acceleration electric field created by a magnetic vortex generated on the back of a target substance through irradiation of the target substance with a beam obtained by concentrating a high-intensity laser (Fig. 17.19). The study demonstrated that the model was capable of generating a high-energy carbon beam with energy of 18 MeV per nucleon and a high-energy oxygen beam with energy of 20 MeV per nucleon. This is a level of energy that is about ten times as high as the level of energy achievable with traditional method-based laser equipment of similar sizes. Because the cluster target used in the study is one that can be supplied on a continuous basis, it can

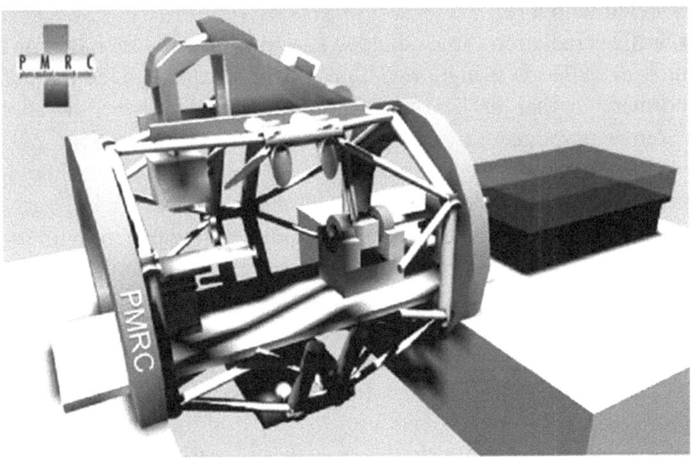

Fig. 17.19 Concept model of laser-driven proton beam-based treatment equipment (Photo by courtesy of the Photomedicine Research Liaison Center)

be said that the model can lead to the realization of a medical accelerator, for which repeated irradiation is a requirement.

This section has presented only two examples of ambitious research and development efforts, but it goes without saying that new attempts are being made continuously from various standpoints in various study areas. In particular, creation of new technologies through integration of technologies that belong to different domains and fields is being pursued in the field of medicine as well.

This chapter has provided a brief overview of the utilization of radiation in the field of medicine. Some of the readers may feel that there is a bias in the selection of the subjects, but we hope that you understand that we have selected the subjects based on the relationship between medicine and radiation as seen from an objective viewpoint of us as people who are not specialists of medicine or medical treatment.

Exercises

1. With reference to the text of this chapter, complete the following sentences concerning the medical use of radiation by selecting the appropriate word to be inserted in parentheses from the following word list:

Utilization of radiation in the field of medicine can be divided into utilization of radiation for [a] and utilization of radiation for [b]. Examples of utilization of radiation for [a] include plain [c] photography and [c] CT (computed tomography), which are technologies for obtaining transmission images through irradiation using a radiation source such as [c] (discovered by Röntgen) located outside the body of the patient, and PET (positron emission tomography) and SPECT (single-photon emission computed tomography), which are technologies for measuring [d] emitted by drugs labeled with a [e].

[b] that utilizes radiation can be divided into methods that are based on the [f] of the patient with radiation using a radiation source located outside the body of the patient and methods that are based on the [f] of the patient with radiation from a radiation source introduced artificially into the body of the patient. Most of the former types of methods are methods that use [c]. Recently, technologies and equipment that enhance the therapeutic effect and minimize unnecessary [g] have been developed. In addition, equipment for irradiating the patient with [d] emitted by [h] and [b] that utilizes the phenomenon wherein a [i] generated by an accelerator such as a synchrotron suddenly releases a large amount of energy immediately before it stops its movement are becoming popular. Radiation sources can be introduced into the body of the patient by using a [j] radiation source such as Iodine-125 (I-125) or Iridium-192 (I-192) or by dosing the patient with a drug labeled with a [e].

Radiation is also being utilized for radiation-based [k] of medical supplies and for producing reagents for examinations and research.

Word List

1. Disinfection
2. Diagnosis
3. Sealed
4. Therapy
5. Particle beam
6. X-ray
7. Cobalt-60 (Co-60)
8. Radionuclide
9. Exposure
10. Gamma ray
11. Irradiation

2. Complete the following sentences about radioisotope-based diagnosis and therapy by selecting the appropriate words that should be inserted in the brackets from the word list below referencing the text of Chap. 17:

Medical examinations that use radioisotopes (RIs) can be divided into [a] examinations and [b] examinations. In terms of the amount of RIs used, most of the medical examinations that use RIs are [a] examinations, and the proportion of [b] examinations among the medical examinations that use RIs is small. The two radionuclides that are far more frequently used for [a] examinations than other radionuclides are [c] for [d] and SPECT and [e] for PET. [e] is dosed to the patient in the form of fluorodeoxyglucose (FDG) labeled with [e], which is an analogue of glucose. PET-based examinations detect, using the simultaneous measurement method, the pair of [g] generated when [f]s emitted by a [f]-emitting radionuclide such as [e] annihilate with electrons. Because the two [g] travel in opposite directions (i.e., in directions 180° apart from each other), the point of generation of the radiation can be determined without using a [h] unlike in the case of SPECT.

Only three RIs (radionuclides) have been approved for use in treatment in Japan. These are [i], Strontium-89 (Sr-89), and [j]. Initially, only [i] was approved as the RI for use in treatment in Japan, and this situation continued for a long period of time. [i] is used not only for treatment of cancers but also for treatment of thyroid disorders. It is also dosed to patients in small amounts for diagnostic purposes, because it emits gamma ray. Sr-89 behaves like calcium in the human body and tends to accumulate in bones. For this reason, Sr-89 is used for treatment aimed at alleviating pains. [j] is a radionuclide that only emits a 2.2 MeV beta ray. It is used to produce a drug that has an immunological therapeutic effect and a radiation therapeutic effect by [j] labeled rituximab (a monoclonal antibody), which is a pharmaceutical for treatment of malignant lymphomas that utilizes the antigen-antibody reaction, with it. The treatment method in which a RI-labeled drug like this is introduced into the body of the patient is called [k].

Word List

1. Positron
2. In vivo
3. Fluorine-18 (F-18)
4. Radionuclide therapy
5. Scintigraphy
6. Iodine-131 (I-131)
7. Collimator
8. In vitro
9. Yttrium-90 (Y-90)
10. Tecnetium-99m (Tc-99m)
11. Annihilation gamma rays

References

1. Murayama H, Kougakkai no ima to korekara pojitoron CT (in Japanese); http://annex.jsap.or.jp/OSJ/50th_cd/main/mokuji/index.htm; Murayama H, "The present and future of the optical society of Japan – Positron CT" (in Japanese); http://annex.jsap.or.jp/OSJ/50th_cd/main/mokuji/index.htm
2. Senda M, Misuta PET no kakuigaku kyoushitu (in Japanese); http://www.asca-co.com/nuclear/; Senda M, "Seminar on nuclear medicine by Mr. PET" (in Japanese) http://www.asca-co.com/nuclear/
3. Gan shinryou ni okeru PET shindan no ichi zuke, Konsensasu, 5 (3) (2006) (in Japanese); "Role of PET-based diagnosis in cancer examinations". Consensus 5(3) (2006) (in Japanese)
4. Nakano Y (2006) Zenshin PET/MRI yuugou gazou ni kansuru kentou, Nihon houshasen gijutu gakkai zasshi, 62(6) (in Japanese); Nakano Y (2006) "A study on combined whole-body PET/MRI images". J Jpn Soc Radiol Technol 62(6) (2006) (in Japanese)
5. Houshasen ryouhou souron, Kokuritu gan senta gan taisaku jouhou senta (in Japanese); http://ganjoho.jp/public/dia_tre/treatment/radiotherapy/radiotherapy.html; "General guide to radiotherapy", Center for Cancer Control and Information Services, National Cancer Center (in Japanese); http://ganjoho.jp/public/dia_tre/treatment/radiotherapy/radiotherapy.html
6. Gan no houshasen chiryou kanja san no ishi kettei ni yakudatu jouhou, Beikoku houshasen shuyou gakkai (Kokuritu gan senta gan jouhou sabisu) (in Japanese); "Radiotherapy for cancer treatment : information that helps patients' decision making". American College of Radiation Oncology (Center for Cancer Control and Information Services, National Cancer Center) (in Japanese)
7. Tsujii, H. : Houshasen o riyou shita "Gan" no hakken to chiryou (in Japanese); http://www.fnca.mext.go.jp/new/images/asia080118_tsujii_ab.pdf; Tsujii, H. : "Radiation-based cancer detection and treatment" (in Japanese); http://www.fnca.mext.go.jp/new/images/asia080118_tsujii_ab.pdf
8. Gan chiryou, (Kabu) Akyusera (in Japanese); http://www.accuthera.com/RT01.html; "Cancer treatment", Accuthera Inc. (in Japanese); http://www.accuthera.com/RT01.html
9. Karasawa K (2008) Houshasen iryou no genjou, Dai 5 kai Genshiryoku iinkai teirei kaigi shiryou (in Japanese); http://www.aec.go.jp/jicst/NC/iinkai/teirei/siryo2008/siryo05/siryo1-2.pdf; Karasawa K (2008) "Current status of radiation therapy", a reference material for the 5th

Regular Meeting of the Japan Atomic Energy Commission (in Japanese); http://www.aec.go.jp/jicst/NC/iinkai/teirei/siryo2008/siryo05/siryo1-2.pdf
10. Nakagawa K (2004) Houshasen chiryou to iryou you riniakku no genjou to tenbou, Dai 1 kai Nihon kasoku ki gakkai nenaki, Dai 29 kai Riniakku gijutu kenkyuukai puroshi deingusu (in Japanese); http://www.pasj.jp/web_publish/pasj1_lam29/WebPublish/4AP9.pdf; Nakagawa K (2004) "Current status and future prospects of radiation therapy and medical linacs". Proceedings of the 1st annual conference of the particle accelerator society of Japan and the 29th meeting of the Linac Technology Study Group (in Japanese); http://www.pasj.jp/web_publish/pasj1_lam29/WebPublish/4AP9.pdf

Answers to Exercises

Chapter 1

1. See the text.
2. For example, consult the home page of the Atomic Energy Commission (AEC) of Japan.

Chapter 2

1. The energies of γ-rays from Co-60 are 1.17 and 1.33 MeV. The dominant process is the Compton effect.
2. Consult the home page of the National Institute of Radiological Science (NIRS), Japan.

Chapter 3

1. Note that G is in mol•l^{-1} and is the number per 100 eV, $G = 100eN_Ac/D$, where e is the electron charge, and N_A is Avogadro's number.
2. $G(Ce^{3+}) = G(H) + 2G(H_2O_2) - G(OH)$.

Chapter 4

1. The hydroxyl radical (•OH) has an unpaired electron and is electrically neutral. The hydroxyl ion (OH^-) does not have an unpaired electron and is electrically negative.
2. It has 12 lines. Two H atoms, bonded to a C atom having an unpaired electron, split into three lines. Further, three H atoms bonded to the adjacent C atom split each line into four. In all, 12 lines will be observed.

Chapter 5

1. See the text. If it was not pulsed, hydrated electrons appearing at different times would have to be observed simultaneously.
2. See the text.

H. Kudo (ed.), *Radiation Applications*, An Advanced Course in Nuclear Engineering 7, https://doi.org/10.1007/978-981-10-7350-2

Chapter 6

1. About 6. Consult a book on radiation chemistry.
2. An example is the pulse radiolysis method. The decay constant of the optical density at different concentrations of the solute is used. For the OH radical, because its absorption is weak and appears at short wavelengths, the competition with potassium thiocyanate (KSCN) is then used.

Chapter 7

1. Water is not appropriate. It is necessary to reduce the oxygen content.
2. The direct effect can be dominant. The absorbed dose can vary depending on the elemental composition and density.

Chapter 8

1. (1) Graft polymerization
   (2) Cross-linking
   (3) Degradation
2. (1) Heat-resistant electrical cable
   (2) Polyethylene foam sheet
   (3) Membrane for button-shaped electric cell
   (4) Teflon^{TM} bearings, sealant, and O-rings
   (5) Wound dressing
   (6) Plant growth promoter
3. (1) Hesitation to use radiation due to misunderstanding:
   Polymer processing techniques by radiation is sometimes not well-known to public.
   It is necessary to properly convey the characteristic of radiation, safety handling, and usefulness of application.
   (2) Investment to install irradiation facilities:
   It is possible to ask the irradiation service centers (companies that irradiate customer's products with charge).
   (3) Less understanding in polymer processing techniques:
   It is possible to apply organizations such as Japan Atomic Energy Agency for technical consultants.
   It is also possible to see the current status of research and development and commercialized products at their pavilions.

Chapter 9

1. Types of effects of irradiation of semiconductor elements with radiation and the types of radiation that cause each of the types of effects:
   (1) Total ionizing dose effect (TID)
   The main cause is gamma rays and electron beams. An example of effective measures against TID is radiation shielding.

(2) Displacement damage dose effect (DDD)
The main cause is electrons and protons captured by the "geomagnetically captured radiation belt" (the "Van Allen belt") and protons from solar flare radiation. An example of effective measures against DDD is radiation shielding.
(3) Single-event effect (SEE)
The main cause is high-energy heavy particles from galactic cosmic rays. Examples of measures against SEE are element structure-based measures and system-based measures (such as software-based ones).

2. (1) CMOS image sensors for monitoring applications (TID)
(2) Semiconductor detectors (DDD)
(3) CPUs and memories used in personal computers, motor controllers, oscilloscopes, FPGAs, etc. used for beam control and experiments (SEE)

Chapter 10

1. Radiation- and discharging-based methods for 10ppm; adsorption-, catalysis-, and burning-based methods for 100ppm; and concentration and burning-based methods for 1000ppm.
2. Approximately 1 mg
3. One reason is the fact that, in the field of exhaust gas treatment, infrastructures and treatment work systems suitable for existing technologies have been developed over many years, and introduction of a new technology sometimes requires a large initial investment because of incompatibility of the technology with the existing infrastructures or involves complexities in ensuring the continuity of treatment work during the transition period for switching technologies.

Chapter 11

1. The thickness of the oxidized layer is expressed by Formula 11-4. To halve the thickness of the oxidized layer, which is represented by the left-hand side of the formula, it is necessary to quadruple the electron beam dose rate to 2 kGy/s or reduce the partial pressure of oxygen to one fourth (2.5×10^4Pa) by diluting the dry air with nitrogen, etc. In the case where the values of two or more parameters can be changed at the same time, the thickness of the oxidized part can also be halved by doubling the dose rate to 1 kGy/s and halving the partial pressure of oxygen to 5×10^4Pa.
2. The elements that are hardest to be radioactivated are H, He, C, O, and Ne, and the elements that are second hardest to be radioactivated are N, F, Si, Ar, Ca, etc. SiC ceramic fibers which are synthesized from precursor polymeric materials, as well as SiC ceramic fiber-based composite materials, have excellent radioactivation resistance because the amounts of elements contained in them (in the form of impurities) are negligible compared to the amounts of their constituent elements, that is, Si and C, which are the constituent elements of all polymeric materials.

Chapter 12

1. The use of the K factor value calculated using Formula 12-2 allows the backscattering energy of the backscattered He particles to be calculated. The calculation result is 0.52 MeV for ^{12}C, 1.12 MeV for ^{27}Al, 1.54 MeV for ^{59}CO, and 1.85 MeV for ^{197}Au. Therefore, the sample is made of gold (Au) and the answer to the question is (4).
2. The mass separability can be improved by increasing the difference in the K factor. Therefore, the mass separability can be improved by [1] increasing the analysis beam acceleration energy or [2] using an analysis beam of heavy particles.

Chapter 13

1. Ion beam-based induction of mutations tends to cause multilocation damage at the truncated ends of DNA and consequently induce mutations that involve large structural changes. This is because ion beams impart large energy to the irradiated object. Induction of mutations using electron beams, gamma ray, etc. tends to cause non-deletion-type point mutations because of overlapping in the base sequence of the truncated ends of DNA.
2. Ion beam-based induction of mutations in plants can be utilized in breeding of many types of plants including crops, flowering plants, orchard, and trees. It can also be utilized in basic research including gene identification-related research. Ion beams are suitable for these applications because they often cause gene deletion etc.
3. Answer omitted.

Chapter 14

1. As described in the text of Chap. 14.
2. Refer to Reference [7] and other literature.
3. Answer omitted.

Chapter 15

1. As described in the text of Chap. 15.
2. Answer omitted.
3. Answer omitted.

Chapter 16

1. [a] (6) Noninvasive
   [b] (3) Tracers
   [c] (1) Positive electrons
   [d] (8) PET
   [e] (10) Proton
   [f] (11) Neutron

[g] (4) Mass
[h] (9) Positive
[i] (7) Energy
[j] (5) Annihilate
[k] (4) Mass
[l] (2) Simultaneous measurement

2. [1] ×
[2] ×
[3] ○
[4] ○
[5] ○

Chapter 17

1. [a] (2) Diagnosis
[b] (4) Treatment
[c] (6) X-ray
[d] (8) Radioisotope
[e] (10) Gamma ray
[f] (11) Irradiation
[g] (9) Exposure
[h] (7) Cobalt-60 (Co-60)
[i] (5) Particle beam
[j] (3) Sealed
[k] (1) Disinfection

2. [a] (2) In vivo
[b] (8) In vitro
[c] (10) Technetium-99m (Tc-99m)
[d] (5) Sintigraphy
[e] (3) Fluorine-18 (F-18)
[f] (1) Positron
[g] (11) Annihilation gamma rays
[h] (7) Collimator
[i] (6) Iodine-131 (I-131)
[j] (9) Yttrium-90 (Y-90)
[k] (4) radionuclide therapy

Index

H. Kudo (ed.), *Radiation Applications*, An Advanced Course in Nuclear Engineering 7, https://doi.org/10.1007/978-981-10-7350-2

D

E

F

Kolibri 360 GmbH
Ferdinand-Jühlke-Straße 7
99095 Erfurt, Deutschland
produktsicherheit@kolibri360.de